Advances in Industrial Control

Other titles published in this series:

Digital Controller Implementation and Fragility
Robert S.H. Istepanian and James F. Whidborne (Eds.)

Optimisation of Industrial Processes at Supervisory Level
Doris Sáez, Aldo Cipriano and Andrzej W. Ordys

Robust Control of Diesel Ship Propulsion
Nikolaos Xiros

Hydraulic Servo-systems
Mohieddine Jelali and Andreas Kroll

Model-based Fault Diagnosis in Dynamic Systems Using Identification Techniques
Silvio Simani, Cesare Fantuzzi and Ron J. Patton

Strategies for Feedback Linearisation
Freddy Garces, Victor M. Becerra, Chandrasekhar Kambhampati and Kevin Warwick

Robust Autonomous Guidance
Alberto Isidori, Lorenzo Marconi and Andrea Serrani

Dynamic Modelling of Gas Turbines
Gennady G. Kulikov and Haydn A. Thompson (Eds.)

Control of Fuel Cell Power Systems
Jay T. Pukrushpan, Anna G. Stefanopoulou and Huei Peng

Fuzzy Logic, Identification and Predictive Control
Jairo Espinosa, Joos Vandewalle and Vincent Wertz

Optimal Real-time Control of Sewer Networks
Magdalene Marinaki and Markos Papageorgiou

Process Modelling for Control
Benoît Codrons

Computational Intelligence in Time Series Forecasting
Ajoy K. Palit and Dobrivoje Popovic

Modelling and Control of Mini-Flying Machines
Pedro Castillo, Rogelio Lozano and Alejandro Dzul

Ship Motion Control
Tristan Perez

Hard Disk Drive Servo Systems (2nd Ed.)
Ben M. Chen, Tong H. Lee, Kemao Peng and Venkatakrishnan Venkataramanan

Measurement, Control, and Communication Using IEEE 1588
John C. Eidson

Piezoelectric Transducers for Vibration Control and Damping
S.O. Reza Moheimani and Andrew J. Fleming

Manufacturing Systems Control Design
Stjepan Bogdan, Frank L. Lewis, Zdenko Kovai and Jos Mireles Jr.

Windup in Control
Peter Hippe

Nonlinear H2/H8 Constrained Feedback Control
Murad Abu-Khalaf, Jie Huang and Frank L. Lewis

Practical Grey-box Process Identification
Torsten Bohlin

Control of Traffic Systems in Buildings
Sandor Markon, Hajime Kita, Hiroshi Kise and Thomas Bartz-Beielstein

Wind Turbine Control Systems
Fernando D. Bianchi, Hernn De Battista and Ricardo J. Mantz

Advanced Fuzzy Logic Technologies in Industrial Applications
Ying Bai, Hanqi Zhuang and Dali Wang (Eds.)

Practical PID Control
Antonio Visioli

(continued after Index)

M. A. A. Shoukat Choudhury • Sirish L. Shah •
Nina F. Thornhill

Diagnosis of Process Nonlinearities and Valve Stiction

Data Driven Approaches

Springer

M. A. A. Shoukat Choudhury
Bangladesh University of
Engineering and Technology (BUET)
Department of Chemical Engineering
Dhaka-1000
Bangladesh

Sirish L. Shah
University of Alberta
Department Chemical & Materials
Engineering
Edmonton AB T6G 2G6
Canada
Sirish.Shah@ualberta.ca

Nina F. Thornhill
Imperial College London
Department of Chemical Engineering
South Kensington Campus
London SW7 2AZ

ISBN: 978-3-540-79223-9 e-ISBN: 978-3-540-79224-6

Library of Congress Control Number: 2008928126

Cover design: eStudio Calamar S.L.

Printed on acid-free paper

9 8 7 6 5 4 3 2 1

springer.com

To all our mentors and students

Preface

Classical control has matured to the point where off-the-shelf control solutions, for linear time-invariant systems, are easily available from most major control vendors. In the relentless drive for improved regulatory performance due to increasing energy costs, higher quality specifications and competition, advanced tools for monitoring closed-loop performance are becoming available. Such measures of closed-loop control performance can now be readily computed in an on-line manner.

Performance metrics, from routine operating data, give an indication of closed-loop performance with respect to user-defined benchmarks, such as minimum variance benchmark, or a historical benchmark of satisfactory performance or a desired settling time benchmark. Conversely, such benchmarks also provide an indication of the difficulty of controlling a process. However, the diagnosis of poor performance is an area with many challenges. Poor closed-system performance may be due to a variety of reasons, for example, poor controller tuning, improper controller configuration, process operating condition changes, process constraints, actuator problems such as valve stiction or saturation and so on. The diagnosis of poor performance is a non-trivial problem.

Access to routine process data is now a norm. The purpose of this book is to introduce novel tools in the analysis of such data to determine the diagnosis of poor controller performance. The main material in the book is concerned with the detection, diagnosis and quantification of process nonlinearities. Process changes such as saturation constraints, sticky valves and nonlinear operating regimes introduce nonlinearities in control signals. The very specific focus of this book is on detecting process nonlinearities and diagnosing and quantification of the source of such nonlinearities. The main tool for the detection of such nonlinearities are higher-order statistics, in particular third-order cumulants and the bispectrum, as the spectral counterpart of the third-order moments of the distribution of a signal. Specifically, this text provides answers to questions such as the following:

- Are the process data well-conditioned for analysis in the first place? Well-conditioned data are an indispensable part of computing any performance metric. It is futile to compute performance metrics and make diagnosis conclusions on poorly conditioned data, as the results may lead to erroneous conclusions. This book dwells on checking the integrity of archived data prior to using such data for computing performance metrics. For example, methods for estimating data compression factors and sensor quantization estimates are considered in this book.

- How to determine a measure of process nonlinearity from process data? Such a metric may indeed give a clue to the degree of difficulty in controlling a given process. If the process is diagnosed to be strongly nonlinear than the suggested remedial action would be to consider a nonlinear control algorithm.
- How to check if the actuator is performing satisfactorily? Statistics from several large-scale industrial studies have confirmed that as many as 40% of the control loops do not perform well because of valve problems. In this respect, an important part of this book is focussed on developing a reliable test for the detection of valve stiction. A significant portion of this monograph is devoted towards data-based modelling of valve dynamics and the detection and diagnosis of valve or actuator problems from routine operating data. Many simulation examples and over a dozen industrial case studies are provided to illustrate the application of higher-order statistical tools developed in this book.
- Are there plant-wide oscillations and if so how to detect loops with common oscillations and diagnose the possible source of the oscillations? It is not uncommon, as evident from the three industrial case studies considered here, for many loops to have common oscillations. Because of the highly integrated nature of modern plants, oscillations that may arise in one loop due to overly tuned controller gains or a sticky valve or an oscillatory disturbance, can easily propagate to many other loops downstream as well as upstream due to the presence of recycles. Newly developed tools for the detection and diagnosis of such problems are presented towards the end of this book.

Most of the results presented here have appeared in the archival literature, albeit in compact forms, over the last several years. One motivation in preparing this monograph was to consolidate all these results in a logical and presentable form in one place. In this respect, the book is likely to be of use for new researchers as a place to access basic information presented in a tutorial format as space is no longer a restriction, as it is in technical papers in archived journals and conference proceedings. However, there are results here that will also appeal to the practising engineer. Portions of this book can also be used in a graduate-level course in process control, signal processing and fault detection and isolation. Many of the chapters do not require knowledge of earlier chapters as prerequisite information and therefore can be read independently.

The work in this book has been the outcome of several years of research efforts by the authors with contributions from many other graduate students at the University of Alberta, Canada, and in the Centre for Process Systems Engineering at University College London and Imperial College London, UK, as well as collaborative work with industry. It is a pleasure to thank the many people who have contributed so generously to the conception and creation of the material in this book. The research environment at U of A, UCL and Imperial College has been very stimulating and superbly conducive towards doing collaborative and cross-disciplinary research and has provided much needed fuel to the efforts that underwent in the work reported here. In this respect, we would particularly like to thank Biao Huang and Fraser Forbes for their ongoing support to our research activity.

Many past as well as current graduate students, research associates and visiting professors have participated in the work reported here, through engaging discussions at seminars, in the class rooms and the labs. These include Salim Ahmed, Ian Alleyne, Margret Bauer, Lamia Benabbas, Bhushan Gopaluni, Enayet Halim, Syed Imtiaz, Mridul Jain, Hailei Jiang, Vinay Kariwala, Weihua Li , Shankar Narasimhan, Sachin Patwardhan, Hari Raghavan and Arun Tangirala. We are indebted to these people who in a way or another have shaped the quality and outcome of our research. The spectral envelope ideas reported in Chaps. 17 and 18 is the contribution owing to the work by Hailei Jiang.

An important driving force in the material presented in this book has been our collaborative partnership with industry. In this respect, Rohit Patwardhan, David Shook, Warren Mitchell and Anand Vishnubhotla of Matrikon Inc. have been instrumental in providing many challenging data sets and practical insights in the evaluation of the stiction detection algorithm. The material in Chaps. 9, 16, 17, 18 and 19 is the outcome of our on-going collaborative work with AT Plastics (Ian Alleyne), Amerene, BP Chemicals (Adrian Meaburn), Celanese Canada Company Corporation (Enbo Feng (now at Petro-Canada)), Eastman Chemical (Mike Paulonis and John Cox), Mitsubishi Chemical (Hisato Douke and Haruo Takada), Teck Cominco Limited (Mike Foley), Cascade Boise Pulp and Paper (Chris McNab) and Suncor Energy (Ramesh Kadali). We are thankful to all of these individuals and their organizations for cooperating so enthusiastically and providing us an opportunity for industrial applications and evaluation of our research results.

We are also grateful to the Natural Sciences and Engineering Research Council (NSERC) of Canada, Matrikon Inc., Suncor Energy, iCORE and the Government of the Province of Alberta for providing us the research funding to carry out the work reported here. The first author would also like to acknowledge support from the Canadian International Development Agency (CIDA) in the form of a scholarship and the Alberta Ingenuity program for an Industrial Fellowship. The support for the first author from Bangladesh University of Engineering and Technology (BUET) is also gratefully acknowledged. The third author acknowledges support of a Global Research Award of the Royal Academy of Engineering, London, UK.

Bangladesh University of Engineering and Technology (BUET), Dhaka, Bangladesh — *M. A. A. Shoukat Choudhury*
U. of Alberta, Edmonton, Canada — *Sirish L. Shah*
Imperial College London, London, UK — *Nina F. Thornhill*
July 2007

Chapter	Contents	Suitable reading for*
	Part I Higher-Order Statistics	
2	Higher-Order Statistics: Preliminaries	B
3	Bispectrum and Bicoherence	B
	Part II Data Quality – Compression and Quantization	
4	Impact of Data Compression and Quantization on Data-Driven Process Analyses	B
	Part III Nonlinearity and Control Performance	
5	Measures of Nonlinearity – A Review	B
6	Linear or Nonlinear? A Bicoherence-Based Measure of Nonlinearity	B
7	A Nonlinearity Measure Based on Surrogate Data Analysis	B
8	Nonlinearities in Control Loops	A and B
9	Diagnosis of Poor Control Performance	A and B
	Part IV Control Valve Stiction – Definition, Modelling, Detection and Quantification	
10	Different Types of Faults in Control Valves	A and B
11	Stiction: Definition and Discussions	A and B
12	Physics-Based Model of Control Valve Stiction	B
13	Data-Driven Model of Valve Stiction	B
14	Describing Function Analysis	B
15	Automatic Detection and Quantification of Valve Stiction	B
16	Industrial Applications of the Stiction Quantification Algorithm	B
17	Confirming Valve Stiction	B
	Part V Plant-wide Oscillations – Detection and Diagnosis	
18	Detection of Plantwide Oscillations	A and B
19	Diagnosis of Plant-wide Oscillations	B

*Material of interest to

Category A	Category B
Senior undergraduate students	Graduate students
Undergraduate-level process control instructors	Graduate-level course instructors
Industrial practitioners	Researchers

Author Biographies

M. A. A. Shoukat Choudhury received his Bachelor and Master degree in Chemical Engineering from Bangladesh University of Engineering and Technology (BUET), Dhaka, Bangladesh. He obtained his Ph.D. in Process Control from the University of Alberta, Canada. For his outstanding research performance, he has been honored with several awards namely University of Alberta Ph.D. Dissertation Fellowship, Andrew Stewart Memorial Prize and ISA Educational Foundation Scholarship. He has published more than 25 refereed papers in international journals and conference proceedings in the area of control valve stiction, control performance assessment, condition based monitoring, data compression, and plantwide oscillations. He has an international patent on "Methods for Detection and Quantification of Control Valve Stiction". He is one of the leading experts in the data-based detection and quantification of control valve stiction. He applied his expertise extensively in industrial practice. His research contributions in control valve health monitoring have enjoyed wide applications in chemical, petrochemical, oil & gas, mineral processing, and pulp & paper industries throughout the world. He is currently working in the Department of Chemical Engineering, BUET. His main research interests include diagnosis of poor control performance, stiction in control valves, data compression, control loop performance assessment, condition-based monitoring, and troubleshooting plant wide oscillations.

Sirish Shah received his B.Sc. degree in Control Engineering from Leeds University in 1971, an M.Sc. degree in Automatic Control from UMIST, Manchester in 1972, and a Ph.D. degree in Process Control (Chemical Engineering) from the University of Alberta in 1976. During 1977, he worked as a computer applications engineer at Esso Chemicals in Sarnia, Ontario. Since 1978 he has been with the University of Alberta, where he currently holds the NSERC-Matrikon-Suncor-iCORE Senior Industrial Research Chair in Computer Process Control. He was the recipient of the Albright & Wilson Americas Award of the Canadian Society for Chemical Engineering (CSChE) in recognition of distinguished contributions to chemical engineering in 1989, the Killam Professor in 2003 and the D.G. Fisher Award of the CSChE for significant contributions in the field of systems and control in 2006. He has held visiting appointments at Oxford University and Balliol College as a SERC fellow in 1985–1986, at Kumamoto University (Japan) as a senior research fellow of the Japan Society for the Promotion of Science (JSPS) in 1994, at the University of Newcastle, Australia in 2004, at IIT-Madras India in 2006 and at the National University of Singapore in 2007. His work involves theoretical investigations as well as industrial applications. The main focus of his current research is process and performance monitoring, system identification, design and implementation of softsensors. He has held consulting appointments with a wide variety of process industries and has also taught many industrial courses.

Nina Thornhill received her B.A. in Physics at Oxford University, an M.Sc. degree in Control Systems from Imperial College London and a PhD from University College London (UCL). She worked for six years in the chemical industry and 23 years at UCL before being appointed to the ABB/RAEng Chair of Process Automation in the Department of Chemical Engineering at Imperial College London in 2007. In 2001, she joined the programme of Professor Sirish Shah's Industrial Research Chair in Process Control at the University of Alberta for eight months as an academic visitor, and in 2005 spent half a year of sabbatical leave with ABB Corporate Research in Norway. She has also held a 24-month Royal Society/EPSRC Industry Fellowship on secondment to BP. Her work includes time series analysis and signal processing, plant-wide controller performance assessment, and applications in oil and gas, chemicals and electricity supply. She finds great satisfaction in transferring academic research into useful industrial applications.

Contents

Part III Nonlinearity and Control Performance

Part IV Control Valve Stiction – Definition, Modelling, Detection and Quantification

Chapter 1
Introduction

Data-driven analysis of chemical processes is an area that has developed rapidly since chart recorders started to be replaced by data loggers and digital computers in the 1970s. The rapid development is due in part to the significant advances in computer hardware. An example of the pace of development is that in the last few years while we have been researching and writing this book, the memory available in a portable hard drive for PC data backup has increased by a factor of more than 20 while the physical dimensions and power consumption have stayed the same, and the cost has decreased. Readily available computing power and storage together with high quality and reliable measurements from process instruments have led to increasingly sophisticated and informative methods of analysis to convert data into insights about operations and to provide decision support for process control, operation and maintenance.

At the same time, the education of students and researchers is expanding and becoming more interdisciplinary. Time series analysis, statistical analysis and frequency domain concepts are crossing the boundaries from their traditional homes in electronic, mechanical and aerospace engineering into graduate courses in process operation and control. The sophistication of the methods being applied in data-driven analysis of processes is increasing both in the published academic literature and in commercial tools.

In this monograph, our goal is to present theory and applications for monitoring the performance of continuous processes in the chemical industry using process data. It brings together many results from the past few years in one place and also illustrates the ideas with a range of industrial case studies in which the methods have yielded valuable information. We hope to bring these exciting and useful ideas to the attention of students and researchers who will take them forward into their academic or industrial workplaces.

This chapter begins in Sect. 1.1 by outlining terminology and basic concepts that will underpin the ideas presented later. Section 1.2 gives a brief introduction to control valves and some of their performance issues, and the chapter concludes with a detailed overview of the ideas presented in the book's chapters.

M. A. A. S. Choudhury et al., *Diagnosis of Process Nonlinearities and Valve Stiction*,

1.1 Concepts in Data-Driven Analysis of Chemical Processes

Typical measurements in a continuous process comprise the outputs of instruments for monitoring basic process variables such as flows of liquid and gas, levels of the contents of tanks, pressures and temperatures and, more rarely, quantities related to product such as composition and quality. A business motivation for making measurements is to monitor throughput and profitability, which usually requires some compression of the measurements such as hourly or daily averages. A technical reason for monitoring is that many process variables are controlled by means of feedback controllers, which require accurate measurements from calibrated instruments that are sampled often. In between these extremes lies the area that is the subject of this book, the use of process measurements to monitor the performance of the process and control valves. Figure 1.1 shows the time series of three process measurements from a continuous stirred tank heater in the research laboratory of the University of Alberta. The three measurements are operating at a constant set point, but the dynamic features manifested in the deviations from the operating point are very different. It is the dynamic nature of such measurements that gives insight into underlying problems with the operation or control of the process.

With today's high levels of automation, each process control engineer at an industrial site may be responsible for plants containing hundreds of control valves, while maintenance staff are also thinly spread. Traditional diagnostic methods such as valve travel tests are time-consuming and disruptive to the economics of the process so it is beneficial if there is some indication of *which* valve to test. Without such an indication, the majority of these disruptive tests will be on valves that are perfectly healthy. The methods in this book achieve exactly this objective because they interpret the dynamic features in the measurements from routine operation to indicate where the causes of faults are likely to be found.

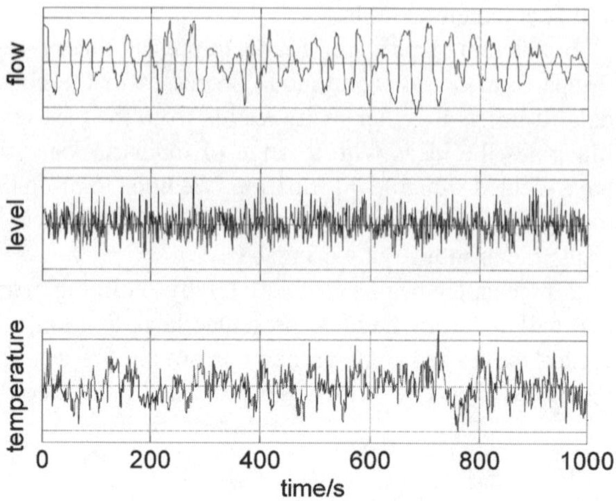

Fig. 1.1 Time series of three process measurements with different dynamic behaviour

1.1.1 Linear and Nonlinear Time Series

The measurements from a process instrument are in the form of a *time series*, a sequence of data samples ordered in time and normally taken at a regular sampling interval. If measurements are taken frequently enough, then the time series reflects the dynamics of the source that generated it, and the current value has a dependence on past values. For instance, if a temperature instrument in a stirred tank is sampled often, then it is possible to determine the thermal time constant of the tank because the time series will capture the first-order response of the temperature in response to a disturbance.

Many operational problems are due to process or control valve nonlinearities. A core theme of the book is that these types of problems are detectable via nonlinearity in the time series of the process measurements. A *nonlinear time series* is a time series generated by a dynamic nonlinear source. Put simply, the current value in a nonlinear time series has a nonlinear dependence on past values. The methods of bicoherence and surrogate data analysis are sensitive to nonlinearity in a time series and are effective tools for the detection of sources of nonlinearity in a plant.

This book deals extensively with the analysis of the time series of process measurements using concepts such as:

- the statistical distribution of the values of the measurement;
- the extent to which the samples are randomly related to each other, or whether there are time-delayed correlations between samples;
- nonlinearity in the time series;
- the frequencies represented in the time series.

1.1.2 Statistics and Randomness

A time series generally has a deterministic dynamic component and a random stochastic component. The values of the samples are characterized by their *probability distribution*, while correlations between one sample and another are characterized by the *autocovariance function*. Both these properties of a times series are used in later chapters.

The upper panel in Fig. 1.2 shows a random time series consisting of Gaussian white noise in which each sample is independent and not correlated with any other sample. The vertical axis is x, the value of the sample, and the horizontal axis is n, the sample number (i.e. it is a time domain plot for a sampled data sequence). The lower panel is a plot of the probability density function (PDF). It shows the frequency of occurrence of each value of x. The horizontal axis is the value of x and the vertical axis is the probability density function. A continuous line is shown representing the true distribution together with a bar chart derived from the values observed in the upper plot.

In the example of Fig. 1.2, the values of the measurements in the time series have a Gaussian distribution that is completely specified by its second-order statistics,

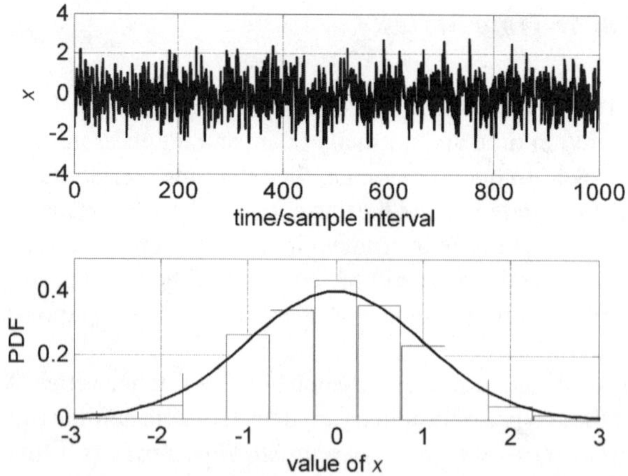

Fig. 1.2 The probability density function (PDF) of a Gaussian time series

i.e. the mean value and variance. There are many other PDFs besides Gaussian, however. For instance, the values in a random time series resulting from of a series of coin flips has a binary distribution because the result can only be heads or tails. Figure 1.3 shows the PDFs of the time series in Fig. 1.1 together with a Gaussian PDF that has the same second-order statistics. From a visual inspection, the level measurement has an approximately Gaussian distribution, while the temperature measurement is slightly flattened and slightly skewed towards smaller values. The distribution of the flow measurement has two peaks, a distribution known as the camel distribution. This distribution arises when a time series has oscillatory behaviour, because an oscillating signal spends more time at the peaks and valleys whereas the transition between the peaks and valleys is relatively fast.

The probability distribution of a nonlinear time series is generally not Gaussian and requires higher-order statistics to describe it fully, a fact which is exploited throughout this book for nonlinearity detection.

Autocovariance is a property of a time series that reveals time-delayed correlations and is given by the following expression:

$$m_2(\tau) = E\left(x(k)x(k+\tau)\right)$$

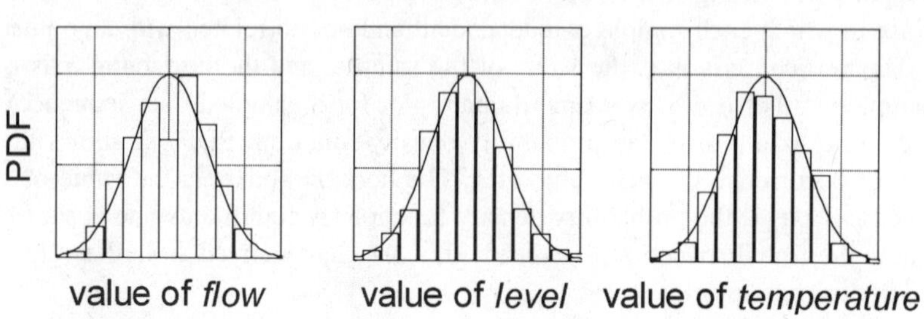

Fig. 1.3 PDFs of the times series shown in Fig. 1.1. The distribution is shown as a histogram and the solid line is a Gaussian distribution with the same mean and standard deviation

where E is the expectation operator and x is a mean-centred variable (i.e. $E(x) = 0$). The expression indicates that the autocovariance m_2 is a function of lag τ and it represents a time-delayed correlation. When $\tau = 0$, the autocovariance is equal to the variance of x. If there is no correlation, as is the case with white noise, then $m_2(\tau)$ is zero for $\tau \neq 0$. The autocovariance of an oscillating signal is also oscillatory with the same period of oscillation as the time series, and this feature is exploited in one of the oscillation-detection methods presented in Part V of the book. A concept related to autocovariance is *autocorrelation*, which applies to time series that have not had the mean value removed. Figure 1.4 shows the autocovariance functions of the time series from Fig. 1.1, where the oscillatory nature of the flow measurement can be clearly seen. The autocovariance at zero lag is equal to 1 in all cases because the time series were scaled to unit standard deviation before the autocovariance calculation. The level measurement has an autocovariance that indicates the noise in the level measurement is white because the autocovariance function is small for lags greater than zero. The temperature measurement has some time-delayed correlation that is different at different lags, which indicates that the temperature time series has some dynamic features.

Autocovariance is a second-order measure and is sufficient to specify fully the dynamic properties of a minimum-phase linear time series. Nonlinear time series, though, have more complicated behaviour which is investigated using higher-order correlations called moments. The third-order moment is:

$$m_3(\tau_1, \tau_2) = E(x(k)x(k+\tau_1)x(k+\tau_2))$$

Third-order moments and quantities called cumulants that are closely related to the moments are needed for the analysis of nonlinear time series and therefore have a prominent role in this book. Parts I and III of the book gives an exposition of higher-order statistics and show how they can be applied to process data.

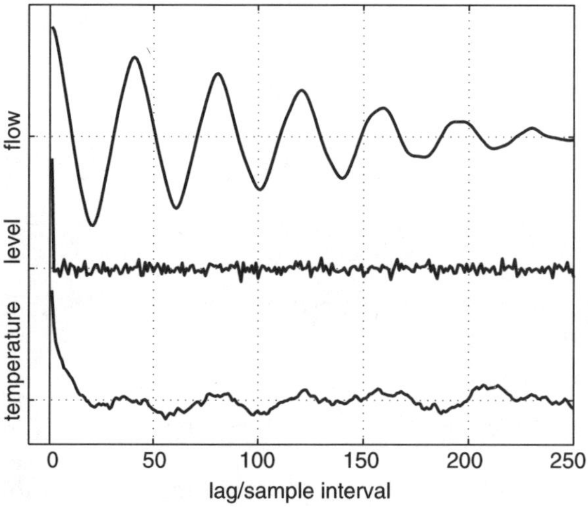

Fig. 1.4 Autocovariance function of the times series shown in Fig. 1.1

1.1.3 Frequency Content and Spectral Methods

The power spectrum is a transform of a time series that reveals its frequency content. Its usefulness in detection and diagnosis is that it provides a characterization of the dynamics of a linear time series because each peak in the spectrum indicates a frequency that is present in the time series. As an example, Fig. 1.5 shows the power spectra of the time series from Fig. 1.1. The power in the flow, level and temperature signals is the same because they are scaled to unit variance before computation of the power spectra. Therefore the total power, which is represented by the area of the power spectrum, is the same in each case. Almost all the power of the flow measurement is concentrated in a peak because the time series contains a narrow band of frequencies. The spectral power of the level measurement is broadband, with spectral content spread evenly across the whole frequency range. This time series has the characteristic of white noise, which has a flat frequency spectrum. The temperature measurement also has broadband spectral content but it is concentrated into a series of peaks towards the low frequency end of the spectrum with less power at high frequency. The ability of power spectra to distinguish between time series has been used extensively in this book. As is demonstrated in Part V, finding the measurement points where the measurements have similar spectra can help greatly in the detection of plant-wide disturbances.

The power spectrum can be determined from the discrete Fourier transform $(X(f))$ of the time series $x(k)$:

$$X(f) = \text{DFT}(x(k)) = \sum_{k=0}^{N-1} x(k)\, \mathrm{e}^{-j2\pi kf/N}$$

where N is the number of samples in the time series. The discrete Fourier transform (DFT) has N frequency channels and is a complex quantity that captures both the amplitude and phase of the frequency components within the time series. The power

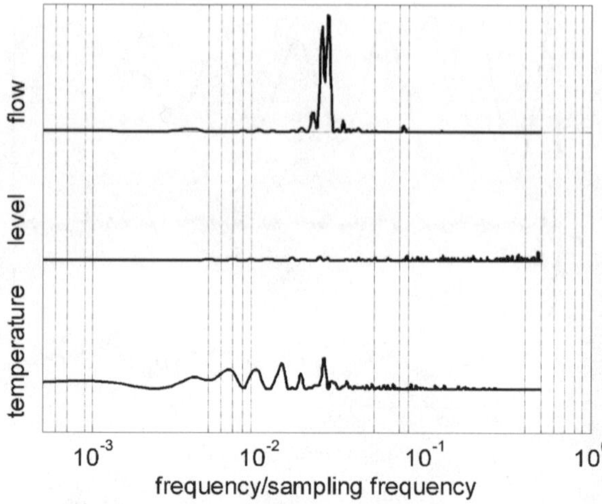

Fig. 1.5 Power spectra of the times series shown in Fig. 1.1

spectrum considers only the amplitudes and is given by $P(f) = |X(f)|^2$, where f specifies the frequency and $P(f)$ is the spectral power at that frequency.

In this book, we use a normalized frequency axis to describe spectral features. Rather than presenting the frequencies in Hertz, we scale the frequency axis by the sampling frequency, f_s. The relationship of the spectral features to the time series then becomes straightforward. For instance, if the power spectrum has a spectral feature at $f/f_s = 0.1$, it means the time series contains a sinusoidal signal whose frequency is one tenth of the sampling frequency. This can be stated in the time domain as a sinusoidal feature having ten samples per cycle.

The point $f/f_s = 0.5$ on the normalized frequency axis is the Nyquist sampling frequency corresponding to a frequency component with two samples per cycle. There is no additional spectral information above the Nyquist frequency because of the phenomenon of aliasing, which means that the power spectrum above the Nyquist frequency is a mirror image of the spectrum below the Nyquist frequency. The reason for this can be seen in Fig. 1.6. The solid line shows a sine wave signal sampled at fewer than two samples per cycle, and the dashed line shows that the same samples also represent a sine wave with a longer period of oscillation and whose frequency is below the Nyquist frequency. The output from a physical instrument should really be filtered before sampling to remove high-frequency content that might be aliased in the manner shown in Fig. 1.6, but in process applications this is rarely necessary because the sampling rates are fast enough by design to capture process effects.

The power spectrum is related to the autocovariance function by the *Wiener–Khinchin theorem*, which states that the power spectrum $P(f)$ is the Fourier transform of the autocovariance function:

$$P(f) = \mathrm{DFT}(m_2(\tau)) = \sum_{\tau=0}^{N-1} m_2(\tau)\, \mathrm{e}^{-j2\pi\tau f/N}$$

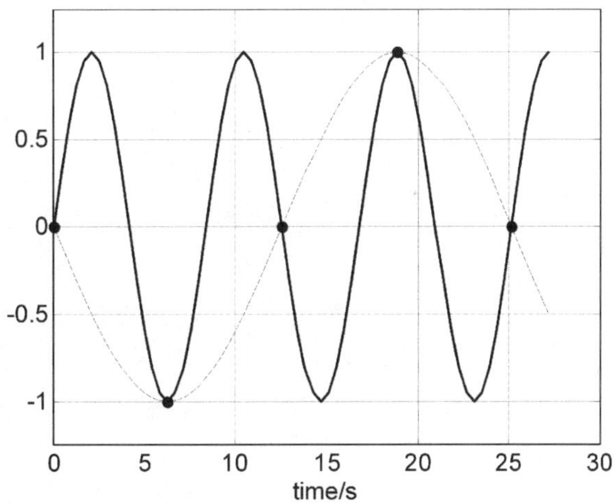

Fig. 1.6 Demonstration of aliasing in an inadequately sampled waveform

This relationship between autocovariance and the power spectrum shows that the power spectrum is also only a second-order description of a time series, and therefore that a generalization is needed for analysis of nonlinear time series. The generalization, which is discussed in detail in Part I of the book, is the bispectrum that is derived from the DFT of the third-order cumulant, a quantity closely related to the third-order moment. The bispectrum of a nonlinear time series shows distinctive, quantifiable features, and these are exploited throughout the book for nonlinearity detection.

1.2 Nonlinearity in Control Valves

A significant theme of this book is nonlinearity in control valves, and Part IV, the most substantial section in the book, is devoted to this topic. The control valve is the actuator for most process control loops and, as the only moving part in the loop, its function is to implement the control action. If the control valve malfunctions, the performance of the control loop is likely to deteriorate, no matter how good the controller is. Commonly encountered control valve problems include nonlinear responses to the demand signal caused by effects such as stiction, deadband or saturation. Because of these problems, the control loop may be oscillatory, which in turn may cause oscillations in many process variables causing a range of operational problems including increased valve wear. Figure 1.7 shows the controlled variable, controller output and flow through the valve in a feedback loop that contains a valve with a deadband, where the controlled variable has considerable dynamic deviations from the set point.

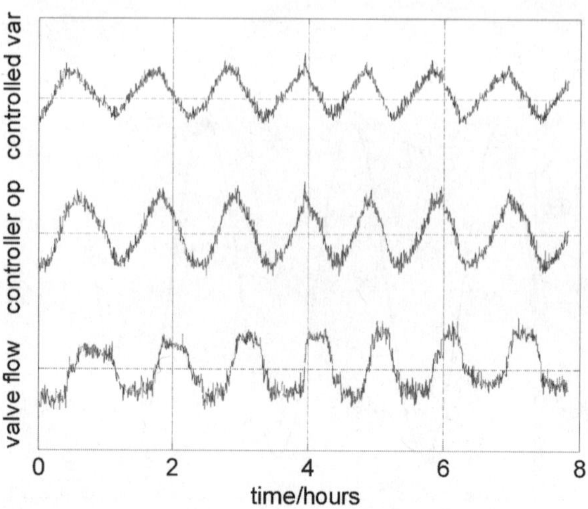

Fig. 1.7 Deviations in variables in a feedback loop containing a valve with a deadband

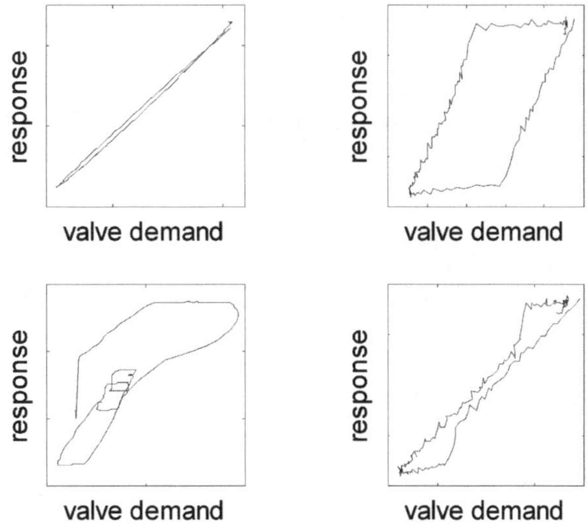

Fig. 1.8 Input–output plots for some real valves

Nonlinearity in control valves is a long-standing and common problem in the process industry. Characteristic patterns in an input–output plot for a control valve show a variety of forms depending upon the nature of the nonlinearity. Figure 1.8 shows plots from some real valves of the valve demand signal (input) versus the valve response (output), in which the plot is a trajectory traced out as time passes. Some of them are far from ideal. In a feedback control loop, the valve demand is the signal from the controller that is sent to the valve, while the response is the flow through the valve. The top left-hand panel shows an ideal linear valve that moves in an almost linear manner and where the travel on closing is the reverse of path on opening. The other panels show behaviour that is highly nonlinear including deadband, where the valve does not move in response to a changing demand, and stiction where the valve jumps once it starts to move. Valve stiction is one of the main causes of oscillation in a control loop.

Part IV of this book gives a detailed coverage of the detection and quantification of stiction in a control valve, with the constraint that the methods should be minimally invasive and based largely on measurements from routine operation. The concepts used include

- mathematical analysis and modelling of control valve nonlinearity;
- mathematical analysis of the impact of control valve nonlinearity in a feedback control loop;
- detecting nonlinearity in the time series of the controlled variable;
- estimation of the extent of the stiction effect from operating data.

Through regular performance monitoring and preventative maintenance of the control valves in many cases, it is possible to avoid unscheduled shutdown of the plant. Every unscheduled shut-off or downtime of a plant takes a costly toll on the maintenance budget of the plant.

The last part of this book, Part V, deals with the detection and diagnosis of plant-wide oscillations. This is not an uncommon problem in modern plants that are highly heat integrated for efficient energy utilization. Oscillations may originate from one source and can quickly propagate downstream as well as upstream, for example due to recycles streams and control loop interactions. The emphasis in these chapters is on

- Spectral methods for identification of all process variables that are oscillating at the same frequency;
- Diagnosis of the root cause variables of plant-wide oscillatons.

1.3 The Layout of the Book

This book is divided into five parts. Most of the sections are independent of each other and can be read in any order according to the reader's preference, or the suggestions of a course instructor. A summary of each part of the book follows:

1.3.1 Part I Higher-Order Statistics

Part I introduces the topic of Higher-Order Statistics (HOS) and motivates their use in the detection and diagnosis of process nonlinearities and valve problems. There are many situations where processes deviate from Gaussianity and linearity, and these type of processes can conveniently be studied using higher order statistics to detect and quantify nonlinearities. First- and second-order statistics include the mean, variance, autocovariance and power spectrum, but analysis of the time series from nonlinear processes is greatly enhanced by the use of the bispectrum, which is the frequency domain representation of the third-order cumulant or moment. The bispectrum and its squared bicoherence function have played a central role in this book because bicoherence is used to examine the presence of nonlinearities in a time series or a signal-generating system.

1.3.2 Part II Data Quality – Compression and Quantization

Part II of the book considers issues of Data Quality. A key requirement of any data-driven method of process performance analysis is the availability of high fidelity data that are sampled sufficiently fast to capture process dynamics. A manifestation of increasing process automation is extensive data logging and process archiving, and historical data can be an invaluable source of information. But the usefulness of historical data is degraded if the data are compressed or manipulated by the data historian. This section of the book illustrates the problems of data quality degradation and shows that compression induces spurious nonlinearity. It also presents an automatic method for detecting and quantifying the degree of compression present in the archived data.

Data quantization, for instance, due to limited resolution in the analogue to digital conversion stage of the instrumentation electronics also has an adverse effect on the data. The section also provides an automated procedure to detect and quantify quantization. If an archived historical data set fails one or the other of the compression and quantization tests, then it is not qualified for analysis by the methods presented in the subsequent sections of this book because it may give spurious or misleading results.

1.3.3 Part III Nonlinearity and Control Performance

Part III, which is presented over five chapters, concerns Nonlinearity and Control Performance. Process nonlinearity poses challenges for process control and can affect the control performance adversely by changing the dynamics of a process at different operating point and rendering the controller tuning non-optimal. Nonlinearities in valves, particularly friction effects such as stiction, may cause oscillations in process variables that can propagate plant wide.

Chapter 5 discusses metrics of nonlinearity, both model- and data-based, where the data-based methods can be classified under two main headings. One approach uses the bicoherence methods that were introduced in Part I, and the other method considered in this book is nonlinear time series analysis based on surrogate data. Chapters 6 and 7 demonstrate the formulation and application of nonlinearity tests based on bicoherence and nonlinear time series analysis. Chapter 6 proposes a total nonlinearity index (TNLI) to quantify nonlinearities together with other tests based on HOS, specifically a non-Gaussianity index (NGI) and a Nonlinearity index (NLI). These indices can be applied to a time series signal and their efficacy is illustrated with worked examples. Chapter 7 describes an algorithm using surrogate data to determine the presence and severity of nonlinearity in the time trends of measurements from a process and demonstrates results of its use in two industrial examples having plant-wide disturbances.

Process and control valve nonlinearities are the subject of Chap. 8. It uses the TNLI from Chap. 6 to quantify nonlinearity in time series data from examples such as the level control of a spherical tank and a nonlinear stirred tank reactor. Control valves with nonlinear opening characteristics are also analysed. The TNLI gives a precise evaluation of the range of deviations from the operating point where the process can safely be considered linear. If normal operation of the process involves excursions larger than the range indicated by TNLI, then it may be necessary to utilize a nonlinear controller.

Chapter 9 applies the NGI and NLI indices from Chap. 6 to the diagnosis of poor control-loop performance. These, together with the specific patterns in plots of process output versus the controller output, have been successfully applied to industrial data sets, which are presented in this chapter. In all cases, the results of the analysis were confirmed by plant engineers at the collaborating companies.

1.3.4 Part IV Control Valve Stiction – Definition, Modelling, Detection and Quantification

Part IV gives a substantial and in-depth discussion over eight chapters of Control Valve Stiction and an analysis of friction in control valves. Stiction is the most common problem in spring-diaphragm-type valves, which are widely used in the process industry. Part IV covers physical and empirical modelling, analysis of the dynamic behaviour of process control loops when stiction is present in the control valve, as well as methods for detecting and quantifying stiction from closed-loop operating data.

Chapter 10 describes the structure of a control valve and explains typical common faults including deadband, hysteresis and backlash. These terms are used inconsistently in the published literature, and there have also been many attempts to understand and define and model the stiction phenomenon. Readers are reminded of the ANSI definitions (from ANSI: ISA-S51.1-1979, Process Instrumentation Terminology) of nonlinearities such as deadband, hysteresis and backlash in an attempt to bring some discipline to the usage of these terms, but to date there is no ANSI definition of stiction. Some practical examples of valve stiction from real-world data are provided to gain insights into the underlying mechanisms. Based on an understanding of the mechanism that causes stiction, Chap. 11 makes a proposal of a new formal definition of stiction in an attempt to fill a much-needed gap.

Chapters 12 and 13 present two models of stiction in control valves. One is physics-based and the other, which has the same dynamic behaviour, is an empirical data-driven model. The physical model is based on Newton's second law and a model of friction forces that captures both Coulomb friction and a static friction that is greater than the moving friction. This model successfully reproduces closed-loop limit cycles observed in industrial process control loops. Although the physical model fulfils the purpose of giving in-depth understanding of the origin of deadband and stiction effects in a control valve, its practical usefulness is limited because the physical parameters such as friction forces and spring constant are not generally known. The data-driven empirical model of Chap. 13 is a means of overcoming this disadvantage. The focus of this chapter is a two-parameter data-driven model of valve stiction, which has been validated using the simulation results of the physics-based stiction model. It successfully duplicates the phenomena seen in control loops with sticking control valves. The major advantage is that the input and output of this model are directly related to measured signals in a control loop, so the model can be easily used to perform simulation in order to develop methodologies for stiction compensation, oscillation diagnosis and loop interaction diagnosis.

The analytical approach to control valve stiction is continued in Chap. 14, which studies limit cycles in feedback control loops using the describing function of the stiction nonlinearity. The aim is to give insights into when a limit cycle will arise and when it will not, and the chapter also provides some understanding of how the magnitude and frequency of the limit cycle are influenced by tuning of the controller.

Chapter 15 meets the challenge of detecting the presence of stiction and quantifying the amount of stiction in a faulty control valve. Invasive methods such as the valve travel test can detect stiction, but are expensive and tedious to apply to the hundreds of valves present in a typical plant. The goal is to detect and also quantify stiction using routine operating data obtained from the process so that the valves in need of maintenance can be identified, isolated and repaired. Industrial case studies have demonstrated the wide applicability and practicality of the developed method. It has been extensively evaluated on simulated as well as industrial data sets and is now in commercial use.

The final two chapters of Part IV present industrial case studies. Chapter 16 evaluates the stiction detection and quantification algorithm from Chap. 15 on a variety of control loops obtained from different types of process industries. If stiction is suspected, then process control engineers would normally like to conduct further confirmatory tests on the suspect valve. An approach that is somewhat less invasive than valve travel testing is presented and analysed using the theoretical insights that were developed in Chap. 14. If a control loop is oscillating, then the test confirms whether or not the oscillation is due to stiction in the control valve. The test is based on observations that changing the controller gain also changes the frequency of the limit cycle oscillations if these have been induced in the control loop because of valve stiction. The method has been validated by describing function analysis and a simulation that gave an accurate prediction of the changes observed in an industrial example.

1.3.5 Part V Plant-wide Oscillations – Detection and Diagnosis

Part V of the book comprises two chapters exploring the impact of process and control valve nonlinearities in causing plant-wide oscillations. It has been widely reported that oscillations caused by nonlinear components in a control loop can propagate through a plant with resulting increases in the variability of the process variables causing inferior quality product, larger rejection rates, increased energy consumption and reduced average throughput. Chapter 18 reviews a number of recently developed methods that can detect plant-wide oscillations.

Chapter 19 discusses a systematic procedure for the detection and diagnosis of plant-wide oscillations and demonstrates its efficacy through several industrial case studies that have appeared in the published literature over the past few years. Oscillation detection methods from Chap. 18 have been used for the detection and grouping of process measurements having common oscillations, which are then diagnosed to find and isolate root causes. In cases where the reason for oscillation is suspected to be nonlinearity, the nonlinearity measures from Part III were applied to detect the probable candidates for the root cause.

1.3.6 References

The book concludes with a comprehensive list of references to articles from the archival literature, books, relevant Standards and useful Internet resources.

1.4 Summary

This introductory chapter has explained the motivation for the use of data-driven methods for detection and diagnosis of process and control valve nonlinearities.

It first presented fundamental concepts from time series analysis that are the basis for the methods presented in later chapters, including the notions of nonlinear time series and higher-order statistical moments. It went on to describe the central role of control valves in a chemical process and to discuss the disruption to smooth running of a process that occurs when a control valve fails to respond properly to the valve demand signal sent by the controller, a situation that is unfortunately all too common. Deadband and stiction are two serious valve nonlinearities that cause poor performance of a control valve. The challenge that is going to be addressed throughout the later chapters is how to detect and quantify control valve problems from routine operating data so as to minimize disruptions to the process due to interventionist procedures such as valve travel tests.

The final section of the introduction gave a walk-through of the topics in the book, which has 19 chapters distributed over five parts. The contents include theory and algorithms together with case studies, which show how the methods have worked with real data from real processes to provide useful insights leading to real process improvements.

Part I
Higher-Order Statistics

Chapter 2
Higher-Order Statistics: Preliminaries

This chapter presents a tutorial introduction to higher-order statistics (HOS) and its relation to conventional second-order statistics (SOS). The first- and second-order statistics, for example, mean, variance, autocorrelation and power spectrum, are popular signal-processing tools and are used extensively in data analysis. Such second-order statistics are sufficient for describing linear and Gaussian processes. In practice, there are many situations where the processes deviate from Gaussianity and linearity. These type of processes can conveniently be studied using Higher-Order Statistics (HOS). There are three main reasons for using HOS: to extract information due to deviations from Gaussianity, to recover the true phase character of the signals and to detect and quantify nonlinearities in time series (Nikias and Petropulu, 1993).

2.1 Introduction

A signal is a variable that carries some kind of information that can be conveyed, displayed or manipulated. Signals can be obtained from numerous sources – for example, speech, sound, music, images, chemical processes, radar, seismic surveys, computer simulations and robots. Useful information is often hidden in a signal. The objective in a signal-processing problem is to process a signal or a finite number of data samples to extract important information that may be hidden therein (Smith, 1998; Ifeachor and Jervis, 1993). In most cases, data are available as variables sampled at regular intervals of time. Digital signal processing continues to play an important role in the industrial applications of high technology. Digital signal processing – combined with the ideas and methodologies from various branches of engineering (computer, chemical, electrical and electronics engineering), statistics, numerical analysis and computer science – has become an indispensable tool for data analysis and is being used in many industrial processes. A digital signal can be analysed using either time domain or frequency domain techniques.

2.2 Time Domain Analysis

This section covers the background of moment generating function because the moments play a key role in time domain statistical data analysis. Frequency domain data analysis has also been developed based on the time domain moments. Cumulants are just another representation of moments. Since the book deals with statistical data analysis, many of the ideas developed here will be used in the subsequent chapters of this book.

A time series is a signal usually obtained by sampling a variable in equal intervals of time. Many statistical measures (e.g. moments, cumulants, autocorrelation, cross-correlation) have been developed to measure the temporal characteristics of time series signal. Data for most of the signals are collected as samples at regular intervals of time. In statistical analysis, it is often assumed that the time series or the signal is stationary. This assumption holds for subsequent definitions and analyses.

2.2.1 *Moments*

A random variable is a quantity whose values are random and to which a probability distribution is assigned. Formally, a random variable is a measurable function from a sample space to the measurable space of possible values of the variable.

A generating function of a random variable is an expected value of a certain transformation of the variable. All generating functions have three important properties:

1. The generating function completely determines the distribution.
2. The generating function of a sum of two independent variables is the product of the independent generating functions.
3. The moments of the random variable distribution can be obtained from the derivatives of the generating function.

As the term implies, the moment generating function should be able to generate all moments. For any random variable x, the moment generating function can be defined as the expectation of the transformation, e^{tx}, where $t \in R$, i.e.

$$M_x(t) = E[\mathrm{e}^{tx}] \tag{2.1}$$

Moments can be obtained from the coefficients of the Taylor's series expansion of the moment generating function about the origin,

$$M_x(t) = M_x(t)\mid_{t=0} + \left.\frac{\partial M_x(t)}{\partial t}\right|_{t=0}(t-0) + \frac{1}{2!}\left.\frac{\partial^2 M_x(t)}{\partial t^2}\right|_{t=0}(t-0)^2 + \frac{1}{3!}\left.\frac{\partial^3 M_x(t)}{\partial t^3}\right|_{t=0}(t-0)^3 + \cdots$$

$$= M_x(t)\,|_{t=0} + \left.\frac{\partial M_x(t)}{\partial t}\right|_{t=0} t + \frac{1}{2!}\left.\frac{\partial^2 M_x(t)}{\partial t^2}\right|_{t=0} t^2 + \frac{1}{3!}\left.\frac{\partial^3 M_x(t)}{\partial t^3}\right|_{t=0} t^3 + \cdots \tag{2.2}$$

From the right side of the Eq. (2.2), the first derivative gives the first-order moment:

$$\begin{aligned} m_1 &= \left.\frac{\partial M_x(t)}{\partial t}\right|_{t=0} \\ &= \left.\frac{\partial E[\mathrm{e}^{tx}]}{\partial t}\right|_{t=0} \\ &= E[x\mathrm{e}^{tx}]\,|_{t=0} \\ &= E[x] \end{aligned} \tag{2.3}$$

The second derivative gives the second-order moment:

$$m_2 = \left.\frac{\partial^2 M_x(t)}{\partial t^2}\right|_{t=0} = \left.\frac{\partial^2 E[\mathrm{e}^{tx}]}{\partial t^2}\right|_{t=0} = E[x^2\mathrm{e}^{tx}]\,|_{t=0} = E[x^2] \tag{2.4}$$

Similarly, the third derivative gives the third-order moment:

$$m_3 = E[x^3] \tag{2.5}$$

and so on. Therefore, we can rewrite the moment generating function as

$$\begin{aligned} M_x(t) &= E[\mathrm{e}^{tx}] \\ &= E[1 + tx + \frac{t^2}{2!}x^2 + \frac{t^3}{3!}x^3 + \cdots] \\ &= 1 + tE[x] + \frac{t^2}{2!}E[x^2] + \frac{t^3}{3!}E[x^3] + \cdots \\ &= 1 + tm_1 + \frac{t^2}{2!}m_2 + \frac{t^3}{3!}m_3 + \cdots \end{aligned} \tag{2.6}$$

The first-order moment is the mean ($\mu = E(x)$) of the data series, $x(k)$, which provides a measure of the location or the centre of gravity of the probability density function (PDF) for an ergodic signal. The second-order moment is the variance of the data series and gives the spread of the PDF, while the third-order moment provides a measure of the skewness of the distribution and the fourth-order moment provides a measure of the flatness of the distribution. Alternatively, for any zero-mean data series, $x(k)$, the 2nd, 3rd, ..., nth order moment of the data series can be defined as:

$$
\begin{aligned}
m_2(\tau_1) &\triangleq E[x(k)x(k+\tau_1)] \\
m_3(\tau_1,\tau_2) &\triangleq E[x(k)x(k+\tau_1)x(k+\tau_2)] \\
\vdots &\quad \vdots \\
m_n(\tau_1,\tau_2,\cdots\tau_{n-1}) &\triangleq E[x(k)x(k+\tau_1)x(k+\tau_2)\cdots x(k+\tau_{n-1})]
\end{aligned} \tag{2.7}
$$

The variance, σ^2, can be obtained from $m_2(0)$; skewness from $m_3(0,0)$ and so on.

2.2.2 Cumulants

Cumulants are another set of statistical measures that can be used instead of moments because of their excellent noise-suppressing properties. The cumulant generating function is defined as the logarithm of the moment generating function. That is, for a random variable x, the cumulant generating function is

$$
C_x(t) \triangleq \ln(M_x(t)) \tag{2.8}
$$

Just as moments are derived from the Taylor's series expansion of the moment generating function, cumulants can be derived from the Taylor's series expansion of the cumulant generating function:

$$
\begin{aligned}
C_x(t) &= C_x(t)\mid_{t=0} + \left.\frac{\partial C_x(t)}{\partial t}\right|_{t=0}(t-0) + \frac{1}{2!}\left.\frac{\partial^2 C_x(t)}{\partial t^2}\right|_{t=0}(t-0)^2 \\
&\quad + \frac{1}{3!}\left.\frac{\partial^3 C_x(t)}{\partial t^3}\right|_{t=0}(t-0)^3 + \cdots \\
&= C_x(t)\mid_{t=0} + \left.\frac{\partial C_x(t)}{\partial t}\right|_{t=0} t + \frac{1}{2!}\left.\frac{\partial^2 C_x(t)}{\partial t^2}\right|_{t=0} t^2 \\
&\quad + \frac{1}{3!}\left.\frac{\partial^3 C_x(t)}{\partial t^3}\right|_{t=0} t^3 + \cdots
\end{aligned} \tag{2.9}
$$

The first-order cumulant is given by:

$$
\begin{aligned}
c_1 &= \left.\frac{\partial C_x(t)}{\partial t}\right|_{t=0} \\
&= \left.\frac{\partial \ln[E[e^{tx}]]}{\partial t}\right|_{t=0} \\
&= \left.\frac{\partial \ln[1 + tm_1 + \frac{t^2}{2!}m_2 + \frac{t^3}{3!}m_3 + \cdots]}{\partial t}\right|_{t=0} \\
&= \left.\frac{1}{[1 + tm_1 + \frac{t^2}{2!}m_2 + \frac{t^3}{3!}m_3 + \cdots]}[m_1 + \frac{2t}{2!}m_2 + \frac{3t^2}{3!}m_3 + \cdots]\right|_{t=0} \\
&= m_1
\end{aligned} \tag{2.10}
$$

The second-order cumulant is:

$$
\begin{aligned}
c_2 &= \left.\frac{\partial^2 C_x(t)}{\partial t^2}\right|_{t=0} \\
&= \left.\frac{\partial^2 \ln[E[\mathrm{e}^{tx}]]}{\partial t^2}\right|_{t=0} \\
&= \left.\frac{\partial^2 \ln[1+tm_1+\frac{t^2}{2!}m_2+\frac{t^3}{3!}m_3+\cdots]}{\partial t^2}\right|_{t=0} \\
&= \left.\frac{\partial}{\partial t}\left[\frac{1}{[1+tm_1+\frac{t^2}{2!}m_2+\frac{t^3}{3!}m_3+\cdots]}[m_1+\frac{2t}{2!}m_2+\frac{3t^2}{3!}m_3+\cdots]\right]\right|_{t=0} \\
&= \left.\left[\frac{-1}{[1+tm_1+\frac{t^2}{2!}m_2+\frac{t^3}{3!}m_3+\cdots]^2}[m_1+\frac{2t}{2!}m_2+\frac{3t^2}{3!}m_3+\cdots]^2\right]\right|_{t=0} \\
&\quad + \left.\left[\frac{1}{[1+tm_1+\frac{t^2}{2!}m_2+\frac{t^3}{3!}m_3+\cdots]}[0+m_2+\frac{6t}{3!}m_3+\cdots]\right]\right|_{t=0} \\
&= -m_1^2+m_2 \\
&= m_2-m_1^2
\end{aligned}
\tag{2.11}
$$

Similarly, the third-order cumulant is:

$$
\begin{aligned}
c_3 &= \left.\frac{\partial^3 C_x(t)}{\partial t^3}\right|_{t=0} \\
&= m_3-3m_2m_1+2m_1^3
\end{aligned}
\tag{2.12}
$$

The fourth-order cumulant

$$
\begin{aligned}
c_4 &= \left.\frac{\partial^4 C_x(t)}{\partial t^4}\right|_{t=0} \\
&= m_4-4m_3m_1-3m_2^2+12m_2m_1^2-6m_1^4
\end{aligned}
\tag{2.13}
$$

Note that all of the above cumulants are about the origin. For any other point (τ), the 2nd, 3rd, ..., nth order cumulant of any random data series, $x(k)$, can be rewritten as:

$$
\begin{aligned}
c_2(\tau_1) &\triangleq \mathrm{cum}[x(k),x(k+\tau_1)] \\
c_3(\tau_1,\tau_2) &\triangleq \mathrm{cum}[x(k),x(k+\tau_1),x(k+\tau_2)] \\
\vdots\quad &\quad \vdots \\
c_n(\tau_1,\tau_2,\cdots\tau_{n-1}) &\triangleq \mathrm{cum}[x(k),x(k+\tau_1),x(k+\tau_2),\ldots,x(k+\tau_{n-1})]
\end{aligned}
$$

where cum represents cumulant. It is important to note that cumulants are closely related to moments (e.g. $c_2 = m_2 - m_1^2$) and can be calculated from the knowledge of moments. The computation of cumulants of order n requires all moments up to order n.

2.2.3 The Relationship Between Moments and Cumulants

Given a set of n random variables, $\{x_1, x_2, \ldots x_n\}$, the general relationship between their joint moments and joint cumulants of order n is given by:

$$\text{cum}[x_1, x_2, \ldots, x_n] = \sum_{p=1}^{n} (-1)^{p-1} \ (p-1)! \ E\{\prod_{i \in s_1} x_i\} \ E\{\prod_{i \in s_2} x_i\} \cdots \ E\{\prod_{i \in s_p} x_i\} \tag{2.14}$$

where the summation extends over all partitions $(s_1, s_2, \ldots, s_p)$; $p = 1, 2, \ldots, n$, of the set of integers $(1, 2, \ldots, n)$ (Rosenblatt and Van Ness, 1965). For example, for the third-order cumulant, the set of integers $(1, 2, 3)$ can be partitioned into

$p = 1$	$s_1 = \{1, 2, 3\}$		
$p = 2$	$s_1 = \{1\}$	$s_2 = \{2, 3\}$	
	$s_1 = \{2\}$	$s_2 = \{1, 3\}$	
	$s_1 = \{3\}$	$s_2 = \{1, 2\}$	
$p = 3$	$s_1 = \{1\}$	$s_2 = \{2\}$	$s_3 = \{3\}$

Therefore, using Eq. (2.14) we can write,

$$\begin{aligned}\text{cum}[x_1, x_2, x_3] &= E\{x_1x_2x_3\} - E\{x_1\}E\{x_2x_3\} - E\{x_2\}E\{x_1x_3\}\cdots \\ &\quad - E\{x_3\}E\{x_1x_2\} + 2E\{x_1\}E\{x_2\}E\{x_3\}\end{aligned} \tag{2.15}$$

If $x_2 = x_3 = x_1 = x$, then

$$\begin{aligned}\text{cum}[x, x, x] &= E\{x^3\} - 3E\{x\}E\{x^2\} + 2\{E\{x\}\}^3 \\ &= m_3 - 3m_1m_2 + 2m_1^3\end{aligned} \tag{2.16}$$

which is the same as the expression for c_3 obtained earlier.

Note that the computation of joint cumulant of order r requires the knowledge of all moments up to order r.

2.2.4 Properties of Moments and Cumulants

1. $\text{mom}[a_1x_1, a_2x_2, \ldots, a_nx_n] = (a_1a_2\ldots,a_n) \ \text{mom}[x_1, x_2, \ldots, x_n]$, and $\text{cum}[a_1x_1, a_2x_2, \ldots, a_nx_n] = (a_1a_2\ldots a_n) \ \text{cum}[x_1, x_2,\ldots,x_n]$, where $(a_1, a_2, \ldots, a_n)$ are constants and *mom* represents moment.

2. Moments and cumulants are symmetric functions in their arguments, e.g. mom $[x_1, x_2, x_3] = \text{mom}[x_1, x_3, x_2] = \text{mom}[x_3, x_2, x_1]$ and so on.
3. If the random variables $(x_1, x_2, \ldots, x_n)$ can be divided into two or more statistically independent groups, then their nth order cumulant is identically zero, i.e. $\text{cum}[x_1, x_2, \ldots, x_n] = 0$, whereas, in general, $\text{mom}[x_1, x_2, \ldots, x_n] \neq 0$.
4. If the sets of random variables $(x_1, x_2, \ldots, x_n)$ and $(y_1, y_2, \ldots, y_n)$ are independent, then $\text{cum}[x_1+y_1, x_2+y_2, \ldots, x_n+y_n] = \text{cum}[(x_1, x_2, \ldots, x_n)] + \text{cum}[y_1, y_2, \ldots, y_n]$, but in case of moments, $\text{mom}[x_1+y_1, x_2+y_2, \ldots, x_n+y_n] \neq \text{mom}[(x_1, x_2, \ldots, x_n)] + \text{mom}[y_1, y_2, \ldots, y_n]$.
5. If the set of random variables $(x_1, x_2, \ldots, x_n)$ is jointly Gaussian, then all information about their distribution is contained in the moments of order $n \leq 2$. Therefore, all moments of order greater than two have no additional information. This leads to the fact that all joint cumulants of order greater than two are identical to zero for Gaussian random vectors.
   The above properties indicate that the cumulants of order greater than two measure the non-Gaussian nature of a time series. This property will be exploited in Sect. 6.3 to develop a practical method for identifying non-Gaussian and nonlinear time series in chemical processes.

Properties 4 and 5 demonstrate the ability of the cumulants to suppress noise in signal processing when the noise is additive Gaussian.

Example 1. Consider the random variables, $z_i = y_i + x_i$, for $i = 1, 2, 3$. If the joint probability density function of (y_1, y_2, y_3) is non-Gaussian and (x_1, x_2, x_3) is jointly Gaussian and independent from (y_1, y_2, y_3) and also $E\{y_i\} \neq 0, E\{x_i\} \neq 0$, then using the fourth and fifth properties of cumulants, we get

$$\text{cum}[(z_1, z_2, z_3)] = \text{cum}[(y_1, y_2, y_3)] \tag{2.17}$$

since $\text{cum}[(x_1, x_2, x_3)] = 0$. In contrast, this is not true for moments. Therefore, cumulants of order greater than two can be conveniently used for analysing signals that contain additive white noise.

Example 2. Consider the random variable, $y = x + e$, where y is a measured signal, x is the true signal and e is the noise, which is independent of x.

Moments: First-order moment of y,

$$m_1 = E[y] = E[x] + E[e] \tag{2.18}$$

Second-order moment of y,

$$m_2 = E[y^2] = E[(x+e)^2] = E[x^2] + 2E[x]\,E[e] + E[e^2] \tag{2.19}$$

Third-order moment of y,

$$\begin{aligned} m_3 &= E\{y^3\} \\ &= E\{(x+e)^3\} \\ &= E\{x^3\} + 3E\{x^2\}E\{e\} + 3E\{x\}E\{e^2\} + E\{e^3\} \end{aligned}$$

$$= \text{mom}[x,x,x] + 3\,\text{mom}[x,x]\ \text{mom}[e] \\ +3\,\text{mom}[x]\ \text{mom}[e,e] + \text{mom}[e,e,e] \tag{2.20}$$

Cumulants: First-order cumulant of y,

$$\text{cum}[y] = E\{y\} = E[x] + E[e] \tag{2.21}$$

Second-order cumulant of y,

$$\begin{aligned} \text{cum}[y,\ y] &= E[y^2] - \{E[y]\}^2 \\ &= E[(x+e)^2] - \{E[x] + E[e]\}^2 \\ &= E[x^2] + 2E[x]\ E[e] + E[e^2] - \{E[x]\}^2 - 2E[x]\ E[e] - \{E[e]\}^2 \\ &= E[x^2] + E[e^2] - \{E[x]\}^2 - \{E[e]\}^2 \end{aligned} \tag{2.22}$$

Third-order cumulant of y,

$$\begin{aligned} \text{cum}[y,\ y,\ y] &= E\{y^3\} - 3E\{y\}E\{y^2\} + 2[E\{y\}]^3 \\ &= E\{(x+e)^3\} - 3E\{(x+e)\}E\{(x+e)^2\} + 2\,[E\{(x+e\}]^3 \\ &= E\{x^3\} - 3E\{x^2\}E\{x\} + 2[E\{x\}]^3 \\ &\quad + E\{e^3\} - 3E\{e^2\}E\{e\} + 2[E\{e\}]^3 \\ &= \text{cum}[x,x,x] + \text{cum}[e,e,e] \end{aligned} \tag{2.23}$$

If the noise is zero-mean and Gaussian distributed with the following properties:

$$\begin{aligned} &\text{the first moment of } e: \ E\{e\} = 0 \\ &\text{the second moment of } e: E\{e^2\} = \sigma_e^2 \\ &\text{the third moment of } e: \ E\{e^3\} = 0 \end{aligned} \tag{2.24}$$

First-order moment of y,

$$m_1 = E[y] = E[x] \tag{2.25}$$

Second-order moment of y,

$$m_2 = E[y^2] = E[x^2] + \sigma_e^2 \tag{2.26}$$

Third-order moment of y,

$$m_3 = \text{mom}[x,x,x] + 3\ E[x]\ \sigma_e^2 \tag{2.27}$$

First-order cumulant of y

$$\text{cum}[y] = E\{y\} = E[x] \tag{2.28}$$

Second-order cumulant of y,

$$\begin{aligned} \text{cum}[y,\ y] &= E[y^2] - \{E[y]\}^2 \\ &= E[x^2] + \sigma_e^2 - \{E[x]\}^2 \end{aligned} \tag{2.29}$$

Third-order cumulant of y,

$$\text{cum}[y,\ y,\ y] = \text{cum}[x,x,x] \tag{2.30}$$

From the above it is clear that cumulants of order greater than 2 are unaffected by the Gaussian noise, but the same order moments are not.

2.2.5 Moments and Cumulants of Stationary Signals

If $x(k),\ k\ =\ \pm 1,\ \pm 2,\ \ldots,$ is a real stationary random signal and its moments up to order n exist, then the nth order moment of x is defined as

$$m_n(\tau_1, \tau_2, \cdots \tau_{n-1}) \triangleq E[x(k)x(k+\tau_1)x(k+\tau_2)\cdots x(k+\tau_{n-1})] \tag{2.31}$$

where, $\tau_i = 0, \pm 1, \pm 2, \ldots$.

Similarly, the nth order cumulant of x is defined as

$$c_n(\tau_1, \tau_2, \cdots \tau_{n-1}) \triangleq \text{cum}[x(k), x(k+\tau_1), x(k+\tau_2), \ldots, x(k+\tau_{n-1})] \tag{2.32}$$

Combining Eqs. (2.32), (2.31) and (2.14), the following relationships of moment and cumulant sequences of $x(k)$ can be obtained:
1st order cumulant:

$$c_1 = m_1 = E[x(k)] \tag{2.33}$$

2nd order cumulant:

$$c_2(\tau_1) = m_2(\tau_1) - m_1^2 = m_2(-\tau_1) - m_1^2 = c_2(-\tau_1) \tag{2.34}$$

3rd order cumulant: combining Eqs. (2.32), (2.31) and (2.15)

$$c_3(\tau_1, \tau_2) = m_3(\tau_1, \tau_2) - m_1[m_2(\tau_1) + m_2(\tau_2) + m_2(\tau_2 - \tau_1)] + 2m_1^3 \tag{2.35}$$

Similarly, the other higher-order cumulants can be obtained.

2.3 Spectral Analysis

Not all the information content of a signal can necessarily be obtained easily from time domain analysis of the data. Transforming a signal from time to frequency domain can expose various characteristics. For example, the periodicity and non-linearities of the signal can be highlighted by means of such a transformation. It can also aid in understanding the signal-generating processes. The discrete Fourier transform (DFT), which plays a key role in modern digital signal processing, is the primary tool for this type of transformation.

For a real strictly stationary signal, $\{x(k)\}$, $k = 0, \pm 1, \pm 2, \ldots$ the nth order cumulant sequence is defined by

$$c_n(\tau_1, \tau_2, \ldots, \tau_{n-1}) = \mathrm{cum}[x(k),\, x(k+\tau_1),\, \ldots,\, x(k+\tau_{n-1})] \tag{2.36}$$

Assuming that the cumulant sequence satisfies the condition

$$\sum_{\tau_1=-\infty}^{\infty} \cdots \sum_{\tau_{n-1}=-\infty}^{\infty} (1+|\tau_i|)\,|c_n(\tau_1, \tau_2, \ldots, \tau_{n-1})| < \infty, \text{ for } i = 1, 2, \ldots, n-1$$

the nth order cumulant spectrum $c_n(f_1, f_2, \ldots, f_{n-1})$ of $\{x(k)\}$ is defined as the $(n-1)$-dimensional Fourier transform of the nth order cumulant sequence, i.e.

$$c_n(f_1, f_2, \ldots, f_{n-1}) = \sum_{\tau_1=-\infty}^{\infty} \cdots \sum_{\tau_{n-1}=-\infty}^{\infty} c_n(\tau_1, \tau_2, \ldots, \tau_{n-1}) \exp\{-j(f_1\tau_1 + f_2\tau_2 + \cdots + f_{n-1}\tau_{n-1})\} \tag{2.37}$$

where $|f_i| \leq \pi$, for $i = 1, 2, \ldots, n-1$ and $|f_1 + f_2 + \cdots + f_n| \leq \pi$.

The power spectrum, bispectrum and trispectrum are special cases of the nth order cumulant spectrum.

2.3.1 *Power Spectrum, n = 2*

The power spectrum is the frequency domain counterpart of the second-order moment or cumulant of a signal, and is usually obtained from a zero-mean time series. For such a time series, say $x(k)$, the second-order moment and the second order cumulant are identical. The power spectrum can be obtained using either of the following two methods:

1. Indirect method: The power spectrum can be calculated from the DFT of the second-order moments or cumulants of the time series. This is known as the indirect method. This method involves the following two steps:

   (a) Calculation of the second-order moment using the following equation:

   $$m_2(\tau) \triangleq E[x(k)x(k+\tau)] \equiv c_2(\tau) \tag{2.38}$$

   (b) If m_2 is absolutely summable, i.e. the summation of the absolute values of m_2 at all lags is a finite number, then the power spectrum is given by:

   $$P(f) = \mathrm{DFT}[m_2(\tau)] \equiv \sum_{\tau=0}^{N-1} m_2(\tau) \mathrm{e}^{-j2\pi\tau f/N} \tag{2.39}$$

2. Direct method: The power spectrum can also be calculated from the direct Fourier transformation of the time series. This is known as the direct method. It also involves two steps:
   (a) Taking the DFT of the data series.

$$X(f) = \sum_{k=0}^{N-1} x(k)e^{-j2\pi kf/N} \tag{2.40}$$

   where j is a complex variable, the square root of -1.
   (b) Calculation of the power spectrum as

$$P(f) = E[X(f)X(-f)] \equiv E[X(f)X^*(f)] \equiv E[|X(f)|^2] \tag{2.41}$$

   where * denotes a complex conjugate.

The power spectrum can be thought of as a decomposition or spread of the signal energy over the frequency channels obtained from the fast Fourier transform. From the expressions used to estimate the power spectrum, it is clear that the power spectrum is phase blind, i.e. the phase information is ignored. The information obtained from the power spectrum or autocorrelation sequence, or more generally up to the second order statistics are sufficient only for complete description of linear processes. In order to study the deviations from Gaussianity and the presence of nonlinearities in signal-generating processes, it is necessary to look beyond the second order statistics (SOS). This brings us to the domain of Higher Order Statistics (HOS). All statistics of order greater than 2 are called HOS.

2.3.2 Bispectrum, $n = 3$

The method used for estimating the second-order spectrum or power spectrum can easily be extended to obtaining frequency domain counterparts of higher-order cumulants. For example, the bispectrum is the frequency domain representation of the third-order cumulant. It is defined as

$$B(f_1, f_2) \triangleq \text{DDFT}[c_3(\tau_1, \tau_2)] \equiv E[X(f_1)X(f_2)X^*(f_1 + f_2)] \tag{2.42}$$

where DDFT stands for double discrete Fourier transformation.

Similary, the trispectrum ($n = 4$) is defined as

$$T(f_1, f_2, f_3) \triangleq \text{TDFT}[c_4(\tau_1, \tau_2, \tau_3)] \equiv E[X(f_1)X(f_2)X(f_3)X^*(f_1 + f_2 + f_3)] \tag{2.43}$$

where TDFT stands for triple discrete Fourier transformation. These higher-order spectra are known as polyspectra.

2.4 Summary

This chapter presented a simplified tutorial introduction of the Higher Order Statistical concepts. The relationship between HOS and classical second-order statistics (SOS) is also explored. Various properties of HOS have been explained with the help of illustrative examples wherever applicable. This book involves an extensive use of the bispectrum. Therefore, for a thorough understanding of bispectrum and bicoherence, the next chapter is devoted towards the description, interpretation, estimation and statistical properties of these estimators.

Chapter 3
Bispectrum and Bicoherence

The bispectrum is the frequency domain representation of the third-order cumulant or moment. Bicoherence is the normalized representation of the bispectrum. Processes that deviate from Gaussianity and linearity can conveniently be studied by using higher order statistics, particularly the bispectrum and bicoherence. This chapter discusses the definition, estimation methods, estimation issues and properties of the bispectrum and bicoherence.

3.1 Bispectrum

The bispectrum is defined as

$$B(f_1, f_2) \triangleq DDFT[c_3(\tau_1, \tau_2)] \equiv E[X(f_1)X(f_2)X^*(f_1+f_2)] \tag{3.1}$$

where DDFT stands for double discrete Fourier transformation.

Equation (3.1) shows that the bispectrum is a complex quantity having both magnitude and phase. It can be plotted against two independent frequency variables, f_1 and f_2, in a three-dimensional figure. Just as the discrete power spectrum has a point of symmetry at the folding frequency, the discrete bispectrum also has 12 planes of symmetries in the (f_1, f_2) plane. For details, readers are referred to Rosenblatt and Van Ness, 1965; Nikias and Raghuveer, 1987; Kim and Powers, 1979; Nikias and Petropulu, 1993. Examination of the bispectrum in only one region gives sufficient information. The other regions of the (f_1, f_2) plane are redundant. The triangular region OAB in Fig. 3.1 shows the non-redundant principal domain of the bispectrum. The non-redundant region is sometimes divided into two parts: the inner triangular (IT) and the outer triangular (OT) regions (regions OAC and ABC, respectively in Fig. 3.1).

Throughout this book frequencies are normalized such that sampling frequency, $f_s = 1$. Therefore, the Nyquist frequency is $f_N = 0.5$. Each point in the bispectrum plot represents the bispectral content of the signal at the bifrequency, (f_1, f_2). In fact, the bispectrum B(f_1, f_2) at point (f_1, f_2) measures the interaction between

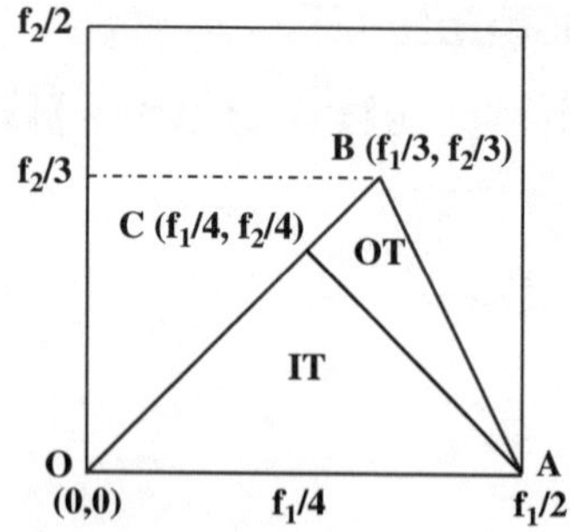

Fig. 3.1 Principal domain of the bispectrum

the signal components at frequencies f_1 and f_2. This interaction can be related to nonlinearities present in the signal-generating systems (Fackrell, 1996) and therein lies the core of its usefulness in the detection and diagnosis of nonlinearities.

3.1.1 Estimation of the Bispectrum

In practice, the higher-order spectra of a signal have to be estimated from a finite set of measurements. The underlying methods for polyspectra estimation are simply extensions of the well-established power spectrum estimation methods. Essentially, there are two broad non-parametric approaches: (1) the indirect method, based on estimating the cumulant functions and then taking the Fourier transform; and (2) the direct method, based on a segment averaging approach. The details of these two approaches are discussed below:

1. **Indirect method:** In this method, the data series are divided into K segments and the third-order cumulants are estimated for each of the segments. The average of the third-order cumulants are taken over the K segments. The bispectrum is then obtained by taking the Fourier transformation of the average third-order cumulants together with the appropriate data window function. The steps are as follows:

   (a) The data series $x(k)$, $k = 0, 1, ..., N-1$, is divided into K segments, $i = 0, 1, ..., K-1$ each of length M. These segments can overlap, so $K \geq N/M$. Let the ith segment of $x(k)$ be $x_i(k)$, $k = 0, 1, ..., M-1$.

   (b) The mean μ_i of the ith segment is calculated and subtracted from each sample in the segment.

$$\mu_i = \frac{1}{M} \sum_{k=0}^{M-1} x_i(k) \tag{3.2}$$

$$x_i'(k) = x_i(k) - \mu_i \tag{3.3}$$

   (c) An estimate of the third-order cumulants for each data segment can be obtained as:

$$c_{3,i}(m,n) = \frac{1}{M} \sum_{k=s_1}^{s_2} x_i'(k) x_i'(k+m) x_i'(k+n) \tag{3.4}$$

where,
$s_1 = max(0, -m, -n)$
$s_2 = min(M-1, M-1-m, M-1-n)$.

(d) The averaging of $c_{3,i}(m,n)$ over all segments gives:

$$\bar{c_3}(m,n) = \frac{1}{K} \sum_{i=1}^{K} c_{3,i}(m,n) \tag{3.5}$$

(e) Then the bispectrum can be estimated as

$$\hat{B}(f_1, f_2) = \sum_{m=-L}^{L} \sum_{n=-L}^{L} \bar{c_3}(m,n)\, w(m,n)\, e^{\{-j(mf_1+nf_2)\}} \tag{3.6}$$

where $L < M-1$ and $w(m,n)$ is a two-dimensional window function. Window functions should possess the same symmetry properties as the third-order cumulants. Detailed descriptions of some window functions are available in the literature (Nikias and Petropulu, 1993; Nikias and Raghuveer, 1987; Mendel, 1991).

2. **Direct method:** The direct method of bispectrum estimation is an extension of the Welch periodogram averaging technique for spectral estimation (Welch, 1967; Fackrell, 1996). The method is depicted schematically in Fig. 3.2 and consists of the following steps:

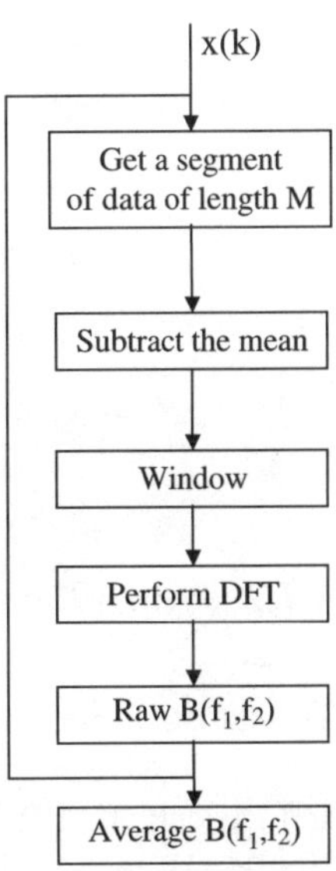

Fig. 3.2 Information flow diagram for the direct method of bispectrum estimation

(a) The data series $x(k),\ k = 0,1,\ldots,N-1$, is divided into K segments, $i = 0,1,\ldots,K-1$ each of length M. These segments can overlap, so that $K \geq N/M$. Let the ith segment of $x(k)$ be $x_i(k),\ k = 0,1,\ldots,M-1$.

(b) The mean μ_i of the ith segment is calculated and subtracted from each sample in the segment.

$$\mu_i = \frac{1}{M}\sum_{k=0}^{M-1} x_i(k) \tag{3.7}$$

$$x_i'(k) = x_i(k) - \mu_i \tag{3.8}$$

(c) The zero-mean-centred segment of the data x_i' is then multiplied by a suitable data window $w(k)$, which provides some control over the spectral leakage. This window function may be a boxcar, Hamming, Hanning or any other window used in ordinary spectral estimation.

$$x_i''(k) = w(k)x_i'(k) \tag{3.9}$$

(d) For each segment, the DFT, $X_i(f)$, is computed as:

$$X_i(f) = \sum_{k=0}^{M-1} x_i''(k)e^{-j2\pi kf/M} \tag{3.10}$$

where f is the discrete frequency. From this DFT, the raw spectral estimates of $P_i(f)$ and the bispectral estimates of $B_i(f_1, f_2)$ can be found.

$$\hat{P}_i(f) = X_i(f)X_i^*(f) \tag{3.11}$$

$$\hat{B}_i(f_1, f_2) = X_i(f_1)X_i(f_2)X_i^*(f_1 + f_2) \tag{3.12}$$

(e) Then the raw estimates from all K segments can be averaged to give the following estimates

$$\hat{P}(f) = \frac{1}{K}\sum_{i=0}^{K-1} \hat{P}_i(f) \tag{3.13}$$

$$\hat{B}(f_1, f_2) = \frac{1}{K}\sum_{i=0}^{K-1} \hat{B}_i(f_1, f_2) \tag{3.14}$$

3.1.2 Properties of Estimators and Asymptotic Behaviour

The statistical properties of the indirect and direct methods of higher-order spectral estimation have been studied extensively (Rosenblatt and Van Ness, 1965; Brillinger and Rosenblatt, 1967a,b; Rao and Gabr, 1984; Nikias and Petropulu, 1993; Riggs, 1999). Assume that $P(f)$ and $B(f_1, f_2)$ are the true power spectrum and bispectrum, respectively, of a stationary zero-mean signal, $x(k)$. Let $\hat{B}(f_1, f_2)$ be the estimate of $B(f_1, f_2)$ using either the indirect method or the direct method and a single realization of the signal, $x(k)$ of length N. How to work with mathematical expectation (E) operator in the case of a single realization of a signal is explained in Eq. (3.27). It has been shown mathematically (Nikias and Mendel, 1993; Kim and Powers, 1979) that for sufficiently large segment size, M, and total length, N, both the direct and indirect methods provide approximately unbiased estimates, namely:

$$E\left\{\hat{B}(f_1,f_2)\right\} \cong B(f_1,f_2) \tag{3.15}$$

with asymptotic variances

$$var\left\{\Re\left[\hat{B}(f_1,f_2)\right]\right\} \cong var\left\{\Im\left[\hat{B}(f_1,f_2)\right]\right\} \cong \frac{1}{2}\sigma^2(f_1,f_2) \tag{3.16}$$

where

$$\sigma^2(f_1,f_2) = \begin{cases} \dfrac{VL^2}{MK}P(f_1)P(f_2)P(f_1+f_2) \text{ (indirect)} \\ \\ \dfrac{N_0^2}{MK}P(f_1)P(f_2)P(f_1+f_2) \text{ (direct)} \end{cases} \tag{3.17}$$

where K is the number of segments, M is the number of samples per segment, V is the total energy of the window used for bispectrum estimation (for a rectangular window V is unity) and L is defined in Step 5 of the indirect method. Because we do not use any frequency domain smoothing, in our case N_0 is same as M. From the above equations, it is apparent that if a rectangular window is used and if $L = N_0$, both methods provide approximately the same estimates.

Brillinger and Rosenblatt (1967a) showed that when M and N are large, the error of the bicoherence estimation is approximately complex Gaussian with zero mean and unity variance:

$$\frac{\hat{B}(f_1,f_2) - B(f_1,f_2)}{\sigma^2(f_1,f_2)} \sim N_c(0,1) \tag{3.18}$$

where N_c denotes a complex normal distribution, i.e. both real and imaginary parts are normally distributed. Another important conclusion that follows from the asymptotic results developed by (Brillinger and Rosenblatt, 1967a,b) is that these statistics can be treated as independent random variables over the grid of principal domain if the grid width is larger than or equal to the bispectrum bandwidth; i.e. $\hat{B}(f_q, f_p)$ and $\hat{B}(f_r, f_s)$ are independent for $q \neq r$ or $p \neq s$ if $|f_{q+1} - f_q| \geq \triangle_0$ or $|f_{r+1} - f_r| \geq \triangle_0$, where

$$\triangle_0 = \begin{cases} \dfrac{1}{L} \quad \text{(indirect)} \\ \\ \dfrac{1}{N_0} \quad \text{(direct)} \end{cases} \tag{3.19}$$

3.1.3 Bicoherence or Normalized Bispectrum

As shown in Eq. (3.17), the bispectral estimates are asymptotically unbiased and the variance of the estimator depends on the second-order spectral properties (Hinich, 1982). That is,

$$var\left(\hat{B}(f_1,f_2)\right) \propto P(f_1)P(f_2)P(f_1+f_2) \tag{3.20}$$

Since the estimate depends directly on signal energies in the bifrequency, the variance of the estimate will be higher at a bifrequency where the energy is high and lower where the energy is low. This causes a serious problem in the estimation. This unsatisfactory property can be resolved in several ways. One way is to prewhiten the signal prior to bispectral analysis (Collis et al., 1998). However, an easier solution is to normalize the bispectrum to get a new measure whose variance is independent of the signal energy. Hinich (1982) has suggested normalization of the bispectrum using the following expression called the skewness function:

$$s^2(f_1,f_2) \triangleq \frac{\left|E\left[B(f_1,f_2)\right]\right|^2}{E[P(f_1)]E[P(f_2)]E[P(f_1+f_2)]} \tag{3.21}$$

where $s^2(f_1,f_2)$ is known as the skewness function. The major drawback of this normalization is that the magnitudes of this function are not bounded. The only reason for dividing the bispectrum with power spectrum is to remove the undesirable variance properties of the estimator. This method of normalization has been extensively used by Hinich (1982) for statistical tests of Gaussianity and linearity of a signal. The higher-order spectral analysis (HOSA) Toolbox in MATLAB® (Swami et al., 1993) has also adopted this normalization. This normalized bispectrum is defined as bicoherence in this toolbox. However, it would be more appropriate to call it 'skewness function'.

The bicoherence is better defined by the following equation:

$$bic^2(f_1,f_2) \triangleq \frac{|B(f_1,f_2)|^2}{E[|X(f_1)X(f_2)|^2]\ E[|X(f_1+f_2)|^2]} \tag{3.22}$$

where *bic* is the bicoherence function. Kim and Powers (1979) have shown that the variance of the bicoherence estimator roughly satisfies the following expression:

$$var\left[\hat{bic}^2(f_1,f_2)\right] \approx \frac{1}{M}\left[1-bic^2(f_1,f_2)\right] \tag{3.23}$$

where M is defined as the number of segments used in the estimation. Note that it is a consistent estimator in the sense that the variance approaches zero as M approaches infinity.

A useful feature of the bicoherence function is that it is bounded between 0 and 1. This can be demonstrated using the Cauchy–Schwartz inequality, which may be expressed as,

$$|E[z_1, z_2]|^2 \leq E[|z_1|^2]E[|z_2|^2] \tag{3.24}$$

Choosing $z_1 = X(f_1)X(f_2)$ and $z_2 = X(f_1 + f_2)$, it can be shown that

$$0 \leq bic^2 \leq 1 \tag{3.25}$$

There are also other normalization methods. Fackrell (1996) gives further details. However, these are not widely used because their properties have not been extensively studied.

3.1.3.1 Estimation of the Squared Bicoherence

To estimate the squared bicoherence from a finite set of measurements, the method of bispectrum estimation can be followed directly. This work has used the direct method throughout, so only the direct method of estimation of bicoherence is discussed here.

1. Using the first four steps of the direct method of bispectrum estimation, obtain the estimates of $\hat{P}_i(f) = X_i(f)X_i^*(f)$ and $\hat{B}_i(f_1, f_2)$ for each segment of the data.
2. Equation (6.2) can be rewritten as

$$bic^2(f_1, f_2) \triangleq \frac{|E[X(f_1)X(f_2)X^*(f_1+f_2)]|^2}{E[|X(f_1)X(f_2)|^2]E[|X(f_1+f_2)|^2]} \tag{3.26}$$

Now, the expectation operator can be replaced with a summation operator over the number of data segments in the following way:

$$\hat{bic}^2(f_1, f_2) = \frac{\left|\frac{1}{M}\Sigma_{i=1}^{M} X_i(f_1)X_i(f_2)X_i^*(f_1+f_2)\right|^2}{\frac{1}{M}\Sigma_{i=1}^{M} |X_i(f_1)X_i(f_2)|^2 \;\; \frac{1}{M}\Sigma_{i=1}^{M} |X_i(f_1+f_2)|^2} \tag{3.27}$$

The squared bicoherence can be estimated using the above equation.

3.1.4 Properties of Bispectrum and Bicoherence

1. *The theoretical bispectrum of a Gaussian signal is zero.*
The moment generating function for a Gaussian signal $x(t)$ is given by (Stuart and Ord, 1987):

$$M_x(t) = exp\left[\mu t + \frac{1}{2}\sigma^2 t^2\right] \tag{3.28}$$

Therefore, the cumulant generating function for x is

$$C_x(t) = \ln M_x(t) = \mu t + \frac{1}{2}\sigma^2 t^2 \tag{3.29}$$

Now, Eq. (2.9) can be rewritten as

$$C_x(t) = c_1 t + \frac{1}{2!}c_2 t^2 + \frac{1}{3!}c_3 t^3 + \ldots + \frac{1}{r!}c_r t^r + \ldots \tag{3.30}$$

comparing Eqs. (3.29) and (3.30), we obtain

$$\begin{aligned} c_1 &= \mu \\ c_2 &= \sigma^2 \\ c_r &= 0, \quad r > 2 \end{aligned} \tag{3.31}$$

Thus we see that for any Gaussian signal, the cumulants at the zeroth lag of order greater than 2 (e.g. $c_3(0,0)$, $c_4(0,0,0)$ and higher) are identically zero. This result can be easily generalized to other non-zero lags cumulants. As for any autocovariance sequence, $c_2(0) > c_2(\tau)$, $\tau \neq 0$, the same is true for other cumulants of higher order. For example, $c_2(0,0) > c_2(\tau_1, \tau_2)$, $\tau_1 = \tau_2 \neq 0$. If $c_2(0,0) = 0$, then $c_2(\tau_1, \tau_2)$, for any τ_1 or τ_2 unequal to zero ($\tau_1 = \tau_2 \neq 0$) must be equal to zero.

The bispectrum is the frequency domain counterpart of the third-order cumulants. Because third-order cumulants for a Gaussian signal are identically zero, the bispectrum will also be zero for a Gaussian signal.

2. *The theoretical bicoherence of a Gaussian signal is identically zero.*
   This follows directly from the Property 1. Because the bicoherence or the skewness function is a scaled bispectrum, a zero bispectrum leads to a zero bicoherence or skewness.
3. *The theoretical bispectrum of a non-Gaussian signal is blind to the additive Gaussian noise.*
   It is shown in Sect. 2.2.4 that the third-order cumulant of a non-Gaussian signal is independent of the additive Gaussian noise. This property follows directly from that result.
4. *The theoretical bicoherence of a signal is not blind to Gaussian noise.*
   Because the bicoherence or skewness function is scaled with the power of the signal, which is not independent of noise, the bicoherence is affected by the additive Gaussian noise.
5. *If a signal is filtered by a linear filter, then the magnitude of the bicoherence is unchanged.*
   Consider a signal, $x(n)$, filtered with a linear causal time-invariant filter, $h(k)$, to obtain the signal, $y(n)$. This can be represented by the following equation:

$$y(n) = \sum_{k=0}^{\infty} h(k)x(n-k) \tag{3.32}$$

In the frequency domain, this can be rewritten as

$$Y(f) = H(f)X(f) \tag{3.33}$$

Now, substituting this in the defining equation of bicoherence (Eq. (6.2)), we obtain

$$\begin{aligned}
bic^2(f_1,f_2) &\triangleq \frac{|E[Y(f_1)Y(f_2)Y^*(f_1+f_2)]|^2}{E\left[|Y(f_1)Y(f_2)|^2\right]E\left[|Y(f_1+f_2)|^2\right]} \\
&= \frac{|E\left[H(f_1)X(f_1)H(f_2)X(f_2)H^*(f_1+f_2)X^*(f_1+f_2)\right]|^2}{E\left[|H(f_1)X(f_1)H(f_2)X(f_2)|^2\right]E\left[|H(f_1+f_2)X(f_1+f_2)|^2\right]} \\
&= \frac{\left|E\left[H(f_1)H(f_2)H^*(f_1+f_2)\right|^2\right]}{E\left[|H(f_1)H(f_2)|^2\ |H(f_1+f_2)|^2\right]} \\
&\quad\times \frac{|E\left[X(f_1)X(f_2)X^*(f_1+f_2)\right]|^2}{E\left[|X(f_1)X(f_2)|^2\right]E\left[|X(f_1+f_2)|^2\right]} \\
&= \frac{|E\left[X(f_1)X(f_2)X^*(f_1+f_2)\right]|^2}{E\left[|X(f_1)X(f_2)|^2\right]E\left[|X(f_1+f_2)|^2\right]}\,1
\end{aligned} \tag{3.34}$$

Note that the above expression only holds when $H(f) > 0,\ \forall f$. Otherwise, $0/0$ may occur in the calculation of the above quantity. This is equivalent to the requirement that the linear filter should not have any zeros on the unit circle (Fackrell, 1996).

This property enables us to use bicoherence to detect the linearity of a signal-generating process.

6. *The theoretical bicoherence of a harmonic signal that can be decomposed into a sum of sinusoids shows peaks if the signal possesses the property of quadratic phase coupling (QPC).*
   A signal is said to have QPC if the signal phases, for example, ϕ_1, ϕ_2 and ϕ_3 at frequencies f_1, f_2 and $f_3 = f_1 + f_2$, respectively, have the relation $\phi_1 + \phi_2 = \phi_3$. This is an indication of a nonlinear signal-generating mechanism. A system that possesses a quadratic type nonlinearity, for example a square function, can yield a QPC signal. This will be illustrated with an example in the following section and also in Sect. 6.5.

3.2 Bispectrum or Bicoherence Estimation Issues

The main estimation parameters that need to be chosen for bispectral analysis are the same parameters as required for second-order spectral analysis. Examples include the choice of window function, data length, data segment length, length of Fourier transform, and overlapping or non-overlapping windows (Fackrell, 1996). In this section, some of these issues will be addressed and discussed using the following illustrative example.

Illustrative Example. This example consists of the following signal.

$$y(t) = sin(2\pi f_1 t + \phi_1) + sin(2\pi f_2 t + \phi_2) + sin(2\pi f_3 t + \phi_3) + sin(2\pi f_4 t + \phi_4) + n(t) \quad (3.35)$$

where the values of f_1, f_2, f_3 and f_4 are 0.12, 0.18. 0.30 and 0.42, respectively; the values of ϕ_1, ϕ_2, ϕ_3 and ϕ_4 are $\frac{\pi}{3}$, $\frac{\pi}{12}$, $\frac{\pi}{4}$ and $\frac{3\pi}{8}$ respectively; $n(t)$ is a zero-mean white noise signal with variance 0.2, and t is the time from 1 to 4096 s. The signal $y(t)$ is a quadratic phase-coupled signal because its frequencies have the relations $f_1 + f_2 = f_3$ and $f_1 + f_3 = f_4$, and its phases have the relations $\phi_1 + \phi_2 = \phi_3$ and $\phi_1 + \phi_3 = \phi_4$. Therefore, these two phase couplings at bifrequencies $(0.12\ ,\ 0.18)$ and $(0.12\ ,\ 0.30)$ appear in the bicoherence plot as two peaks at those frequencies as shown in Fig. 3.3.

3.2.1 Choice of Window Function

In spectral analysis, the use of a window function is very common. The main reason for using a window is to solve the problem of spectral leakage that occurs between neighbouring frequency channels of a peak. Spectral leakage is the term used to describe the loss of power at a given frequency to other frequency bins in the DFT. Spectral leakage can be visualized from the spread of the frequency components (see Fig. 7.5). Each frequency component of a signal should contribute only to one single frequency of the Fourier transform called a FFT bin, but spectral leakage causes the energy to be spread to the neighbouring frequencies. The window function controls the spreading. The contribution from any real frequency component to a given FFT bin is weighted by the amplitude of the window function's frequency spectrum centred at the FFT bin.

Theories and issues related to the use of various types of windows in bispectral estimation have been addressed in Fackrell (1996), Nikias and Petropulu (1993) and Chandaran and Elgar (1991). The performances of three window functions have been compared in Fackrell (1996) and it has been shown that the Hamming window was the most successful among them to best resolve the peak. The Hanning window stands next to the Hamming window in terms of peak resolution. The box-car window function is the worst for resolving the peaks. Figure 3.3 shows the bicoherence plot for the time series of Eq. (6.6). The figure shows that the box-car function is the worst among the three windows in resolving peaks. This is because the kernel of the rectangular window has approximately two times wider sidelobes than that of the Hamming or Hanning window (Smith, 1998; Ifeachor and Jervis, 1993). A careful observation shows that the peaks in Figure 3.3c are better resolved than those in Figure 3.3b.

Fig. 3.3 These plots show bicoherence for the QPC example. The horizontal axes are both frequency axes and the vertical axis represents the bicoherence. Two phase couplings at bifrequencies (0.12, 0.18) and (0.12, 0.30) appear in the bicoherence plot as two peaks at those frequencies

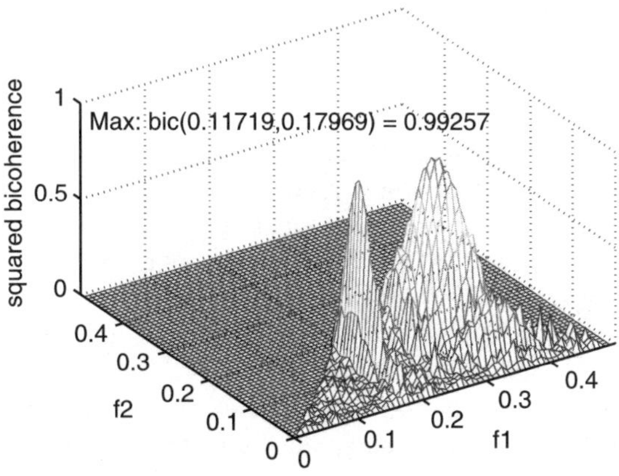

(a) Bicoherence using box-car window

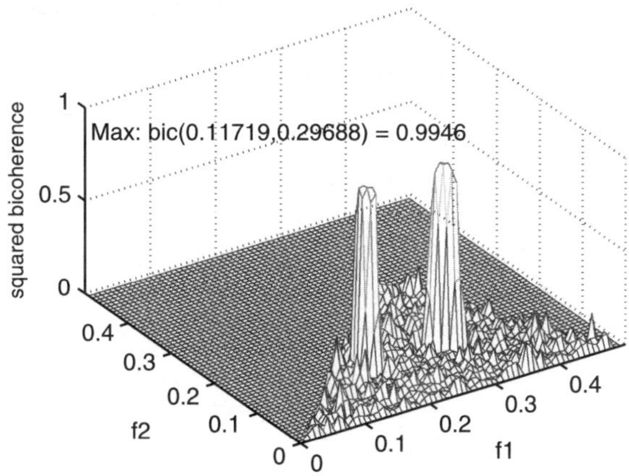

(b) Bicoherence using Hanning wimdow

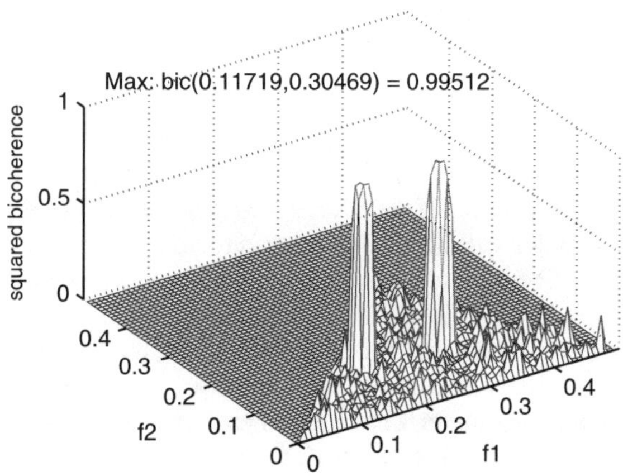

(c) Bicoherence using Hamming window

3.2.2 Choice of Data Length, Segment Length and Fourier Transform Length

It is well known that bispectral estimates generally have higher variance than power spectral estimates for a given data length. The data length that is sufficient for a reliable power spectrum estimation may not be sufficient for a good bispectral estimation. Elgar and Guza (1988) reported empirical results for mean and variance of bicoherence estimates. Hinich (1982) suggested that if no frequency domain smoothing is used, the data should be segmented in a way such that the number of segments of data should be at least as large as the DFT size, i.e. $K \geq M$. However, this may not be achievable and therefore it may be impractical to implement bicoherence in real-life applications. The bottom line is that the number of data segments used in the bicoherence should be sufficient to have asymptotically unbiased and consistent estimation as well as to keep a good frequency resolution. Jamsek et al. (2003) reported that for a real signal with a finite number of data points, the compromise between the frequency resolution and statistical asymptotic properties may be expected at around $K = 30$. The authors also have had similar experience.

In practical bispectral analysis, the length of data required depends on signal-to-noise ratio. Fackrell (1996) presented a useful result for the dependency of theoretical bicoherence of a quadratically coupled sinusoids on the and DFT length. He stated the following: *If the effects of leakage are ignored, then the peak bicoherence $b^2(f_1, f_2)$ corresponds to the coupled components at frequencies f_1, f_2, and $f_1 + f_2$ of a quadratically coupled sinusoids consisting of three equal-amplitude coupled harmonics in variable levels of additive white Gaussian noise with a signal-to-noise ratio of SNR is given by:*

$$bic^2(f_1, f_2) = \frac{1}{1 + \frac{18}{M} 10^{-SNR/10} + \frac{72}{M^2} 10^{-2SNR/10}} \tag{3.36}$$

where, SNR is defined as

$$SNR \triangleq 10 \log_{10} \frac{\sigma_x^2}{\sigma_n^2} \tag{3.37}$$

Equation (3.36) indicates how the bicoherence peak changes with SNR and segment length, M. If the SNR is very high, then the second and third terms in the denominator will be very close to zero and the bicoherence will be close to unity. As the SNR decreases, the bicoherence value also decreases. This can be presented in a three-dimensional plot as shown in Fig. 3.4. The size of the peak of the squared bicoherence is dependent on M because, as the DFT size increases, a better frequency resolution is obtained. Therefore, it is desirable to have the DFT size as large as possible. However, this requirement conflicts with the requirement of having a large number of data segments, K, to obtain reliable estimates. It also increases the computational load. The DFT length is usually chosen to be the same as each segment length. In order to increase the number of data segments for a better estimate, a

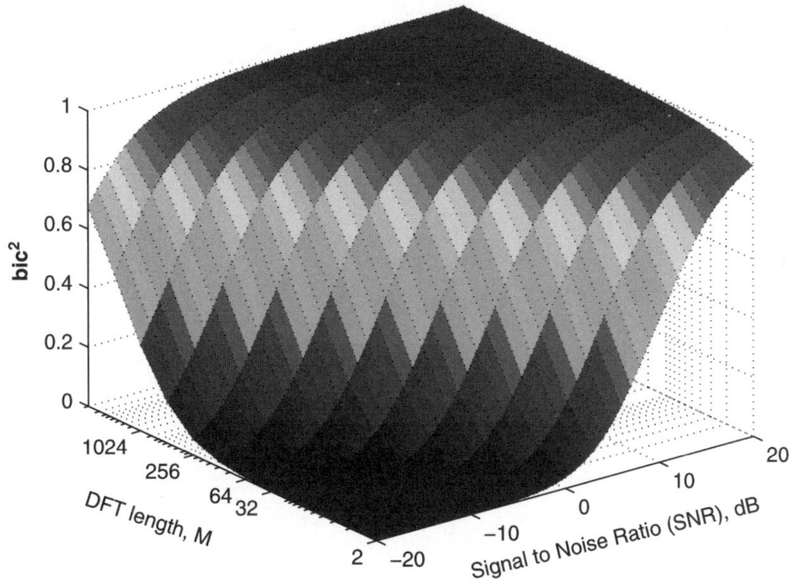

Fig. 3.4 The effect of signal-to-noise ratio and DFT length on bispectral estimation for a quadratically coupled harmonic signal. Larger DFT length and higher *SNR* provide better estimates

certain amount of overlapping of data segments (e.g. 50% or less) can be allowed for data sets having short lengths.

3.3 Summary

This chapter has discussed the definition, interpretation, estimation procedures and statistical properties of bispectrum and bicoherence. Because the bispectrum and the squared bicoherence function have played a central role in this book, these have been introduced with a greater degree of detail in this chapter. As shown in the illustrative example in Sect. 3.2, bicoherence can be used to examine the presence of nonlinearities in a time series or a signal-generating system. In the data-driven approach of detecting nonlinearities, bispectrum and bicoherence are found to be useful. A more detailed discussion is presented in Chap. 6.

Part II
Data Quality – Compression and Quantization

Chapter 4
Impact of Data Compression and Quantization on Data-Driven Process Analyses

Modern chemical plants are extensively instrumented and automated. One manifestation of increasing automation and accessibility is easy availability of process information on the desktop. Information access is typically available as a result of extensive data logging and process archiving. Historical data are an invaluable source of information, but in chemical industrial practice, data are often compressed using various techniques, e.g. box car, backward slope, swinging door, PLOT, wavelet transform, etc. before storing them in a historian. Compression degrades data quality and induces nonlinearity. This chapter focusses on the problems of data quality degradation and nonlinearity induction due to compression of the process data. It also presents an automatic method for detecting and quantifying the degree of compression present in the archived data. Finally, the problem of quantization in the process data is discussed, and an automatic procedure to detect and quantify quantization is presented.

4.1 Introduction

Motivations for data compression discussed in Thornhill et al. (2004) included reduction of the costs of storage of historical data and reduction of cost of transmission of process data through a telecommunications link. The same paper discussed the hidden costs of data compression, in particular that the data often become unsuitable for their intended purposes. Desired end uses of the data surveyed by Kennedy (1993) include:

- Calculation of daily statistics such as daily means and daily standard deviations;
- Averaging for data reconciliation and mass balancing;
- Archiving of data trends for subsequent high fidelity reconstruction;
- Data smoothing by removal of high-frequency noise;
- Feature extraction and recovery of events.

Once the data have been compressed, however, they lose information and the reconstructed trends are deficient in various ways compared to the originals. Figure 4.1 shows an example of a data set that has been reconstructed after compression in a

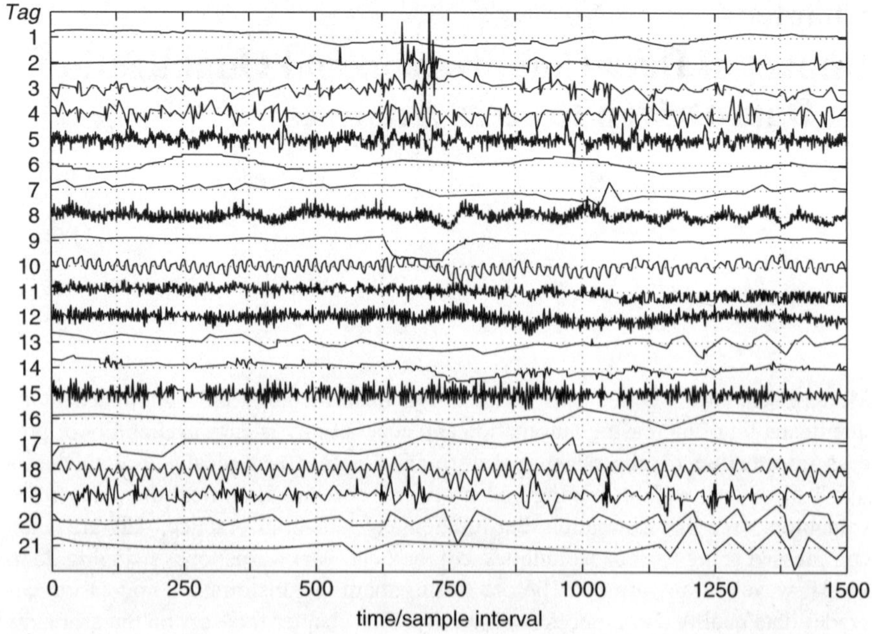

Fig. 4.1 An industrial data set with compression in some tags. Time trends are mean-centred and normalized

data historian. The straight line segments that are characteristic of industrial data compression can be seen in many of the time trends. The original uncompressed data were lost forever when they were compressed and archived, and it is now not possible to determine what features have been lost.

This chapter first describes methods in common use for data compression and shows some examples of the adverse impact of data compression on data-driven plant performance analysis. It also outlines the compression detection algorithm that was first presented in Thornhill et al. (2004) and gives a recommendation about how much compression can be tolerated. Application of the algorithm during the data-preprocessing phase of a plant audit means less time wasted in the evaluation of unsuitable data.

Quantization may also have an impact on the quality of process data. Quantization is often observed in the outputs of process instruments even though 10 bit A/D conversion provides 1024 quantization levels because a measurement controlled to a steady value is likely to sample just a few of the available quantizer levels. The effects of quantization can be enough to give misleading results in a data-driven analysis. An algorithm for the detection of data quantization is discussed in Sect. 4.9.

4.2 Data Compression Methods

Compression using piecewise linear trending is widely used in industrial data historians because it can be applied in real time to spot data. For instance, AspenTech described an adaptive method based upon the box-car/backward-slope method (Aspentech, 2001), while OSI state that their PI data historian uses a type of swinging door compression algorithm (OSI Software Inc., 2002).

4.2.1 Overview of Data Compression

Comparative reviews of various compression methods include Mah et al. (1995), who compared piecewise linear trending methods and introduced a new method called PLOT, and Watson et al. (1998) who also looked at methods utilizing wavelet and Fourier transforms.

A direct method makes the archiving decision in real time, concurrently with the capture of data from the process. Various types of algorithms are used for data compression in process industries. Examples include box car (BC), backward slope (BS), combined box car and backward slope (BCBS), and Swinging door (sdoor). These algorithms (Hale and Sellars, 1981; Bristol, 1990) use heuristic rules to decide whether to archive a spot value. Those rules are tuned to achieve the capture of exceptions and linear trends. They reconstruct a data trend as a series of linear segments connecting the archived spot values of the data. Brief descriptions of these algorithms are given below.

4.2.2 Box-Car (BC) Algorithm

In this algorithm, the current data point is compared to the past recorded data point. If the two values differ by more than or equal to the recording limit or the compression deviation set for this variable, the point immediately prior to the current data point is archived. The details of the algorithm are shown in Fig. 4.2. This method is helpful only for steady processes. If there is a ramp-type change (e.g. the level of a tank as it is being filled, or the process shifts frequently from one operating point to another), then this method does not provide a high level of data compression.

4.2.3 Backward-Slope (BS) Algorithm

As the name suggests, the recording limit or the compression deviation boundary lines run parallel to the projected slope calculated from the last two recorded data points. If the current data point falls outside the deviation limit boundaries, the

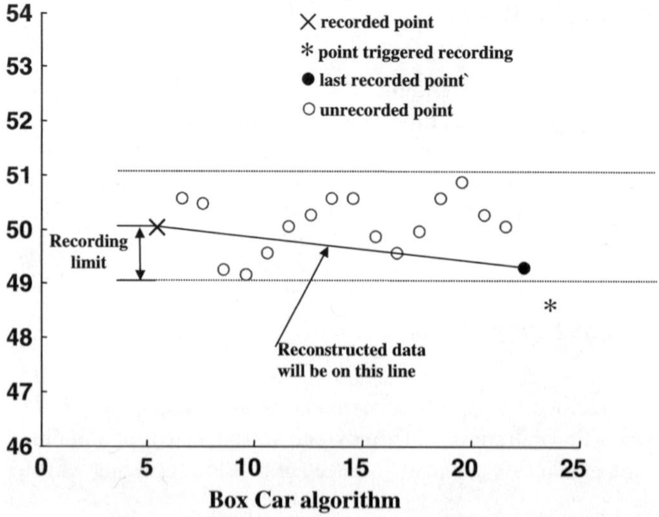

Fig. 4.2 Box-car algorithm for data compression (from Hale and Sellars (1981))

prior point is recorded. Figure 4.3 demonstrates the method. In case of a very noisy variable, the backward-slope algorithm neither produces good results nor gives a high data compression, because the slope is not a meaningful measure in such a case. Therefore, both methods were combined to give the combined box-car and backward-slope (BCBS) method.

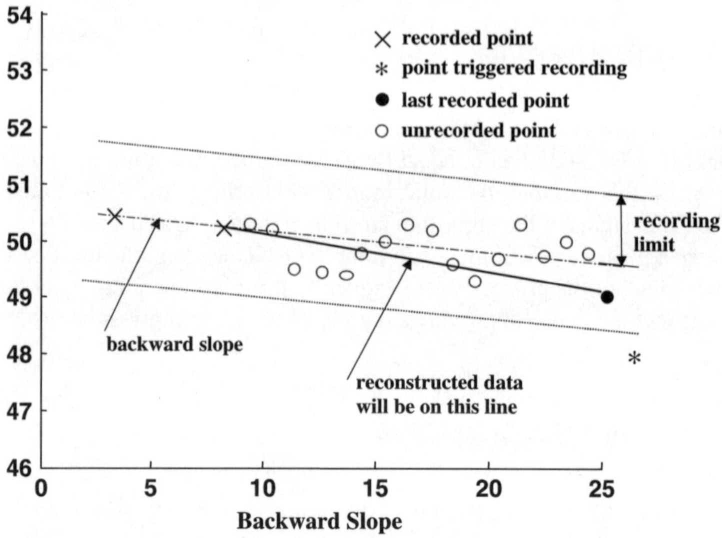

Fig. 4.3 Backward-slope compression algorithm (Hale and Sellars (1981))

4.2.4 Combined Box-Car and Backward-Slope (BCBS) Method

In this algorithm, the previous two methods are applied simultaneously. When the present data point fails to meet the criteria of both box-car and backward-slope method, the point prior to the present value is stored. See Fig. 4.4 for the detailed algorithm.

4.2.5 Swinging Door Compression Algorithm

The details of this method are shown in Fig. 4.5. In Fig. 4.5, point 'a' is obtained from the last recorded point, 'c', plus the compression deviation, while point 'b' appears from the deduction of compression deviation from point 'c'. Points 'a' and 'b' are called pivot points. As new spot values arrive, lines are drawn from the pivot points to form a triangular envelope that includes all the spot values since 'c'. The sides of the triangle are the 'doors'. For instance, in Fig. 4.5, all points up to point 'd' can be enveloped in a triangle. However, the next point, 'e' cannot be included in a triangle because, as shown by the dotted line, the upper and lower doors have opened wider than parallel. This signifies that a new trend has started at point 'd'. Point 'd' is archived and the procedure is repeated from this point.

4.2.6 The Compression Factor

After compression, the time trends are reconstructed from the archived spot values by linear interpolation between archived points at the original sampling instants.

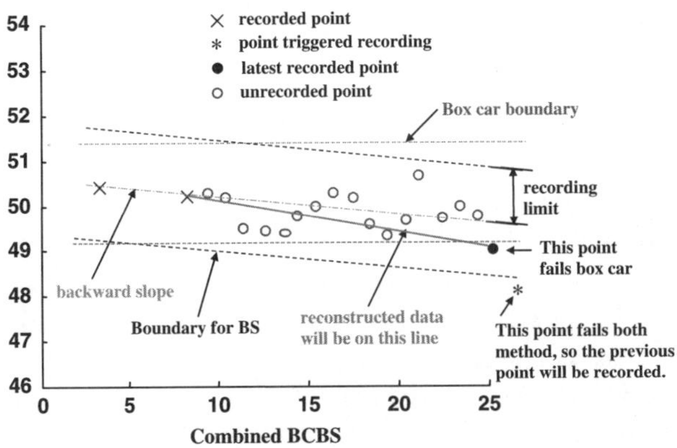

Fig. 4.4 Combined Box-car and backward-slope data compression algorithm (from Hale and Sellars (1981))

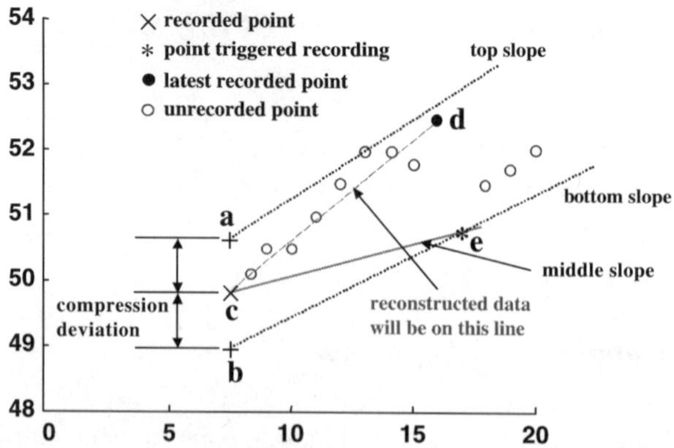

Fig. 4.5 Swinging door algorithm for data compression

The compression factor (CF) is defined as the ratio between the storage requirement of the original data set and that of the archived data. If the original data set had 1,000 observations and 1,000 time tags, a direct method with CF = 10 would yield 100 observations, 100 time tags and 99 linear segments. The compression factor is not specified explicitly in the piecewise linear methods of compression. Instead, the parameter that has to be set is the deviation threshold or recording limit deviation in engineering units. This means that users have no direct control over the compression factor.

4.3 Measures of Data Quality

This chapter will now examine the impact of data compression on industrial process data. First, some measures of data quality are introduced. These include statistical quantities such as mean and variance and the Harris index for control loop performance assessment in the formulation proposed in Desborough and Harris (1992). The nonlinearity index of Chap. 6 is also used as a measure of data quality.

4.3.1 Statistical Measures

As discussed in Thornhill et al. (2004), archived process data may be used for steady-state assessments such as monitoring of plant production rates, data reconciliation and mass balancing. Other uses include variability assessments as process variability is known to have an impact on profit (Martin et al., 1991; Shunta, 1995).

The following measures are introduced to enable the assessment of the effects of compression:

1. The percentage difference between the mean values (PDM) scaled by the standard deviation of the original data:

$$PDM = 100\frac{mean\,(y) - mean\,(\hat{y})}{\sigma_y}$$

2. The ratios between the variance of the reconstructed data ($\sigma_{\hat{y}}^2$) and the variance of the original data (σ_y^2) (RVC), and between σ_y^2 and the variance of the reconstruction error σ_e^2, where $e_i = y_i - \hat{y}_i$ (RVE). The measures are:

$$RVC = \sigma_{\hat{y}}^2 / \sigma_y^2 \quad \text{and} \quad RVE = \sigma_e^2 / \sigma_y^2$$

4.3.2 Nonlinearity Measures

Nonlinearity measures can be used as a diagnostic tool for troubleshooting of hardware faults that may be present in the control loops (Choudhury et al., 2004b; Choudhury et al., 2002) and to make decisions about the type of model needed in inferential sensing (Barnard et al., 2001). Therefore, it is useful to determine how the use of reconstructed data would influence nonlinearity assessment.

The nonlinearity test applied here uses the bicoherence measure to assess nonlinearity. The non-Gaussianity index (NGI) and nonlinearity index (NLI) developed in Chap. 6 will be used to investigate compression-induced nonlinearity in the reconstructed process data.

4.3.3 Performance Index (Harris) Measures

Desborough and Harris (1992) developed a performance index based on minimum variance benchmark of control loop performance. Significant industrial implementations of this index have been reported (Fedenczuk et al., 1999; Paulonis and Cox, 2003; Desborough and Miller, 2002). It is known that the use of data compression influences the Harris index (Thornhill et al., 1999). Consequently, an issue for practitioners is to know whether compressed archived data can be used for the purposes of a minimum variance benchmark calculation.

The Harris indices for the three data sets were calculated using the method described by Desborough and Harris (1992) with an estimated time delay of five samples. The index is determined from the residuals between the measured controller error denoted by y and a b-step ahead prediction, $\hat{y}$.

$$r(i) = y(i) - \hat{y}(i) \tag{4.1}$$

The model for $\hat{y}$ employed 30 autoregressive terms (i.e. $m = 30$) as discussed in Thornhill et al., (1999), and in this case, the prediction horizon was $b = 5$ as the time delay was estimated to be five sample intervals.

$$\hat{y}(i+b) = a_o + a_1 y(i) + a_2 y(i-1) + \cdots\cdots + a_m y(i-m+1) \tag{4.2}$$

The minimum variance benchmark is:

$$1 - \frac{\sigma_r^2}{mse(y_i^2)}, \tag{4.3}$$

where σ_r^2 is the variance of the residuals r and $mse(y_i^2)$ the mean square value of the controller error. An index of 0 represents the minimum variance control, while an index of 1 represents poor control (in which $y \approx \hat{y}$ and r is negligible). In the latter case, the controller is failing to deal with predictable components such as steady offsets or a predictable oscillatory disturbance.

The reconstructed data are more predictable and thus have a worse (larger) Harris Index than the original because compression removes noise and produces piecewise linear segments that have high local predictability. Thus, there is a danger that unnecessary maintenance effort may be spent on the repair of control loops wrongly identified as performing poorly.

4.4 Process Data for Compression Comparison

4.4.1 Industrial Example 1

Industrial example 1 was originally presented in Thornhill et al. (2004), and the main results are outlined here. Three contrasting time series variables were chosen for the evaluation of the impact of compression, courtesy of BP. They are uncompressed liquid flow trends from continuous processes operating at steady state. Each data set comprised nearly 3 h of 10-s samples representing deviations of flow in a process stream from the mean value. Figure 4.6 shows portions of the time trends while the dotted lines in Fig. 4.7 show their power spectra. A detailed description of the data is available in Thornhill et al. (2004). The main results found in Thornhill et al. (2004) are summarized below.

4.4.1.1 Visual Observations

Data set 1 shows a persistent oscillation characterized by an average of about 22 samples per cycle. The challenge for high fidelity compression and reconstruction is to retain the spectral peak in the frequency domain and the oscillatory features in the time domain. Swinging door compression did not follow all the oscillations in

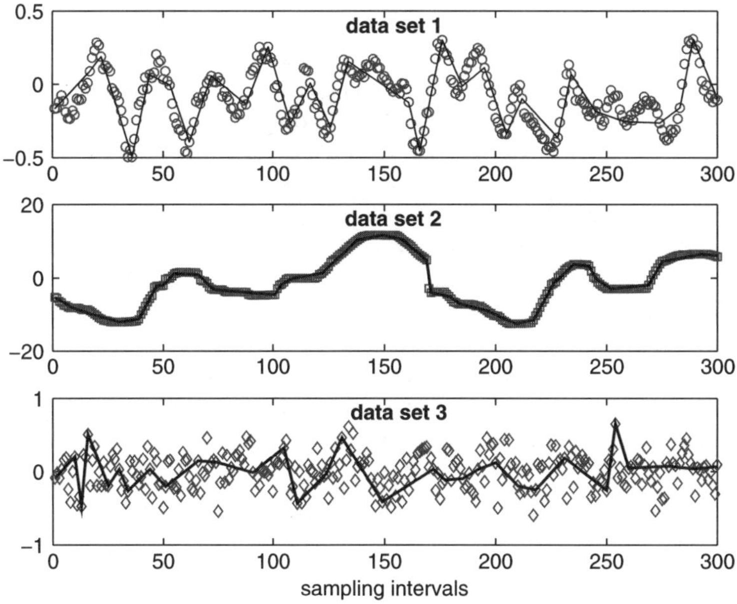

Fig. 4.6 Time trends of original data (*open circles* for data set 1, *squares* for data set 2, and *diamonds* for data set 3) and reconstructed data with compression factor 10 (*lines*)

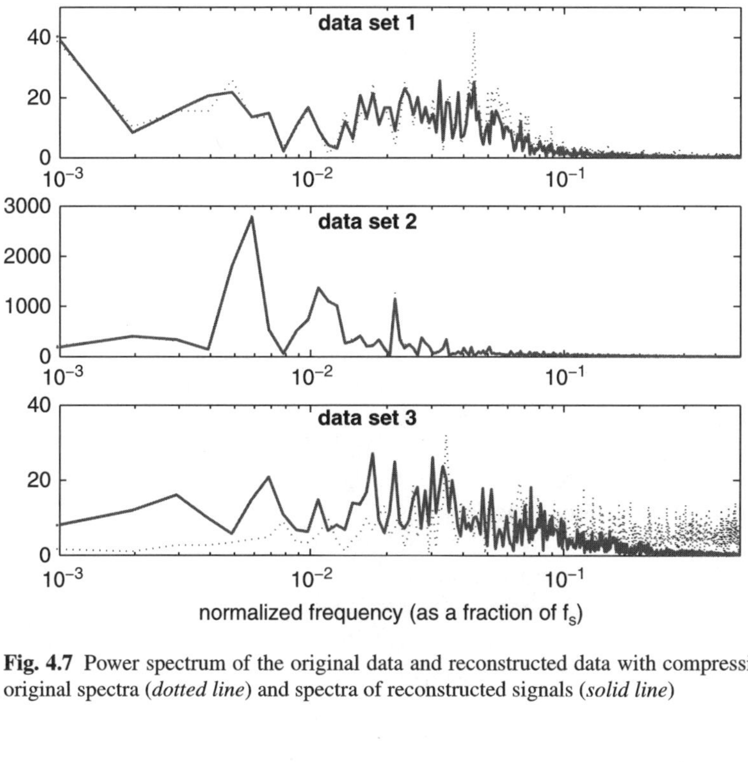

Fig. 4.7 Power spectrum of the original data and reconstructed data with compression factor 10, original spectra (*dotted line*) and spectra of reconstructed signals (*solid line*)

data set 1 because, with CF = 10, the average duration of each linear segment was longer than half of the oscillation period. The reconstructed data had errors at low frequency and a non-zero spectral error at $f = 0$. Therefore, the signal reconstructed after compression had a different mean value than the original.

Data set 2 has a tendency to stay at a constant value for a given time and then to move rapidly to a new level. These data are from a control loop that has a limit cycle caused by a sticking valve. The signal is predictable for long periods, and its spectrum shows very low-frequency features because the period of oscillation is long while a series of harmonics highlights the non-sinusoidal nature of the waveform. High fidelity compression was possible with data set 2 because the data set was predictable in nature.

Data set 3 has little predictability and has spectral features at all frequencies. This signal is dominated by random noise and is from a well-tuned loop operating close to minimum variance. The reconstructed data for this loop lost much of the randomness.

4.4.1.2 Statistical Properties

In Thornhill et al. (2004), it was concluded that data compression provides misleading information about basic statistical properties of data. Compression alters both the mean and variance. The changes in means are only a small percentage of the standard deviation. However, the purpose of data reconciliation is often to find small shifts in the mean value that may be indicative of problems such as leaks. The shift in the mean due to data compression may therefore be wrongly interpreted as evidence of a leak. The decisions used in statistical process control (Wetherill and Brown, 1991) may also be erroneous if the warning and alarm limits have been based upon a statistical distribution determined from compressed archived data.

4.4.1.3 Nonlinearity

The use of compressed archived data to assess nonlinearity, for instance, in an audit of control valves, may be misleading. Time may be wasted in inspection and testing of valves that are in fact operating normally.

4.4.1.4 Control Performance Index

In the case of data sets 1 and 3, it was concluded that compression increases the predictability of the signal and thus affects the Harris index. Data set 2 was inherently predictable (see Sect. 4.4.1.1). The Harris index for data set 2 therefore indicated poor performance even in the uncompressed case and did not change as much on compression as for the other two data sets.

4.4.2 Industrial Example 2

The purpose of choosing this example is to study the effect of compression in fast and slow control loop data from the dynamics point of view. Also, the results of this example will validate some of the results of Example 1. In this data set, data from three different types of control loops were chosen, courtesy of Saudi ARAMCO. They are uncompressed flow, temperature and pressure trends from continuous processes operating at steady state. Each data tag comprised nearly 4 h of 15 s samples representing deviations of flow in a process stream from the mean value. Figure 4.8 shows portions of the time trends, while the dotted lines in Fig. 4.9 show their power spectra.

Flow loop data show a good control loop operating close to the minimum variance benchmark and have little predictability. They have spectral features at all frequencies. Temperature loop data have some irregular oscillations. The oscillations are not always present. Pressure loop data have low-frequency predictable features.

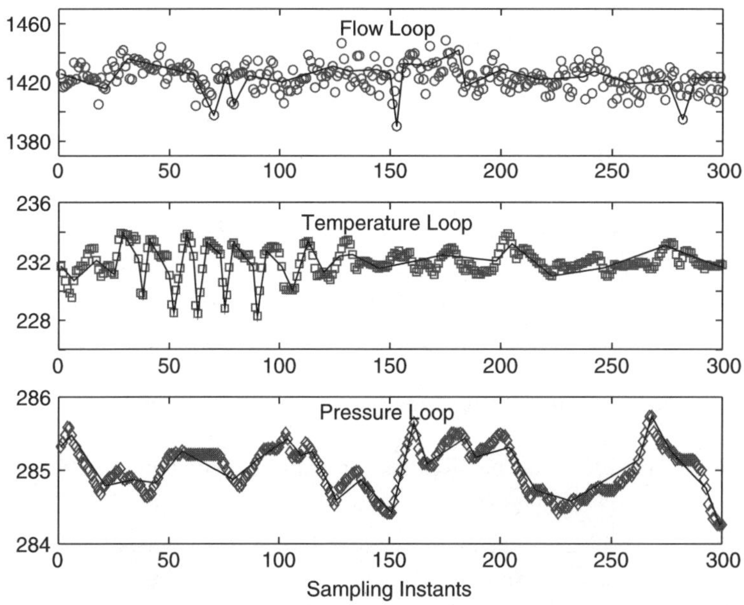

Fig. 4.8 Time trends of original data (*open circles* for flow loop data, *squares* for temperature loop data, and *diamonds* for pressure loop data) and reconstructed data with compression factor 10 (*lines*)

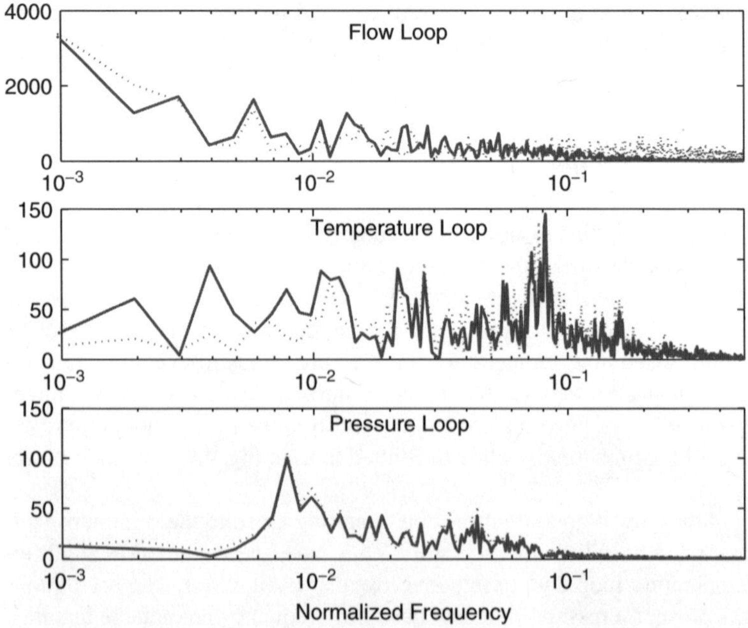

Fig. 4.9 Power spectrum of the original data and reconstructed data with compression factor 10, original spectra (*dotted line*) and spectra of reconstructed signals (*solid line*)

4.5 Results and Discussions for Industrial Example 2

4.5.1 Visual Observations

The panels in Fig. 4.8 show close-up portions of the original data (as points) and reconstructions (solid line) with a compression factor of 10 for data sets 1–3. Each data set had 1024 samples. Features to note are:

- In the reconstructed flow loop data set, much of the randomness of the signal was lost.
- Oscillatory portion of the temperature trend was reconstructed well but the other portion was not.
- High fidelity compression was possible with pressure loop data because it contains mainly low-frequency components.

Figure 4.9 shows reconstruction in the frequency domain. The power spectra of the original signal are shown as dotted lines, while the spectra of the reconstructed signals with CF = 10 are shown as solid lines. When the two are not identical, a reconstruction error exists.

- There are errors in the low- and high-frequency spectral features in the reconstructed flow loop data.

- Flow- and temperature loop data have reconstruction errors at low frequency and a non-zero spectral error at $f = 0$. Therefore, the signal reconstructed after compression had a mean value that differed from the original (see the first panel of Fig. 4.10).
- The pressure loop data was reconstructed well.

4.5.2 Statistical Properties

Figure 4.10 shows the behavior of the mean value and variance measures as a function of compression factor. Similar results were observed in Example 1.

- The mean of the signal reconstructed from the archives differs from the mean of the original. However, the deviation is negligible up to a compression factor (CF) of 3.
- The variances of the reconstructed data are smaller than the variance of the original signal. Ideally, the value of RVC should be 1.
- The variance measures at a given compression factor do not sum to 1.

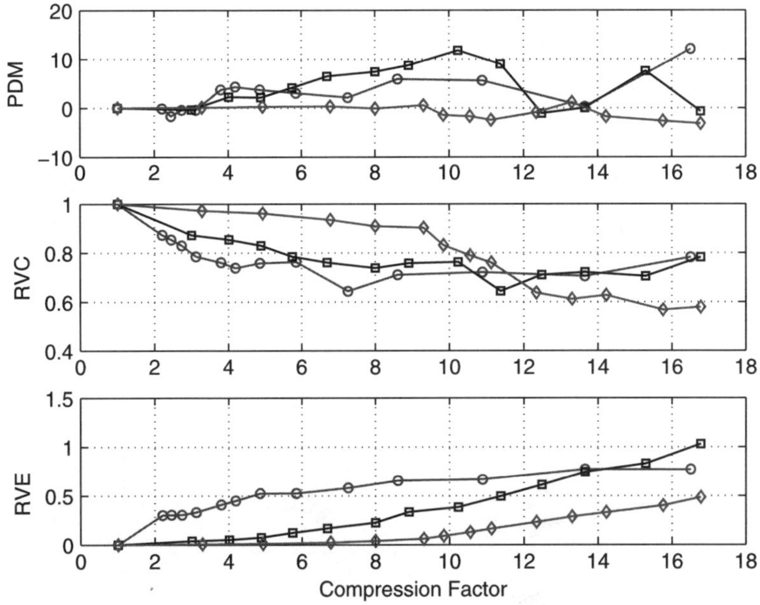

Fig. 4.10 Statistical measures as a function of CF for flow- (*circles*), temperature- (*squares*) and pressure loop data (*diamonds*)

4.5.3 Nonlinearity Assessment

The first and second rows of Fig. 4.11 depict results from nonlinearity assessment of the three data sets. Compression induces non-Gaussianity and nonlinearity in signals when compressed by a factor of more than 3 because uncompressed flow and pressure signals were originally Gaussian and linear. Non-Gaussianity indices increases monotonically for all three data sets, while the nonlinearity indices do not show any specific pattern – indicating the alteration of nonlinear structure of the signal with compression.

4.5.4 Performance (Harris) Index

The third row of Fig. 4.11 depicts the Harris index results. Again, the index increases with compression for all data sets.

In the case of flow- and pressure loop data, it is concluded that compression increases the predictability of signals and thus affects the Harris index. The temperature loop shows poor performance even in the uncompressed case and did not change as much on compression as for the other two data sets.

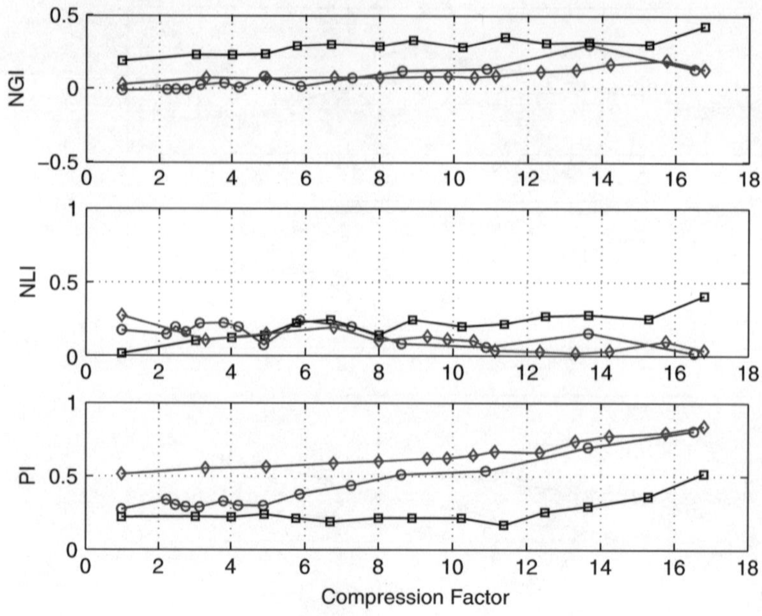

Fig. 4.11 Non-Gaussianity Index (NGI), Nonlinearity Index (NLI) and Harris Performance Index (PI) as a function of compression factor for flow- (*circles*), temperature- (*squares*) and pressure loop data (*diamonds*)

4.6 Summary of Data Quality Measures

From the results and discussion of the industrial examples, the following remarks can be made for real-world industrial applications.

- Data compression changes the statistical properties of data.
- Averaging, data reconciliation and mass-balancing applications should calculate the required quantities directly from the original data because data archived with swinging door compression have a different mean value after reconstruction. This could have serious implications, for example, if the reconstructed data represent oil flow from an off-shore facility being monitored for taxation purposes.
- High fidelity reconstruction requires that the statistical properties of the reconstructed signal be similar to those of the original. Minimum variance and nonlinearity assessment are two procedures that require high fidelity data. Swinging door compression alters these measures significantly.
- Data smoothing, feature extraction and reconstruction of events require that the events and features of interest be retained during compression.
- The control performance measure should be calculated from uncompressed data. It may provide erroneous results if calculated from highly compressed data.
- The nonlinearity index should be estimated from uncompressed data.

Therefore, a recommendation from Thornhill et al. (2004) was:

Data having $CF \leq 3$ may be used with caution for data-driven process analyses.

The new results from the industrial data set 2 presented in this chapter also support this recommendation.

It is noted, however, that certain types of process trends may allow higher compression factors because their intended use is to record constant values such as set points, targets and high and low limits.

4.7 Automated Detection of Compression

4.7.1 Motivation

As shown in an earlier section, compression induces changes to many of the quantities commonly used in data-driven process analyses. If archived data are to be used for an automated analysis, it is first necessary to test for the presence of compression. If the number of spot values in the compressed archive and the original sampling rate are known, then the compression factor may be determined by calculating the ratio between the expected number of observations and the number of archived observations. However, such information is not always available, and it may be necessary to estimate the compression factor from the reconstructed data only. An automated method for the detection of piecewise linear compression that was originally presented in Thornhill et al. (2004) is outlined in this section.

4.7.2 Compression Detection Procedure

Compression detection exploits the fact that the reconstructed data set is piecewise linear, and its second derivative is therefore zero everywhere apart from at the places where the linear segments join. The presence of the characteristic linear segments can be detected by counting the zero-valued second differences $\Delta(\Delta\hat{y})$ calculated from:

$$\Delta(\Delta\hat{y})_i = \frac{(\hat{y}_{i+1}-\hat{y}_i)/h-(\hat{y}_i-\hat{y}_{i-1})/h}{h} = \frac{\hat{y}_{i+1}-2\hat{y}_i+\hat{y}_{i-1}}{h^2},$$

where $\hat{y}$ is the reconstructed signal and h the sampling interval. The index i ranges from 2 to $N-1$, where N is the number of samples. Suppose the original data set had N values and after compression there are m archived spot values and $m-1$ linear segments. If the reconstructed data are differenced twice, there will be $n = N-m$ second differences whose values are zero. Therefore, the compression factor can be determined from:

$$CF_{est} = \frac{N}{m},$$

where $m = N-n$. For example, with ten data points compressed to four archived values and three linear segments, there are $10-4=6$ differences whose values are zero.

Figure 4.12 shows results for data sets 1, 2 and 3 in Example 1 while Fig. 4.13 compares the estimated compression factor with actual compression factor for

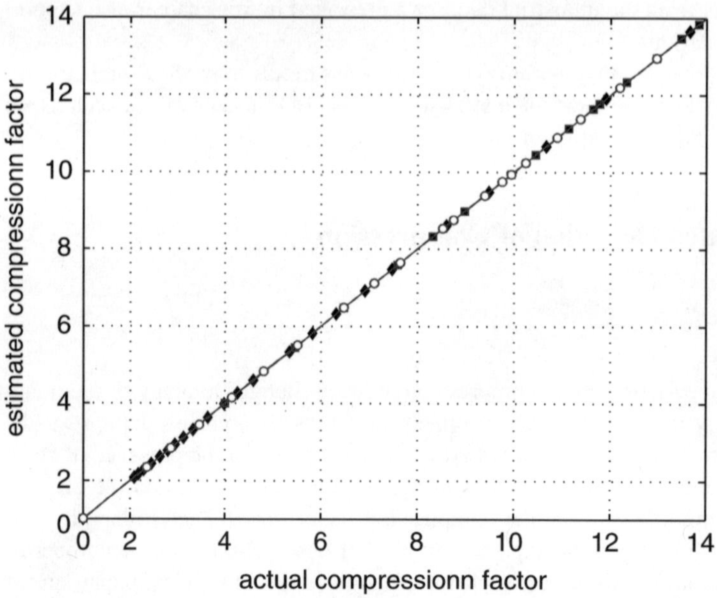

Fig. 4.12 Results from the compression estimation algorithm for data sets 1–3: data set 1 (*circles*), data set 2 (*squares*) and data set 3 (*diamonds*)

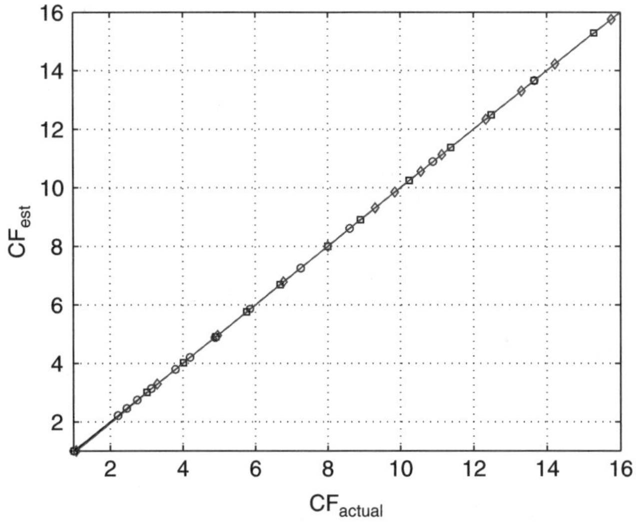

Fig. 4.13 Results from the compression estimation algorithm: flow loop (*circles*), temperature loop (*squares*) and pressure loop (*diamonds*)

Example 2. Results for both examples show that the compression factor derived from counting the zero second differences gives a good estimate of the true compression factor.

4.7.3 Implementation Considerations

Enhancements are needed to the basic algorithm for industrial implementation for the following reasons:

- The sampling interval of the reconstructed signal may be larger than the original; e.g. the compression algorithm may have used 10 s samples but the reconstruction may use 1 min samples.
- The effects of finite precision arithmetic mean that some computed second differences may not be exactly zero.

Suggestions for handling these cases were discussed in detail in Thornhill et al. (2004).

The effects of finite precision arithmetic can be explained with the help of the flow loop of industrial Example 2. The second derivative of the reconstructed compressed flow trend with a compression factor of 10 resembles the plot shown in the left panel of Fig. 4.14. The distribution of the second derivative (right panel of Fig. 4.14) shows that all the zero second derivatives fall in the central bin, whose centre is at zero. Therefore, the size of the central bin provides a good estimate of the number of zero second derivatives. The challenge in this approach is to choose the

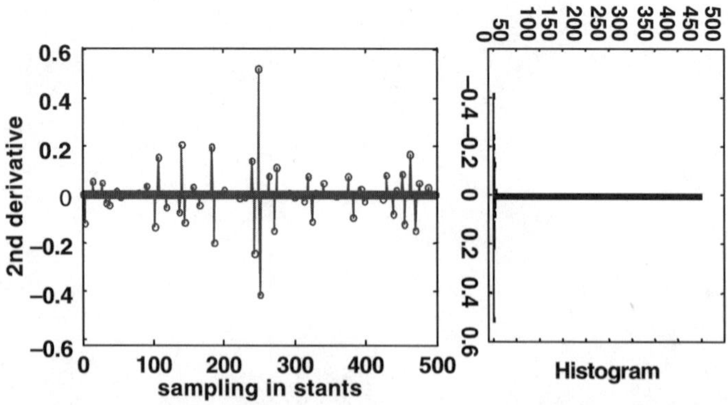

Fig. 4.14 *Left panel* shows the second derivative of the flow trend, *Right panel* shows the distribution of the second derivative

correct number of bins for the histogram. In order to solve this problem, an investigation on how the central bin size varies with the number of bins (Fig. 4.15) reveals that the size of the central bin does not change much beyond a certain number of bins. This was observed for many other compressed data. Thus, the number of bins can be chosen to be greater than the number after the elbow joint in Fig. 4.15. For an automatic compression detection algorithm, the number of bins for the histogram calculation can be fixed anywhere between 300 and 400 for a data set whose length is larger than 1,000 samples.

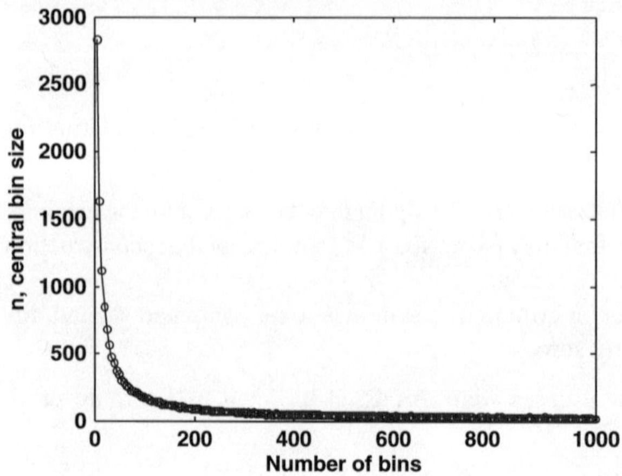

Fig. 4.15 The variation of central bin size with number of bins

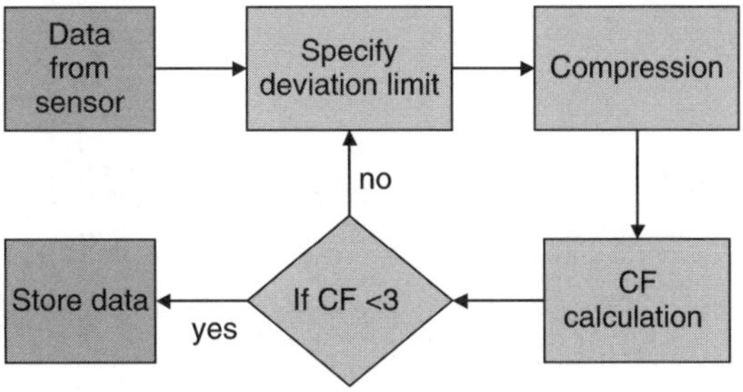

Fig. 4.16 Logic flow diagram for harmless storing of process data

4.8 A Recommendation for Harmless Storing of Data

Ideally, data should not be compressed. If compression is absolutely necessary, they should not be compressed more than a factor of 3. In order to ensure this, an algorithm such as one explained in Fig. 4.16 can be implemented in a data historian before storing the data. From the experience of the authors, if compression is an absolute necessity, then it is recommended that wavelet compression–reconstruction method be used because wavelet transform performs superbly in data reconstruction compared to any other method (Imtiaz et al., 2007).

4.9 Quantization

Modern control systems work with the help of computers and digitization. Because computers cannot read analogue signals, all analogue signals must be converted to digital signals. For digitization, two steps are required: sampling and quantization. Sampling is only the first phase in acquiring data into a computer. Computational processing further requires that the samples be quantized – analogue values are converted into digital form. In short, one needs to perform analogue-to-digital (A/D) conversion. The number of bits in the A/D converter determines the precision of the digitized data. Old A/D converters have fewer number of bits, i.e. low resolution. Those may introduce a significant amount of quantization errors. Significant quantization errors produce oscillations (Horch, 2000) in process variables. Sometimes the quantization errors are too large to use the data for any practical analysis. A method of estimating quantization in process data is now outlined.

Figure 4.17 shows an example of heavily quantized process data. The quantized data usually resembles a staircases and quantization is easy to detect through visualization of the data. However, to check quantization in hundreds of variables in the data- preprocessing stage, an automatic method of detecting quantization is required. We define a quantization factor as:

$$QF \triangleq \frac{Q_{level}}{\sigma}, \tag{4.4}$$

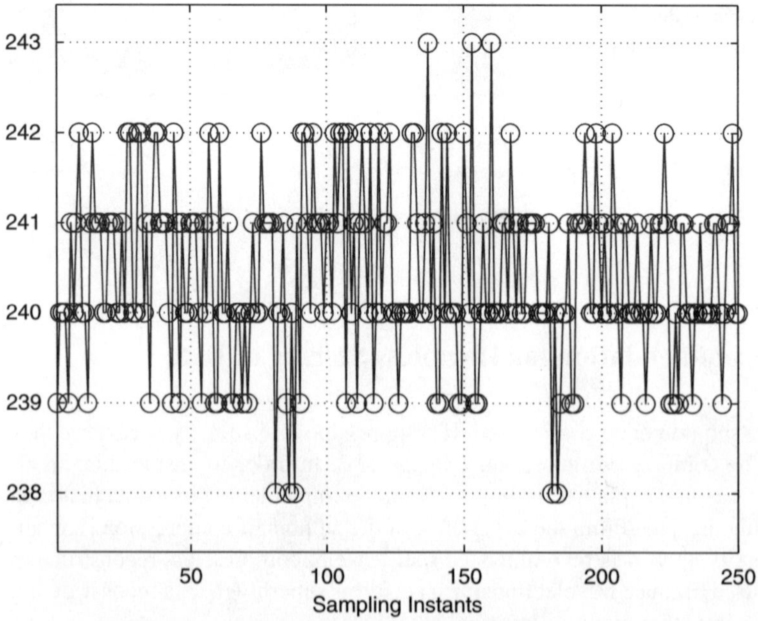

Fig. 4.17 An example of a heavily quantized process variable

where Q_{level} is the quantization level present in the data and can be obtained from the minimum value of the non-zero first difference of the data, σ is the standard deviation and QF is the quantization factor. For data that have no quantization problems, the magnitude of QF should be close to zero. In contrast, for a heavily quantized data set, the magnitude of QF will be closer to 1.

Figure 4.18 depicts an industrial example of heavily quantized data. This data set represents vibration monitoring data of a large pump. The right side of Fig. 4.18 shows the estimated quantization factor for each data tag. Clearly, tags 3, 6, 9, 12 and 13 are heavily quantized. The example shown in Fig. 4.17 represents tag 3 of this industrial data set, for which the estimated quantization factor was 1. For most plants that implement modern control systems, quantization is not a problem because they use high resolution A/D converters. However, many plants still use old control systems and sensors, where quantization remains a problem. This automatic quantization-detection algorithm can be used in the data-preprocessing stage to check for quantization and may serve as a useful diagnostic tool to decide which variables are not worth further analysis.

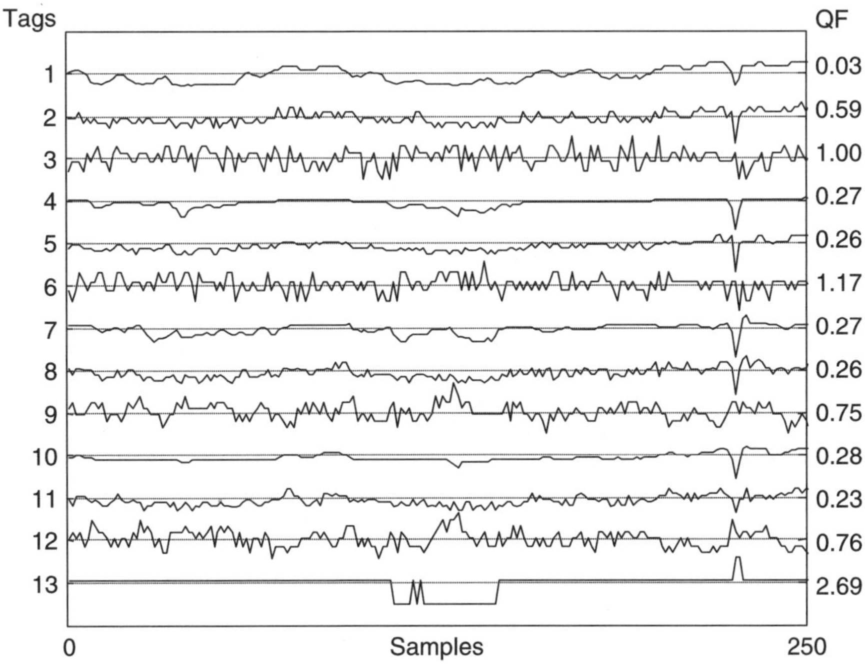

Fig. 4.18 An industrial data set where some variables were heavily quantized

4.10 Summary

The chapter has discussed the impact of compression and quantization on the quality of process data stored in a historical data archive. Examples using real industrial data from continuous processes showed that piecewise linear compression alters key statistical features of the data set such as the mean and standard deviation. Other data-driven analyses were also influenced.

A strong recommendation from this chapter is that operating companies should avoid using data compression when archiving process data, and that archived data are useless for many types of analysis if the compression factor is greater than 3. The algorithm from Thornhill et al. (2004) for the automated detection of compression was reviewed and its effectiveness demonstrated with industrial data sets. It seems essential to run such an algorithm before embarking on any data-driven analysis using archived data. If an archived measurement fails the compression test, it is not qualified for many data-driven analyses and may give spurious or misleading results.

The chapter also considered the influence of quantization. Quantization is often observed in the outputs of process instruments even with a high-resolution analogue to digital converter because a measurement controlled to a steady value may only sample a few of the available quantizer levels. A method for detection of quantization was presented and demonstrated.

Part III
Nonlinearity and Control Performance

Chapter 5
Measures of Nonlinearity – A Review

Nonlinearity plays a significant role in the closed loop performance of a process. If a process is highly nonlinear, a linear controller may not perform well. The presence of nonlinearities in a process can affect its control performance adversely. Because of the nonlinear nature of a process (e.g. in the case of pH control), changes in operating conditions may render the current controller tuning as *non-optimal*. Nonlinearities in valves, e.g. stiction, may cause oscillations in process variables. Such oscillations from one control loop can easily propagate to different units of the plant, thus resulting in plant-wide oscillations. The detection and quantification of process nonlinearity can help decide whether to implement a nonlinear controller or not. It is also useful in diagnosing the root-cause of plant-wide oscillations. The degree of nonlinearity of a process depends on the operating region of the process and the input excitation signals to the process. It is important to be able to measure the degree of nonlinearity of a process under various input excitation signals or operating conditions. This chapter briefly discusses various measures of the degree of nonlinearity.

5.1 Definition of Nonlinear Systems

Nonlinear systems are of interest to researchers because most physical systems in nature are nonlinear. However, nonlinearity is not defined as a rigorous property, rather it has been defined by a contradiction. If a system does not follow the 'linearity theorem', it is termed as nonlinear. There is a formal definition of linear systems, and the systems that do not follow this definition are referred as nonlinear systems. Mathematically, a system or function $f(x)$ is linear if it has the following two properties:

1. Additivity: $f(x+y) = f(x) + f(y)$
2. Homogeneity: $f(\alpha x) = \alpha f(x)$, where α is any constant.

A linear system should satisfy both properties. These two properties in combination are known as *the principle of superposition*. Any system that does not follow the principle of superposition is called a nonlinear system. A simple example of a nonlinear system or function is a quadratic or square function. Neither of the above rules are satisfied by a *square* function. A nonlinear system may exhibit properties such as chaos, bifurcation, limit cycles and multiple isolated equilibrium points. For a sinusoidal input to a nonlinear system, the output may contain many harmonics and subharmonics with various amplitudes and phases. On the other hand, the output of a linear system to a sinusoidal input will have the same frequency with or without a phase lag or a lead.

5.2 Nonlinearity in Process Time Trends

A *linear* time series has a linear dynamic model such as the Box–Jenkins model with constant coefficients driven by Gaussian white noise. A nonlinear time series is generated by a nonlinear process, for instance, a nonlinear feedback function:

$$\begin{aligned} x(n) &= \phi\,(x(n-1), x(n-2), \ldots., u(n-1)) + w(n) \\ y(n) &= h(x(n)) + v(n) \\ u(n) &= g(y(n)), \end{aligned}$$

where $\phi\,(x(n-1), x(n-2), \ldots., u(n-1))$ represents linear dynamics, $x(n)$ is an internal state, $y(n)$ is a measurement from the process, $h(x(n))$ is a measurement function that may be linear or nonlinear, $g(x(n))$ is a nonlinear feedback function and $w(n)$ and $v(n)$ are process and measurement noise, respectively. An example of a nonlinear feedback function is the on–off control of a directly injected steam-heated tank in which the steam valve switches on when the temperature drops to a low limit and switches off when the temperature reaches a high limit. The nonlinear characteristic g in that case is a relay with deadband. The temperature does not remain constant in such a system, rather it cycles in periodic pattern. Control loops having valves with nonlinear friction characteristics and/or nonlinear instrumentation faults can lead to similar behaviour and are described by the same model structure.

The waveform of a limit cycle is periodic but non-sinusoidal and therefore has harmonics. A distinctive characteristic of a nonlinear time series is the presence of phase coupling that creates coherence between frequency bands. The presence of harmonics and phase coupling are therefore indicators of the presence of a nonlinearity.

5.3 Various Measures of Nonlinearity

In recent years, several studies have been undertaken to test the nonlinearity of a system or to obtain a measure of the nonlinearity of a process. There are two broad approaches to measure the nonlinearity of a process. One is based on the input–output relationship or the model of the system, and another is based on the output time series of the process.

5.3.1 *Model-Based Measures of Nonlinearity*

Model-based approaches require a process model for the quantification of nonlinearity. The subject of this book is data-driven approaches, but the main model-based approaches are reviewed here for completeness.

5.3.1.1 Nonlinearity Measures Based on Best Linear Approximation

Helbig et al. (2000) proposed a definition of a nonlinearity measure based on the best linear approximation of a nonlinear input–output system. The nonlinearity of a dynamic system is defined as the normalized largest difference between the nonlinear process and a best linear approximation of the process. It provides a nonlinearity index whose magnitude varies between 0 and 1. However, this method is computationally expensive and requires a fairly accurate model of the process and solution of a complex min–max–min optimization problem. The success of the method depends on the fidelity of the approximation of the nonlinear process by a linear process.

5.3.1.2 Curvature-Based Nonlinearity Measures

Guay (1996) introduced a nonlinearity measure based on the evaluation of the induced local curvature of the process response assuming a process can be represented by a twice-differentiable process map. Guay has used this method to quantify both steady-state and dynamic nonlinearity of chemical processes and demonstrated the potential of the method using simulation example of a bioreactor.

5.3.2 *Time Series-Based Measures of Nonlinearity*

The time series approach requires only the output time series of a process to assess nonlinearity. It does not require identification of the process model.

5.3.2.1 Bicoherence-Based Nonlinearity Measures

Bispectrum and bicoherence metrics have been used to detect the presence of nonlinearity in process data (Emara-Shabaik et al., 1996; Choudhury et al., 2004b). Two indices for testing the 'non-Gaussianity' and 'nonlinearity' of a time series have been defined in Choudhury et al., (2004b, 2006a) and also in Chap. 6 of this book. Zang and Howell (2007) have investigated the types of limit cycles that are amendable to bispectrum analysis.

As discussed in Chaps. 2 and 3, bispectrum is the frequency domain counterpart of the third-order moments and is defined as

$$B(f_1,f_2) \triangleq E\left[X(f_1)X(f_2)X^*(f_1+f_2)\right], \tag{5.1}$$

where, $X(f)$ is the Fourier transformation of the data series $x(t)$. Equation (5.1) shows that it is a complex quantity having both magnitude and phase. It can be plotted against two independent frequency variables, f_1 and f_2 in a three-dimensional plot. Each point in the plot represents the bispectral energy content of the signal at the bifrequency, (f_1,f_2). In fact, the bispectrum, $B(f_1,f_2)$, at point (f_1,f_2) measures the nonlinear interaction between signal components at frequencies f_1 and f_2 (Nikias and Petropulu, 1993). This interaction can be related to the nonlinearities present in the signal-generating systems. These properties are exploited in the detection and diagnosis of nonlinearities.

The bispectrum is normalized in the following way to give a measure called bicoherence whose magnitude is bounded between 0 and 1:

$$bic^2(f_1,f_2) \triangleq \frac{|E\left[B(f_1,f_2)\right]|^2}{E\left[|X(f_1)X(f_2)|^2\right]E\left[|X(f_1+f_2)|^2\right]}, \tag{5.2}$$

where 'bic' is known as the 'bicoherence' function.

A stationary time series, $x(k)$, is called linear or can be treated as the output of a linear system, if it can be represented by

$$x(k) = \sum_{n=0}^{M-1} h(n)e(k-n) \tag{5.3}$$

where $e(k)$ is a sequence of independent identically distributed random variables with $E[e(k)] = 0$, $\sigma_e^2 = E[e^2(k)]$ and $\mu_3 = E[e^3(k)]$.

It can be shown that for a linear time series

$$bic^2(f_1,f_2) = \frac{\mu_3^2}{\sigma_e^6}. \tag{5.4}$$

Equation (5.4) shows that for any linear signal, x, the squared bicoherence will be independent of the bifrequencies, i.e. a constant in the bifrequency plane. If the squared bicoherence is zero, the signal x is Gaussian because the skewness or μ_3 is also zero in such a case. For a nonlinear signal, the bicoherence will not be zero or a non-zero constant. In other words, for a nonlinear time series, the bicoherence will no longer be frequency independent. A detailed discussion on the bicoherence-based nonlinearity measures is provided in Chap. 6.

5.3.2.2 Surrogate Data-Based Nonlinearity Measure

A distinctive characteristic of a nonlinear time series is the presence of phase coupling that creates coherence between different frequency bands in the spectrum. In the case of a limit cycle, the phase structure is what gives the waveforms its distinctive non-sinusoidal waveform. The purpose of surrogate data methods is to create synthetic data sets called surrogate time series having some of the same properties as the time series under test, for instance the same power spectrum, but with the phase coupling removed by randomization. A key property of the test time series is then compared to that of its surrogates, and nonlinearity is diagnosed if the property is significantly different in the test time series (Theiler et al., 1992; Kantz and Schreiber, 1997).

A basic surrogate time series with the same power spectrum as the test time series is created by the procedure outlined in the following pseudo-code, which is for the case when the test time series has an even number of samples, N. The quantity ϕ_k, $k = 1,\ldots,(N/2-1)$, is a random phase in the range 0–2π while FFT and IFFT are the forward and inverse discrete Fourier transforms.

$$z = FFT\,(\text{test time series})$$

$$z_{surr} = \begin{cases} z[i] & i = 1 \\ z[i]e^{j\phi_{i-1}} & i = 2,\ldots,N/2 \\ z[i] & i = N/2+1 \\ z[i]e^{-j\phi_{N-i+1}} & i = (N/2+2),\ldots,N \end{cases}$$

$$\text{surrogate data} = IFFT\,(z_{surr})$$

The assignment of values to z_{surr} in the above expression takes account of the zero and Nyquist frequencies in frequency channels 1 and $N/2+1$, and assigns a complex conjugate phase to the mirror-imaged aliased frequency components above the Nyquist sampling frequency. The TISEAN project (Hegger et al., 1999; Hegger et al., 2000) and the text book of Kantz and Schreiber (1997) present detailed methods and codes for the generation of surrogate data and for testing of time series for evidence of nonlinearity.

In the area of process modelling, Aldrich and Barkhuizen (2003) detected nonlinearity in process data by comparing the singular values of a data matrix comprising time-shifted vectors of the test data against those from surrogate data, while Barnard et al. (2001) and Theron and Aldrich (2004) explored system identification using surrogate methods to detect when the data are the outputs of a nonlinear process, as well as to validate the models. In this monograph, the use of surrogate data to determine nonlinear root causes of plant-wide disturbances is discussed in greater detail in Chaps. 7 and 19, based on the exposition of Thornhill (2005a).

5.3.2.3 Harmonic Analysis

Early studies used the presence of prominent harmonics as an indicator of nonlinearity on the grounds that the waveform in a limit cycle is periodic but generally is non-sinusoidal and therefore has harmonics. The presence of harmonics in a time series has been used successfully for the diagnosis of SISO control-loop faults

(Paulonis and Cox, 2003; Ruel and Gerry, 1998; Thornhill and Hägglund, 1997). Finding harmonics requires signal processing to isolate the spectral frequencies of interest and further inspection to confirm that the frequencies are integer multiples of a fundamental.

It is not always true, however, that the time trend with the largest harmonic content is the root cause because the action of a control loop may split the harmonics of an incoming disturbance between the manipulated variable and the controlled variable. Figure 5.1 shows an example whose root cause is a tightly tuned pressure control loop that is limit cycling due to saturation, as can be seen in the manipulated variable PC.MV. The spectra of PC.PV and PC.MV comprises a fundamental oscillation and a second harmonic. These pressure variations disturbed a downstream-level control loop. The manipulated variable for the level control loop (LC.MV) contains both the fundamental oscillation and a second harmonic while the controlled variable (LC.PV) has a strong second harmonic with little sign of the fundamental oscillation. These spectra show that the level controller is diverting the fundamental component of the oscillating disturbance away from the controlled variable. Harmonic analysis would (wrongly) suggest the controlled variable in the level loop LC.PV as the root cause because of its strong second harmonic. Non-linearity assessment, by contrast, correctly identifies the time trends PC.PV and PC.MV to be the most nonlinear element (Matsuo et al., 2004). Further examples are discussed and analysed in (Zang and Howell, 2005).

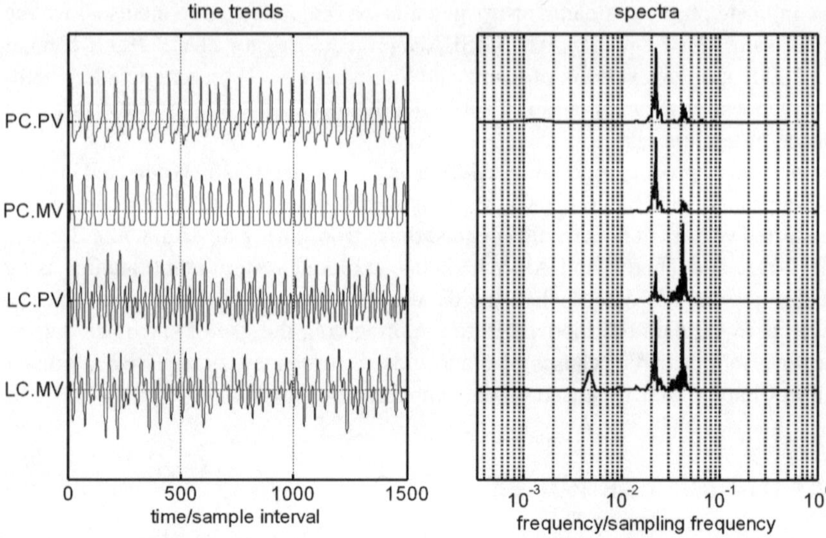

Fig. 5.1 An illustration of a false result from harmonic analysis. Harmonic analysis wrongly suggests LC.PV as the root cause because of its strong second harmonic. Nonlinearity assessment correctly points to PC.MV from an upstream pressure controller as the root cause of the nonlinearity

5.3.2.4 Methods from Nonlinear Time Series Analysis

Several other measures of nonlinearity have been proposed and compared in the nonlinear time series literature. For example, Schreiber and Schmitz (1997) compared the discrimination power of five different nonlinear measures such as two different correlation dimensions, nonlinear predictive error, third-order cumulant and time reversibility.

Correlation dimension and maximal Lyapunov exponent have also been examined for the diagnosis of nonlinearity in chemical processes by Zang and Howell (2004). These are promising because their numerical values are often different for the primary disturbance compared to secondary propagated disturbances. The same authors (Zang and Howell, 2005) compared these and several other nonlinearity assessment methods finding that most of the approaches gave the correct diagnoses in an industrial example.

5.4 Summary

This chapter has discussed various measures or metrics of nonlinearity. There are two broad approaches to measure nonlinearity – model based and data based. Model-based techniques are useful in the area of control relevant nonlinearity under various input excitation signals while the data-based methods find applications in detecting and diagnosing the root causes for poor control performance. Data-based methods were classified in the chapter under two main headings: those using bicoherence methods and those using methods from nonlinear time series analysis. The following chapters demonstrate the formulation and application of nonlinearity tests based on the bicoherence method and the nonlinear time series analysis method.

Chapter 6
Linear or Nonlinear? A Bicoherence-Based Measure of Nonlinearity

This chapter discusses bicoherence-based measures of nonlinearity and non-Gaussianity. First, an improved method for estimating bicoherence with fewer number of spurious peaks is presented. Then, two indices – non-Gaussianity index (NGI) and nonlinearity index (NLI) – are developed to test signals or time series data for the possible presence of non-Gaussianity and nonlinearity. Finally a new index called the total nonlinearity index (TNLI) is defined to quantify nonlinearities. The efficacy of the proposed indices is demonstrated using illustrative examples.

6.1 Introduction

Classical signal-processing tools utilize only the first- and second-order moments, i.e. the mean, variance, covariance and correlation. Such tools are mainly useful for analysing signals from linear processes. The distribution of signals from nonlinear processes is often skewed and non-Gaussian. This necessitates the use of higher-order statistical tools. The third- and fourth-order moments and their frequency domain counterparts (bispectrum and trispectrum) are found to be useful in analysing nonlinearities in communication signals and mechanical machine-condition monitoring (Nikias and Petropulu, 1993; Collis et al., 1998). These higher-order statistical techniques have also been used to detect and diagnose nonlinearities in control valves used in the process industries (Choudhury et al., 2004b; Choudhury, 2004).

In the estimation of the bicoherence, many spurious peaks arise due to the occurrence of small magnitudes in the denominator used to normalize the bispectrum. Collis et al. (1998) suggested an addition of a small constant to the denominator to remove these spurious peaks. However, the use of a fixed or stationary constant does not produce reliable results. Here we describe a method to choose this constant dynamically, depending on the noise level of the time series. Choosing the constant dynamically helps to obtain a better estimate of the bicoherence with a significant reduction of spurious peaks.

6.2 Bispectrum and Bicoherence

Industrial processes generally deviate from Gaussianity and linearity and exhibit nonlinear behaviour. These processes can conveniently be studied using higher-order statistics (HOS) (Nikias and Petropulu, 1993). The bispectrum is the simplest of the various frequency domain HOS measures. It is the frequency domain counterpart of the third-order moments and is defined as

$$B(f_1,f_2) \triangleq E[X(f_1)X(f_2)X^*(f_1+f_2)], \tag{6.1}$$

where $X(f)$ is the Fourier transform of the data series $x(t)$. Equation (6.1) shows that it is a complex quantity having both magnitude and phase. It can be plotted against two independent frequency variables, f_1 and f_2, in a three-dimensional plot. Each point in the plot represents the bispectral energy content of the signal at the bifrequency, (f_1, f_2). In fact, the bispectrum, $B(f_1, f_2)$, at point (f_1, f_2) measures the nonlinear interaction between frequencies f_1 and f_2 (Nikias and Petropulu, 1993). This interaction between frequencies can be related to the nonlinearities present in the signal-generating systems and therein lies its usefulness in the detection and diagnosis of nonlinearities.

The bispectrum is normalized in the following way to give a measure called bicoherence whose magnitude is bounded between 0 and 1:

$$bic^2(f_1,f_2) \triangleq \frac{|E[B(f_1,f_2)]|^2}{E\left[|X(f_1)X(f_2)|^2\right]E\left[|X(f_1+f_2)|^2\right]}, \tag{6.2}$$

where 'bic' is the bicoherence function. Equation (6.2) can be rewritten as

$$bic^2(f_1,f_2) \triangleq \frac{|E[X(f_1)X(f_2)X^*(f_1+f_2)]|^2}{E\left[|X(f_1)X(f_2)|^2\right]E\left[|X(f_1+f_2)|^2\right]} \tag{6.3}$$

If Welch's periodogram method is used to estimate the bicoherence, the expectation operator can be replaced with a summation operator over the number of data segments using the assumption of ergodicity:

$$\hat{bic}^2(f_1,f_2) = \frac{\left|\frac{1}{M}\sum_{i=1}^{M} X_i(f_1)X_i(f_2)X_i^*(f_1+f_2)\right|^2}{\frac{1}{M}\sum_{i=1}^{M} |X_i(f_1)X_i(f_2)|^2 \cdot \frac{1}{M}\sum_{i=1}^{M} |X_i(f_1+f_2)|^2} \tag{6.4}$$

The squared bicoherence is usually estimated using this equation.

6.2.1 Spurious Peaks in the Estimated Bicoherence

Often, there are small spurious peaks in the bicoherence plot, which make the interpretation of the peaks difficult. An example can be seen in the left panel (both top and bottom) of Fig. 6.2.

Collis et al. (1998) reported that these spurious peaks are due to the small values occurring in the denominator of the bicoherence (Eq. (6.4)), which cause the estimate to become ill-conditioned. To mitigate such effects, they suggest adding a small constant, ε, to the denominator of Eq. (6.2):

$$bic^2(f_1,f_2) \triangleq \frac{|E[B(f_1,f_2)]|^2}{E[|X(f_1)X(f_2)|^2]E[|X(f_1+f_2)|^2]+\varepsilon}. \tag{6.5}$$

It is also reported in their paper that the disadvantage of using this correction technique is to introduce a negative bias error in the bicoherence estimator. One needs to select ε such that it is large enough to remove spurious artifacts, but not so large as to cause significant bias effects. However, they did not suggest a suitable numerical value or any method to choose a correct value of ε. In our experience of working with many industrial time series data, a static numerical value of ε does not provide a good estimate of the bicoherence for all cases (see the middle panel in Fig. 6.2). Also, it may introduce bias errors large enough to mask the significant peaks. Therefore, the magnitude of ε needs to be changed dynamically with the calculation of bicoherence for each time series. This point will be illustrated using the following example.

6.2.2 *Illustrative Example 1*

This example illustrates the effect of ε on the estimation of bicoherence of a quadratically phase-coupled (QPC) signal under the influence of various noise levels. Let a QPC signal be constructed as follows:

$$\begin{aligned} y(t) &= sin(2\pi f_1 t+\phi_1)+sin(2\pi f_2 t+\phi_2) \\ &\quad + 0.1\ sin(2\pi f_3 t+\phi_3)+n(t), \end{aligned} \tag{6.6}$$

where the values of f_1, f_2 and f_3 are 0.12, 0.18 and 0.30, respectively; the values of ϕ_1, ϕ_2 and ϕ_3 are $\frac{\pi}{3}$, $\frac{\pi}{12}$ and $\frac{5\pi}{12}$, respectively; $n(t)$ is a zero-mean white noise signal and t corresponds to discrete sample instants from 1 to 4096 s. The signal $y(t)$ is a quadratic phase-coupled signal because its frequencies have the relation $f_1+f_2=f_3$ and its phases have the relation $\phi_1+\phi_2=\phi_3$. Therefore, the phase coupling at bifrequency $(0.12,0.18)$ should appear as a single peak in the bicoherence plot. Let the signal-to-noise ratio (*SNR*) be defined as:

$$\mathrm{SNR} = \frac{variance\ of\ signal}{variance\ of\ noise}. \tag{6.7}$$

All subplots in Fig. 6.2 show that there is one peak at the bifrequency $(0.12,0.18)$. The top panel of Fig. 6.2 shows the case for $SNR=3$ while the bottom panel is for $\mathrm{SNR}=10$. As evident from this figure, if $\varepsilon=0$ is used, there are many small spurious peaks in the bicoherence plot. As expected, the scenario is worse for the case of low signal-to-noise ratio. From the top-left plot of Fig. 6.2, it is hard to find the true peak. For a static value of ε (in this case, chosen as $\varepsilon = 0.00001$), the spurious peaks are absent, but the bias effect on the bicoherence magnitude is very significant. Also, it is difficult to choose a proper static value for ε. In fact, one needs to adopt a trial and error procedure. For the known signals whose characteristics are known, a proper value for ε using a trial and error technique can be chosen without the fear of loosing the significant peaks. But such is not the case for the practical usage of the bicoherence, where the peaks of bicoherence are unknown and uncertain. Therefore, a better method for choosing ε is sought. The following section suggests a new method of choosing ε dynamically.

6.2.3 How to Choose ε?

Let us denote the denominator of Eq. (6.4) as D. The terms in the denominator depend on the power spectrum. The behaviour of the denominator D is qualitatively similar to that of the power spectrum of the original signal. If the signal has one or more spectral peaks at specific frequencies then the denominator also shows features at the same frequencies. Figure 6.1 shows D plotted against f_1 for $0 \leq f_1 \leq 0.5$. The multiple lines in the plot represent various values of f_2 since D depends on both f_1 and f_2. This figure shows that there are only a few peaks and the rest are small mainly due to noise. The noise level is denoted as a dotted line in Fig 6.1. Choosing the magnitude of ε at this level helps to get rid of most of the spurious peaks. In order to obtain the value of ε automatically, it can be chosen as the maximum of the Pth percentiles of the calculated values of $D(f_1,f_2)$. If it is assumed that 25% of the values in $D(f_1,f_2)$ constitute the peaks, then P can be chosen as the 75th percentile. The dotted line in Fig. 6.1 is drawn at the maximum of the 75th percentiles of the calculated values of $D(f_1,f_2)$.

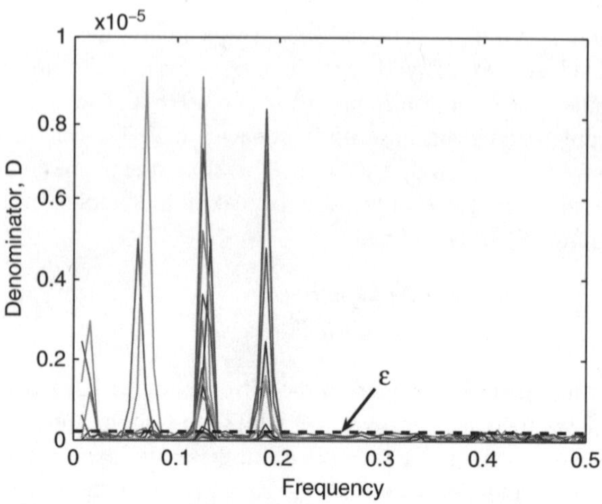

Fig. 6.1 Denominator of Eq. (6.2) for illustrative example 1

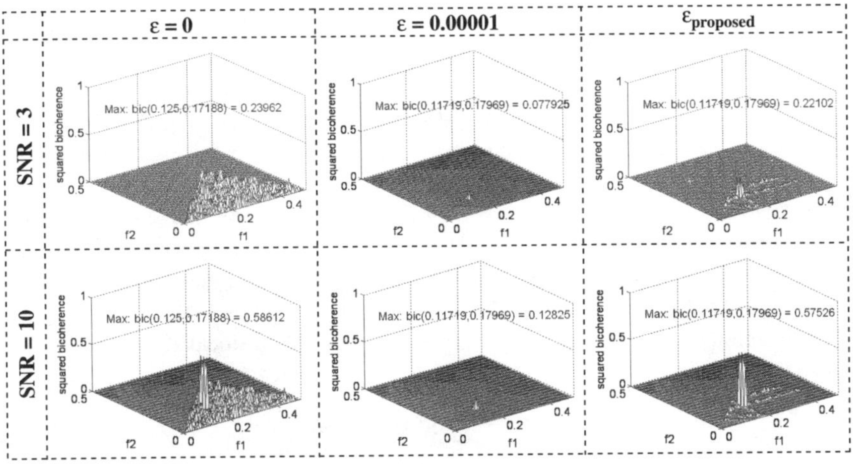

Fig. 6.2 Effect of ε in bicoherence estimation

Two plots in the right column of Fig. 6.2 show the improvement of the bicoherence estimation using the proposed ε. The use of the proposed dynamic ε could clearly remove most of the spurious peaks without introducing a significant negative bias effect on the bicoherence peak, which can be observed by comparing the maximum peak magnitudes for the cases of $\varepsilon = 0$ (left column plots) and $\varepsilon_{proposed}$ (right column plots).

6.3 Test of Gaussianity and Linearity of a Signal

Testing for non-Gaussianity and nonlinearity in a system can serve as a useful diagnostic tool to analyse poor performance of a control loop. Over the last few decades, many researchers have used the notion of bispectrum to test Gaussianity and linearity of a time series (Hinich, 1982; Rao and Gabr, 1980, 1984; Collis et al., 1998; Terdik and Máth, 1998; Yuan, 1999). Hinich (1982) and Yuan (1999) constructed statistical hypothesis test using sample interquartile range of the skewness function for testing the linearity of a time series. Rao and Gabr (1980) used bispectral density function (bispectrum) to construct a Hotelling T^2-test for the same purpose. The problem of using bispectrum or skewness function for testing the linearity of time series is that the magnitudes of these functions are not bounded. It is difficult to compare magnitudes of an unbounded function when applied to multiple time series or signals. Terdik and Máth (1998) also used bispectrum to test the null hypothesis that a predictor of a time series is linear against the alternative that the predictor is quadratic. This is a narrow and less powerful test because, often, the type of nonlinearity of a signal is unknown. Collis et al. (1998) used the bicoherence function whose magnitude is bounded between 0 and 1 to check the linearity of a signal or time series. However, they did not construct any statistical test. In this work, bicoherence is used to construct a chi-square test for examining the nonlinearity of a time series or signal. The hypothesis tests are incorporated in two new indices – the Non-Gaussianity Index (NGI) and the Nonlinearity Index (NLI). The magnitudes of the new indices are always bounded between -1 and 1 because bicoherence, whose magnitude is also bounded between 0 and 1, is used to define them. The definition and development of these indices using statistical tests on a bounded function (bicoherence) allow the users to apply them for comparing multiple time series or control loops. The derivation and development of these indices are described below.

A discrete ergodic stationary time series, $x(k)$, is called linear if it can be represented by

$$x(k) = \sum_{n=0}^{M-1} h(n)e(k-n), \tag{6.8}$$

where $e(k)$ is a sequence of independent identically distributed random variables with $E[e(k)] = 0$, $\sigma_e^2 = E[e^2(k)]$ and $\mu_3 = E[e^3(k)]$. For this case, the following frequency domain relationships can be obtained.

$$\text{The power spectrum}: \quad P(f) = \sigma_e^2 \mid H(f) \mid^2 \equiv |X(f)X^*(f)| \tag{6.9}$$

$$\text{and the bispectrum}: \quad B(f_1, f_2) = \mu_3 H(f_1)H(f_2)H^*(f_1+f_2), \tag{6.10}$$

where $H(f) = \sum_{n=0}^{M-1} h(n)\exp(-2\pi i f n)$. Equation (6.2) can be rewritten as

$$\begin{aligned} bic^2(f_1,f_2) &\triangleq \frac{|B(f_1,f_2)|^2}{E[|X(f_1)X^*(f_1)||X(f_2)X^*(f_2)|] \;\; E[|X(f_1+f_2)X^*(f_1+f_2)|]} \\ &\equiv \frac{|B(f_1,f_2)|^2}{E[|P(f_1)||P(f_2)|]E[|P(f_1+f_2)|]} \end{aligned} \tag{6.11}$$

For a linear time series, by substituting expressions from Eqs. (6.9) and (6.10), it can be shown that

$$bic^2(f_1,f_2) = \frac{\mu_3^2}{\sigma_e^6} \tag{6.12}$$

Equation (6.12) shows that for any linear signal, x, the squared bicoherence will be independent of the bifrequencies, i.e. a constant in the bifrequency plane. If the squared bicoherence is zero, the signal x is Gaussian because the skewness or μ_3 is also zero in such a case. Strictly speaking, such a signal should be called non-skewed with a symmetric distribution rather than Gaussian. However, in most of the HOS literature (Nikias and Petropulu, 1993; Hinich, 1982; Rao and Gabr, 1980; Kim and Powers, 1979; Collis et al., 1998; Fackrell, 1996), the two terms – nonskewed and Gaussian – have been used interchangeably. To check whether the squared bicoherence is constant, two tests are required. One is to determine whether the squared bicoherence is zero, which would show that the signal is Gaussian and thereby the signal-generating process is linear. The other is to test for a non-zero constant squared bicoherence, which would show that the signal is non-Gaussian but the signal- generating process is linear.

The bicoherence is a complex quantity with real and imaginary parts. The square of the magnitude of the bicoherence can be obtained as

$$bic^2 = \Re(bic)^2 + \Im(bic)^2, \tag{6.13}$$

where $\Re$ and $\Im$ are real and imaginary parts, respectively. It is well established in the HOS literature that bicoherence is a complex normal variable, i.e. both the estimates of real and imaginary parts of the bicoherence are normally distributed (Hinich, 1982) and asymptotically independent, i.e. the estimate at a particular bifrequency is independent of the estimates of its neighbouring bifrequencies (Fackrell, 1996). Therefore, for linear signals, the squared bicoherence at each frequency is a central chi-squared (χ^2) distributed variable with 2 degrees of freedom. Hinich (1982) showed that the signal of interest is Gaussian if the skewness function is asymptotically centrally χ^2 distributed with 2 degrees of freedom. He used this information to construct statistical hypothesis tests for examining bicoherence at each frequency in the principal domain. A disadvantage of this test is that while applying to each of the bifrequencies in the principal domain of squared bicoherence plot, the probability of false detection accumulates owing to a large number of bifrequencies in the principal domain. Thus it overestimates the number of bifrequencies in which the bicoherence magnitude is significant. Therefore, a test based on the average squared bicoherence over the principal domain has been formulated in Choudhury et al. (2004b). However, this approach resulted in the loss of frequency resolution. At the time when the results of the 2004 paper by Choudhury et al. (2004b) were published, the improved method of bicoherence estimation was not available. Subsequent to the publication of this paper our practical evaluation of the method motivated us to improve the bicoherence estimation method to get rid of most of the spurious peaks during the estimation. With this refinement, the statistical test to check the significance of bicoherence can be applied at each bifrequency with a lower probability of false 'positives'.

The following statistical test is suggested to check for the significance of bicoherence magnitude at each individual bifrequency:

$$P\{2K\, bic^2(f_1, f_2) > c_\alpha^{\chi^2}\} = \alpha \tag{6.14}$$

or,

$$P\{bic^2(f_1, f_2) > \frac{c_\alpha^{\chi^2}}{2K}\} = \alpha, \tag{6.15}$$

where K is the number of data segments used in bicoherence estimation and $c_\alpha^{\chi^2}$ the critical value calculated from the central χ^2 distribution table for a significance level of α with two degrees of freedom. For example, for $\alpha = 0.05$, the value of $c_{0.05}^{\chi^2}$ is 5.99.

Often the principal domain of the bicoherence plot contains more than a hundred bifrequencies. The hypothesis test results for this large number of bifrequencies can be conveniently summarized into the following Non-Gaussianity Index (NGI):

$$\mathrm{NGI} \triangleq \frac{\sum bic^2_{significant}}{L} - \frac{c_\alpha^{\chi^2}}{2KL}, \tag{6.16}$$

where $bic^2_{significant}$ are those bicoherence values which satisfy Eq. (6.15), i.e. $bic^2(f_1, f_2) > \frac{c_\alpha^{\chi^2}}{2K}$, and L is the number of $bic^2_{significant}$. Therefore, the following rule-based decision is suggested:

- if NGI ≤ 0, the signal is GAUSSIAN
- if NGI > 0, the signal is NON-GAUSSIAN

Thus, a signal is non-skewed or Gaussian at a confidence level of α if the NGI is ≤ 0. Thus NGI facilitates the automation of this statistical hypothesis test.

If a signal is found to be Gaussian, the signal-generating process is linear (Rao and Gabr, 1980). In the case of a non-Gaussian signal, the signal generating process should be tested for its linearity. As shown in Eq. (6.12), if the signal is non-Gaussian and linear, the magnitude of the squared bicoherence should be a non-zero constant at all bifrequencies in the principal domain because μ is a non-zero constant in such a case.

A simple way to confirm the constancy of squared bicoherence is to examine the three-dimensional squared bicoherence plot and observe the flatness of the plot. However, this can be tedious and cumbersome when the method is applied to a large number of time series. Alternatively, if the squared bicoherence is of a constant magnitude at all bifrequencies in the principal domain, the variance of the estimated bicoherence should be zero. To check the flatness of the plot or the constancy of the squared bicoherence, a nonlinearity index has been reported in Choudhury et al. (2004b) in which the maximum squared bicoherence is compared with the average squared bicoherence plus two times the standard deviation of the estimated squared bicoherence. The disadvantage of using this index is that the presence of a few large peaks significantly bias the standard deviation and mean of the estimated bicoherence, which leads to some false negatives. In order to avoid these limitations, the nonlinearity index defined in (Choudhury et al., 2004b) was modified in (Choudhury et al., 2006a) and can be stated as:

$$\mathrm{NLI} \triangleq \hat{bic}^2_{max} - \left(\overline{\hat{bic}^2_{robust}} + 2\sigma_{\hat{bic}^2,\, robust} \right), \tag{6.17}$$

where $\overline{\hat{bic}^2_{robust}}$ and $\sigma_{\hat{bic}^2,\, robust}$ are, respectively, the robust mean and the robust standard deviation of the estimated squared bicoherence. They are calculated by excluding the largest and smallest $Q\%$ of the bicoherence. A good value of Q may be chosen as 10. There are similar concepts used in the statistical literature, for example, robust mean, trimmed mean and interquartile mean.

Therefore, it can be concluded that:

- if NLI ≤ 0, the signal generating process is LINEAR
- if NLI > 0, the signal generating process is NONLINEAR

Since the squared bicoherence is bounded between 0 and 1, the Nonlinearity Index (NLI) is also bounded between -1 and 1.

6.3.1 Total Nonlinearity Index (TNLI)

It is important to measure nonlinearity in terms of a metric especially when there is a need to compare the extent of nonlinearities in various time series. A limitation of using NLI to measure the extent of nonlinearity is that it only uses information from the maximum bicoherence peak. Therefore, it will only capture the extent of nonlinearity induced by the interaction of two frequencies corresponding to the location of the maximum bicoherence peak. In order to capture all nonlinear interactions among the frequency channels, a new index called the 'Total Nonlinearity Index (TNLI)' is introduced. If a time series is detected as non-Gaussian and nonlinear by the above-mentioned NGI and NLI indices, then the total nonlinearity present in the time series can be quantified using the following new index:

$$\text{TNLI} \triangleq \sum bic^2_{significant}, \tag{6.18}$$

where TNLI is the 'Total Nonlinearity Index', $bic^2_{significant}$ are those bicoherence values which satisfy Eq. (6.15), i.e. $2K\ bic^2(f_1, f_2) > c_\alpha^{\chi^2}$. The TNLI is bounded between 0 and L, where L is the number of $bic^2_{significant}$.

6.4 Illustrative Example 2: Bicoherence of a Linear and a Nonlinear Signal

Two signals, y_{linear} and $y_{nonlinear}$, were generated using the following equations.

$$x(k) = 3F(q^{-1})d_1(k)$$

$$y_{linear} = x(k) + d_2(k) \tag{6.19}$$

$$y_{nonlinear} = x(k) + 0.1x(k)^2 + d_2(k), \tag{6.20}$$

where $d_1(k)$ and $d_2(k)$ are zero-mean white noise sequences with variance 1 and 0.001, respectively, and $F(q^{-1})$ is a third-order narrow pass Butterworth filter with a frequency range 0.095 – 0.105 in a 0 – 0.5 normalized frequency scale, such that $f_s = 1$ is the sampling frequency. Figure 6.3 shows the frequency response of the Butterworth filter.

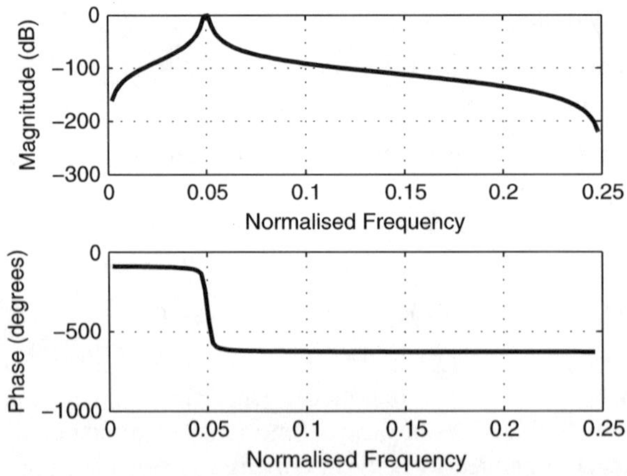

Fig. 6.3 Frequency response of the designed Butterworth filter

The purpose of this example is to demonstrate the power of the bicoherence in the detection of nonlinearity. By merely looking at the time trend of the signals (the left panel of Fig. 6.4), it is not possible to differentiate between them. Also, the power spectra (the middle panel of Fig. 6.4) or the second-order moments look alike and are unable to distinguish the two signals. The right panel of Fig. 6.4 shows the three-dimensional bicoherence plots. For y_{linear}, the test result is NGI = 0. The NGI indicates that the probability distribution of the signal is symmetrical. Therefore, the nonlinearity test result is not required here. In contrast, for $y_{nonlinear}$, the NGI equals 0.53, thereby detecting that the probability density function of the signal is

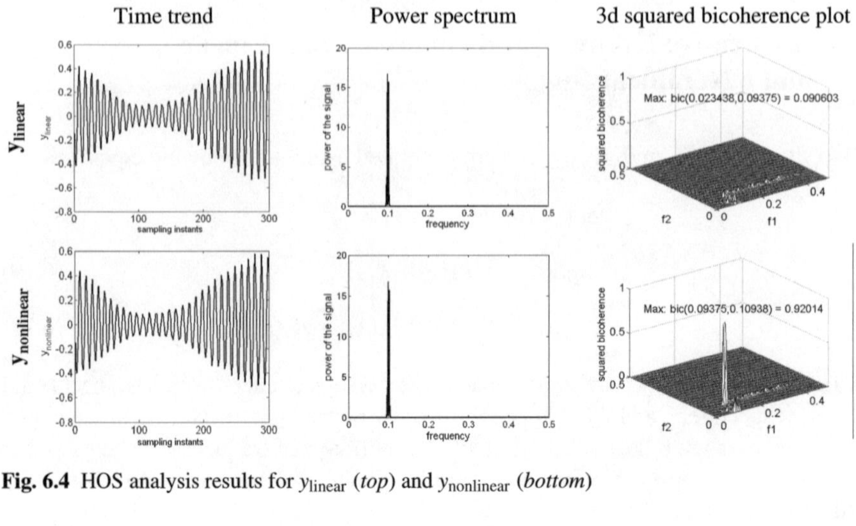

Fig. 6.4 HOS analysis results for y_{linear} (*top*) and $y_{\text{nonlinear}}$ (*bottom*)

not symmetrical, which is an indication that it may be nonlinear. The nonlinearity test gives NLI $= 0.91$, which clearly indicates the presence of nonlinearity in this signal. The total nonlinearity index for this signal is 3.71. From the bicoherence plot, the peak position in the principal domain is approximately at the (0.1,0.1) bifrequency. This means that the nonlinearity in the signal is due to the interaction of these two frequencies. Examination of the signal-generating system reveals that the band-pass filtered signal has the frequency range [0.095–0.105]. This signal was squared to introduce nonlinearity. Thus, the nonlinearity is due to the multiplication of two signals, each having a frequency of approximately 0.1. The bicoherence plot (right plot on the second row) shows that the bifrequency of the identified peak is (0.1,0.1). Therefore, the HOS-based method is capable of not only detecting the nonlinearity but it can also correctly identify the frequencies of nonlinear interactions.

6.5 Illustrative Example 3: Bicoherence of a Nonlinear Sinusoid Signal with Noise

An input signal was constructed by adding two sinusoids, each having a different frequency and phase. That is,

$$\begin{aligned} x'(k) &= sin(2\pi f_1 k + \phi_1) + sin(2\pi f_2 k + \phi_2) \\ x(k) &= x'(k) + d(k) \\ y(k) &= x'(k) + n_l x'(k)^2 + d(k), \end{aligned} \tag{6.21}$$

where $f_1 = 0.12$, $f_2 = 0.30$, $\phi_1 = \pi/3$, $\phi_2 = \pi/8$, n_l is a multiplication factor employed to represent the contribution of the nonlinear component of the signal and $d(k)$ is a white noise sequence with variance 0.04. Again, frequencies are normalized such that the sampling frequency is 1.

The quadratic term in Eq. (6.21) will introduce phase coupling in the output signal, y. It can be better understood by rewriting Eq. (6.21) in the following form:

$$\begin{aligned} y(k) = {} & \sin(2\pi f_1 k + \phi_1) + \sin(2\pi f_2 k + \phi_2) + 0.25[1 - \cos(2(2\pi f_1 k + \phi_1)) \\ & - \cos(2(2\pi f_2 k + \phi_2)) + \cos(2\pi(f_2 - f_1)k + \phi_2 - \phi_1) \\ & - \cos(2\pi(f_1 + f_2)k + \phi_1 + \phi_2)] + d(k). \end{aligned} \tag{6.22}$$

The quadratic nonlinearity generates the sinusoidal components in Eq. (6.22) with frequencies $2f_1$, $2f_2$, $f_2 - f_1$ and $f_2 + f_1$. Six possible frequency couplings are presented in Table 6.1.

Table 6.1 QPC relations for the output signal, y

	Frequency relations	Phase relations	Peak locations
1	$f_1 + f_1 = 2f_1$	$\phi_1 + \phi_1 = 2\phi_1$	(f_1, f_1)
2	$f_2 + f_2 = 2f_2$	$\phi_2 + \phi_2 = 2\phi_2$	(f_2, f_2)
3	$f_1 + f_2 = f_1 + f_2$	$\phi_1 + \phi_2 = \phi_1 + \phi_2$	(f_1, f_2)
4	$f_1 + (f_2 - f_1) = f_2$	$\phi_1 + (\phi_2 - \phi_1) = \phi_2$	$(f_1, f_2 - f_1)$
5	$(f_2 - f_1) + (f_1 + f_2) = 2f_2$	$\phi_1 + \phi_2 + (\phi_2 - \phi_1) = 2\phi_2$	$(f_2 - f_1, f_1 + f_2)$
6	$(f_2 - f_1) + 2f_1 = f_1 + f_2$	$2\phi_1 + (\phi_2 - \phi_1) = \phi_1 + \phi_2$	$(f_2 - f_1, 2f_1)$

6.5.1 Mild Nonlinearity (n_l=0.05)

The left panel of Fig. 6.5 shows the time series while the middle panel depicts the power spectrum of the signal x and y, respectively. Neither of these plots helps in distinguishing the two signals. However, the use of higher-order statistics can successfully detect the nonlinearities present in y. The right panel of Fig. 6.5 shows the three-dimensional squared bicoherence plots of x and y, respectively. For the signal x, NGI $= 0$, which indicates that the signal is "Gaussian", i.e. that the probability density function is symmetrical, and linear.

In contrast, in the case of y, the test results are NGI $= 0.32$ and NLI $= 0.62$. Thus the nonlinearity present in y is correctly detected. This example also shows the sensitivity of the proposed indices to the presence of nonlinearity in the signal. The presence of as little as 5% of the nonlinear square term in the noisy signal y has been detected. For this case, the squared bicoherence plot shows peaks at (0.12,0.12), (0.12,0.18), (0.30,0.30) and (0.12,0.30) bifrequencies. These

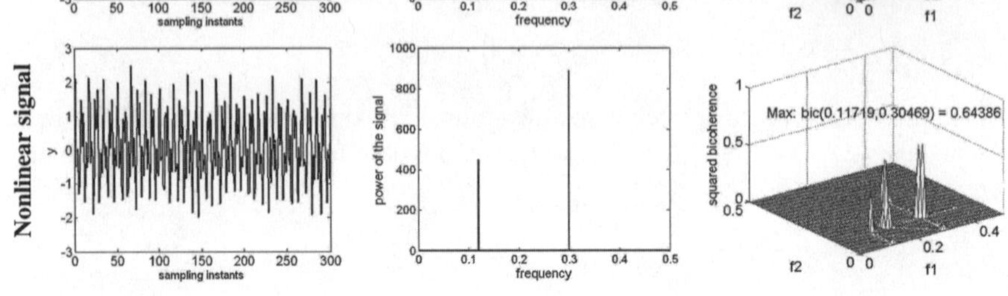

Fig. 6.5 HOS analysis results for the linear and nonlinear sinusoid signals, mild nonlinearity

bifrequencies correspond to (f_1, f_1),$(f_1, f_2 - f_1)$, (f_2, f_2) and (f_1, f_2), respectively. The other two peaks are not visible in the bicoherence plot owing to their small sizes. This is because only 5% of the nonlinear term was added.

6.5.2 Strong Nonlinearity (n_l=0.25)

For this case, the signal y was produced using the same Eq. (6.21) but with a larger multiplying factor for the nonlinear quadratic term in order to observe all the peaks in the bicoherence plot resulting from the theoretical analysis in Table 7.3. The magnitude of n_l was chosen as 0.25. Figure 6.6(a) shows the time trend of y, while the power spectrum of y is shown in Fig. 6.6(b). This time, the power spectrum shows extra peaks of small magnitudes at frequencies 0.18, 0.24 and 0.42. The use of higher-order statistics can successfully detect the nonlinearities present in y. The magnitudes of NGI and NLI are 0.59 and 0.85, respectively, clearly detecting the nonlinearity of the signal. Figure 6.6(c) depicts the three-dimensional squared bicoherence plots for y. This plot clearly shows the location of all six peaks at

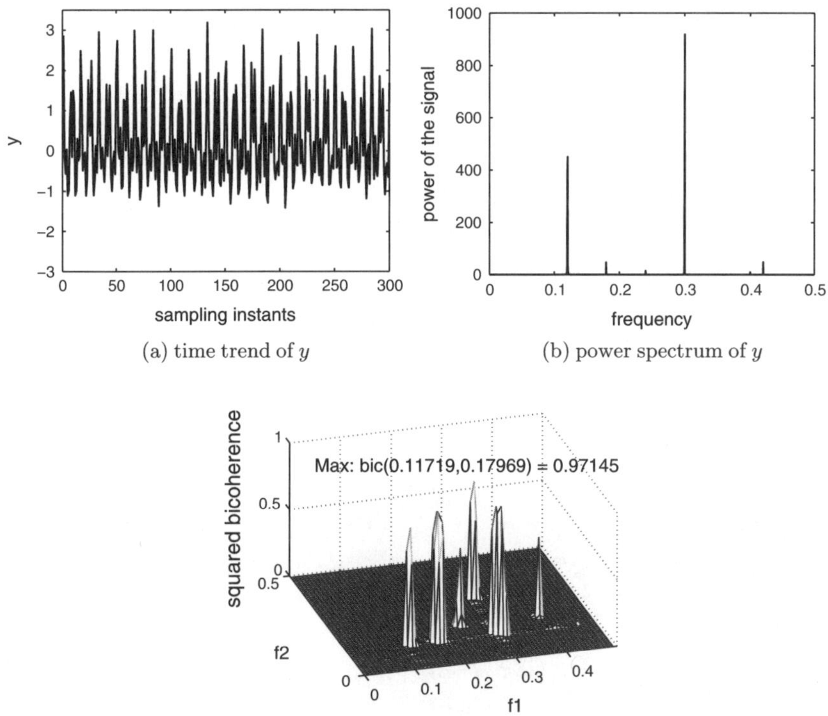

(a) time trend of y (b) power spectrum of y

(c) bicoherence plot

Fig. 6.6 HOS analysis results for the linear and nonlinear sinusoid signals, strong nonlinearity

(0.12,0.12), (0.30,0.30), (0.12,0.30), (0.12,0.18), (0.42,0.18) and (0.18,0.24) bifrequencies. These bifrequencies correspond to (f_1, f_1), (f_2, f_2), (f_1, f_2), $(f_1, f_2 - f_1)$, $(f_1 + f_2, f_2 - f_1)$ and $(f_2 - f_1, 2f_1)$, respectively. All the peaks were visible for this case because 25% of the nonlinear terms was added. Therefore, the bicoherence plot correctly identifies the frequency interactions resulting from the presence of nonlinearity in the signal.

6.5.3 Extent of Nonlinearity and Effect of Noise

The purpose of this example is to demonstrate the efficacy of the proposed indices in the presence of varying noise and degree of nonlinearity in a signal.

Figure 6.7 shows the non-Gaussianity index (NGI) and the nonlinearity index (NLI) plotted against n_l for varying cases of signal-to-noise (SNR) ratio. The dotted lines represent the NGI and the solid lines show the NLI. It is clear from the figure that both indices increase with the increase of nonlinearity in the signal. It is also evident that the indices work properly in a highly noise-corrupted environment (SNR $= 2$). The total nonlinearity index shown in Fig. 6.8 represents the extent of nonlinearity in the signal. For $n_l = 0$, there is no nonlinearity. Therefore, TNLI $= 0$. With an increase of n_l, the TNLI also increases. The amount of noise influences all indices to some extent because the denominator of the bicoherence estimator is affected by the noise.

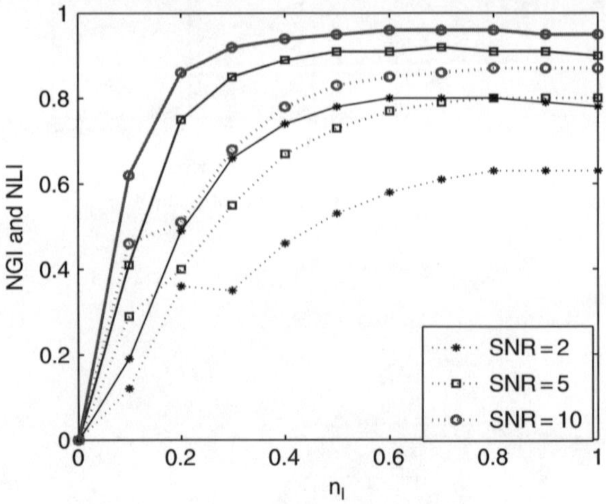

Fig. 6.7 NGI (*dotted lines*) and NLI (*solid lines*)

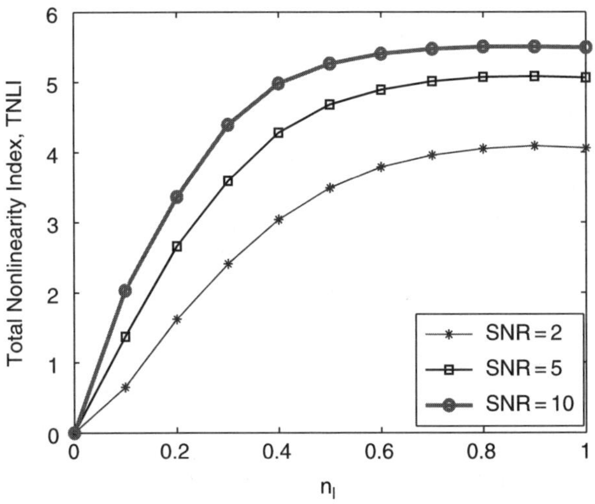

Fig. 6.8 TNLI increases with the increase of the nonlinearity of the signal

6.6 Summary

This chapter discusses an improved method for bicoherence estimation and a bicoherence-based method for detecting the non-Gaussianity and nonlinearity of time series signals. Two indices, NGI and NLI, have been developed and evaluated. A new index, total nonlinearity index (TNLI), has been defined and its performance in quantifying signal nonlinearity is demonstrated using illustrative simulation examples. These indices can be applied to any time series signal and do not require a sinusoidal excitation of a system.

Chapter 7
A Nonlinearity Measure Based on Surrogate Data Analysis

The previous chapter (Sect. 5.3.2.2) gave an introduction to the concept of surrogate time series for the purposes of nonlinearity detection. This chapter discusses in detail a measure based on surrogate data to determine the presence and severity of nonlinearity in the time trends of measurements from a process.

7.1 Surrogate Time Series

7.1.1 Nonlinearity Detection Using Surrogates

A nonlinearity test using surrogate data determines whether a time series could be the output of a linear system driven by Gaussian white noise, or whether its properties can only be explained as the output of a nonlinear system. A nonlinear time series with phase coupling is more structured and more predictable than a similar time series known as a *surrogate* having the same power spectrum but with random phases (see Sect. 5.3.2.2). The spread of values of some statistical property of a group of surrogate time series provides a reference distribution against which the properties of the test time series can be evaluated.

Theiler et al., (1992) gave an early exposition of the use of surrogates in nonlinearity testing. The review paper by Schreiber and Schmitz (2000) explains several methods for producing surrogate time series and gives worked examples, while the text book by Kantz and Schreiber (1997) covers surrogates and many other approaches to nonlinear time series analysis.

7.1.2 Predictability in Nonlinear Time Series

The basis of the test is a comparison of the predictability of the time trend compared with that of its surrogates. Figure 7.1 illustrates the concept. The top panel is an oscillatory time trend of a steam flow measurement from a refinery. It has a clearly defined pattern and it is possible to make a good prediction of where the trend will

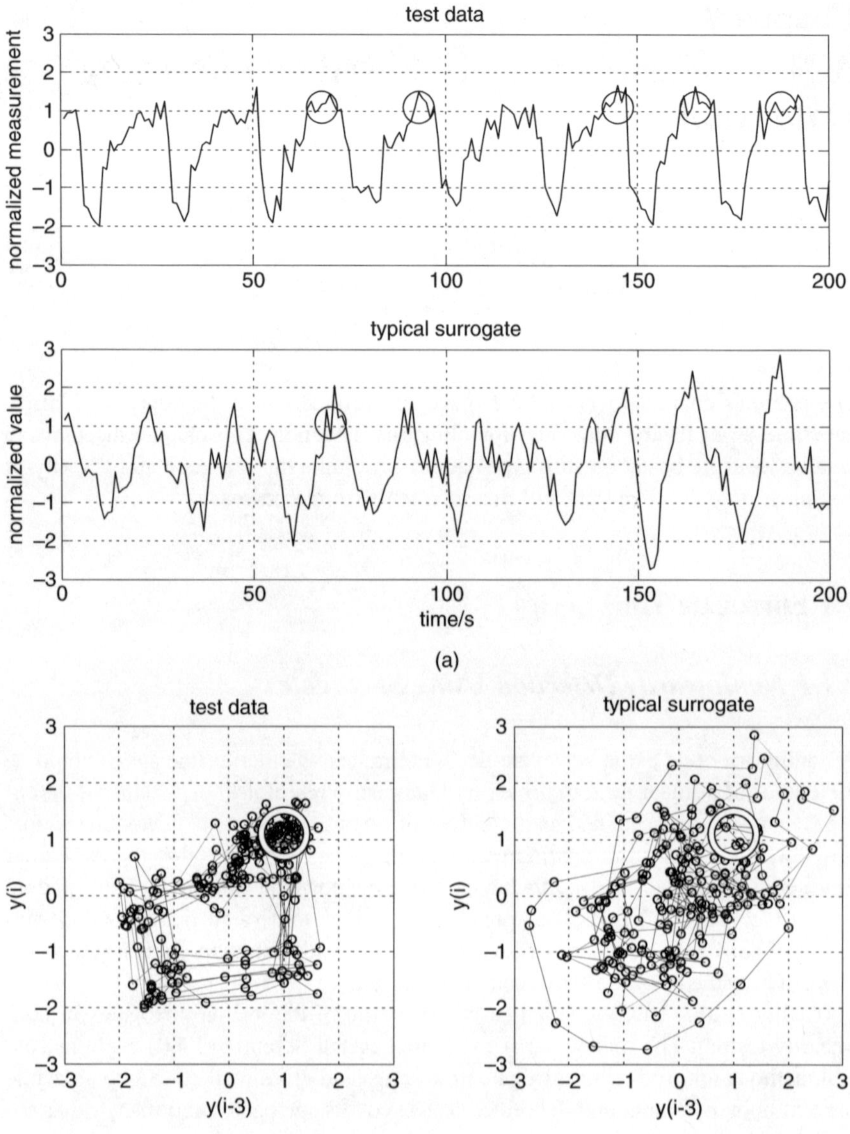

Fig. 7.1 **(a)** A limit cycle oscillation and a typical surrogate represented as time trends. **(b)** Limit cycle and surrogate as a two-dimensional embedding. The white rings in the embedded plots indicate the positions of the peaks circled in the time trends

go after reaching a given position. For example at one of ringed peaks, the prediction of the trend can be achieved by finding similar peaks in the time trend and observing where the trend went next on those occasions.

The middle panel shows a surrogate of the time trend. In contrast to the original time trend, the surrogate lacks structure even though it has the same power spectrum. The removal of phase coherence has upset the regular pattern of peaks. For instance, it is hard to anticipate where the trajectory will go next after emerging from the region highlighted with a circle because there are no other similar peaks on which to base the prediction.

The lower two panels introduce the concept known as *embedding*, a trajectory of the time trend plotted against a time-delayed version of the same signal. The lower left plot confirms the predictable and structured nature of the steam-flow time trend. There are large regions of the embedded space that are never visited by the trajectory. The white circle corresponds to the peak positions and it is easy to predict where the trajectory will go next after visiting the circle because it always does the large vertical downward jump on the right-hand side. The embedded plot in the lower right panel is for the surrogate and it is much less structured. It is not possible to predict how the trajectory will enter or leave the white circle because the entry and exit points on previous occasions were random. The time delay of three samples in these embedded plots was selected for the purposes of demonstration. Both the dimensions and the delays used in the embedded space are key parameters in the nonlinearity test and will be discussed in Sect. 7.3.

Predictability of the time trend relative to the surrogate gives the basis of a nonlinearity measure. Prediction errors for the surrogates define a reference probability distribution under the null hypothesis. A nonlinear time series is more predictable than its surrogates. A prediction error for the test time series which is smaller than the mean of the reference distribution by more than three standard deviations, suggests the time trend is nonlinear.

7.2 Algorithm for Nonlinearity Diagnosis

7.2.1 Construction of the Data Matrix for Nonlinear Prediction

As explained above, testing for nonlinearity is based on a comparison of the predictability of the time trend and its surrogates. Nonlinear prediction of time series was originally described by Sugihara and May (1990) to distinguish determinism from random noise, and the field of nonlinear time series analysis and prediction has been reviewed by Schreiber (1999). Rhodes and Morari (1998) gave an early process application of nonlinear prediction where the emphasis was on modelling of nonlinear systems when noise corrupts a deterministic signal.

Nonlinear prediction uses a data matrix known as embedding, having E columns each of which is a copy of the original data set delayed by a suitable sampling interval. For instance, a data matrix with a delay of one sampling interval and $E = 3$ is:

$$\mathbf{Y} = \begin{pmatrix} y(1) & y(2) & y(3) \\ y(2) & y(3) & y(4) \\ y(3) & y(4) & y(5) \\ \cdots & \cdots & \cdots \\ y(\ell-2) & y(\ell-1) & y(\ell) \end{pmatrix}$$

Rows of the matrix **Y** represent time trajectories that are segments of the original trend. As the original data formed a continuous time trend, the trajectories in adjacent rows are similar. They are called *near-in-time neighbours*. Also, if the time trend is periodic, then the trajectories in later rows of **Y** will be similar to the earlier rows after one or more complete cycles of oscillation. For instance, if the period of oscillation is 50 samples per cycle, then $\|\mathbf{y}_{51} - \mathbf{y}_1\|$ will be small, where $\mathbf{y}_{51}$ is the 51st row vector of **Y** and $\mathbf{y}_1$ is the first. Those rows are called *near neighbours*.

7.2.2 Calculation of Prediction Error

Predictions are generated from near neighbours with near-in-time neighbours excluded so that the neighbours are only selected from other parts of the time trend that are similar. When k nearest neighbours have been identified then those near neighbours are used to make an H step-ahead prediction. For instance, if row vector $\mathbf{y}_1$ were identified as a near neighbour of $\mathbf{y}_{51}$ and if H were 3, then $y(4)$ would give a prediction of $y(54)$. A sequence of prediction errors can thus be created by subtracting the average of the predictions of the k nearest neighbours from the observed value. The overall prediction error is the r.m.s. value of the prediction error sequence.

The analysis is non-causal, and any element in the time series may be predicted from both earlier and later values. Figure 7.2 illustrates the principle using a time series from the SE Asia refinery case study (see Chap. 19, Sect. 19.3), where the embedding dimension E is 16 and the prediction is made 16 steps ahead. The upper panel shows the 100th row of the data matrix **Y**, which is a full cycle starting at sample 100, marked with a heavy line. The five nearest neighbours are the rows of **Y** that begin at samples 67, 133, 166, 199 and 232 and are also shown as heavy lines in the lower panel. The average of the points marked o in the lower panel is then used as a prediction for the value marked **x** in the upper panel.

7.2.3 Calculation of Surrogate Data

There are some rather important data-preprocessing steps prior to the calculation of surrogate time series, which are itemized in Sect. 7.4.3. Once these have been completed, surrogates of the test times series can be calculated as described below. The method presented here uses a class of surrogates called Fourier surrogates.

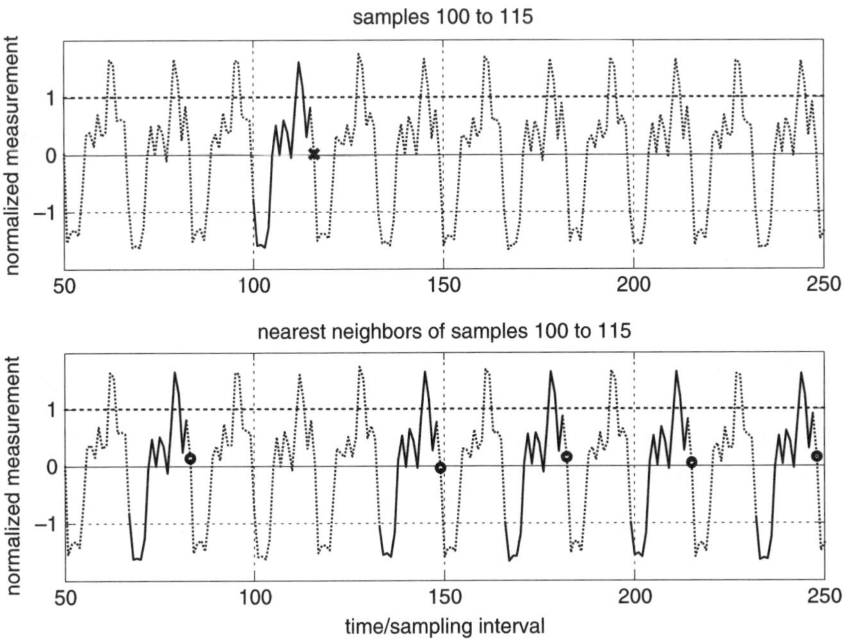

Fig. 7.2 Illustration of the nearest neighbour concept. The highlighted cycles in the lower plot are the five nearest neighbours of the cycle in the upper plot. The average of the points marked **o** gives a prediction for the point marked **x**

Fourier surrogates have the same power spectrum as the time trend under test. The magnitudes of the discrete Fourier transform (DFT) are the same in both cases but the arguments of the DFT of the surrogate data set are randomized. Thus if the DFT in frequency channel i is:

$$|Y(j\omega_i)|\, e^{j\angle Y(j\omega_i)}$$

then the DFT of the surrogate is:

$$|Y(j\omega_i)|\, e^{j(\angle Y(j\omega_i)+\phi_i)}$$

where ϕ_i is a phase angle sampled from a uniform random distribution in the range $-\pi < \phi_i \leq +\pi$. The aliased frequency channels above the Nyquist sampling frequency have the opposite phase added. If the number of samples ℓ is even and if the frequency channels are indexed as $i = 1$ to ℓ, the Nyquist frequency is in channel $\ell/2+1$ and the alias of the ith frequency channel is channel $\ell - i + 2$. Then, $\phi_1 = 0$, $\phi_{\ell/2+1} = 0$ and $\phi_{\ell-i+2} = -\phi_i$ ($i = 2$ to $\ell/2$).

If ℓ is odd, $\phi_1 = 0$ and $\phi_{\ell-i+2} = -\phi_i$ ($i = 2$ to $ceil\left(\ell/2\right)$), where $ceil\left(\ell/2\right)$ is the rounded-up integer value of $\ell/2$.

Finally, the surrogate data set is created from the inverse Fourier transform of the phase randomized DFT.

7.2.4 Statistical Testing

The nonlinearity test requires the determination of mean square prediction errors of M surrogates. The statistical distribution of those errors gives a reference distribution. If the test data prediction error lies on the lower tail of the reference distribution, then the test signal is more predictable and nonlinearity is diagnosed using the following statistic based on a three-sigma test:

$$N = \frac{\bar{\Gamma}_{surr} - \Gamma_{test}}{3\sigma_{\Gamma_{surr}}}$$

where Γ_{test} is the mean square error of the test data, $\bar{\Gamma}_{surr}$ is the mean of the reference distribution and $\sigma_{\Gamma_{surr}}$ its standard deviation. If $N > 1$ then nonlinearity is inferred in the time series. Larger values of N are interpreted as meaning that the time series has more nonlinearity, those with $N < 1$ are taken to be linear.

It is possible for the test to give small negative values of N. Negative values in the range $-1 \leq N < 0$ are not statistically significant and arise from the stochastic nature of the test. Results giving $N < -1$ do not arise at all because the surrogate sequences that have no phase coherence are always less predictable than a nonlinear time series with phase coherence.

7.2.5 Algorithm Summary

Step 1: Form the embedded matrix for the test data $y(1)$ .... $y(\ell)$:

$$\mathbf{Y} = \begin{pmatrix} y(1) & y(2) & \ldots & y(E) \\ y(2) & y(3) & \ldots & y(E+1) \\ y(3) & y(4) & \ldots & y(E+2) \\ \ldots & \ldots & \ldots & \ldots \\ y(\ell-E+1) & y(\ell-E+2) & \ldots & y(\ell) \end{pmatrix}$$

Step 2: For each row $\underline{\mathbf{y}}_i$ of $\mathbf{Y}$ find the indexes $j_p \quad (p = 1 \ldots k)$ of k nearest neighbour rows $\underline{\mathbf{y}}_{j_p}$ having the k smallest values of $\left\|\underline{\mathbf{y}}_{j_p} - \underline{\mathbf{y}}_i\right\|$ subject to a near-in-time neighbour exclusion constraint $\left|j_p - i\right| > E/2$.

Step 3: Find the sum of squared prediction errors for the test data:

$$\Gamma_{test} = \sum_{i=1}^{\ell-H} \left(y(i+H) - \frac{1}{k} \sum_{p=1}^{k} y(j_p + H) \right)^2$$

Step 4: Create M surrogate prediction errors Γ_{surr} by applying Steps 1 through 3 to M surrogate data sets.

Step 5: Calculate the nonlinearity from

$$N = \frac{\bar{\Gamma}_{surr} - \Gamma_{test}}{3\sigma_{\Gamma_{surr}}}$$

7.3 Selection of the Parameter Values

7.3.1 Recommended Default Parameter Values

Empirical studies have been conducted to ascertain the sensitivity of the nonlinearity index to the parameters of the algorithm and are reported in detail in Thornhill (2005). Reliable results have been achieved using the default values shown in Table 7.1. The *floor* function in the third row indicates that for non-integer values of S, E is set to the rounded-down integer value of S.

Table 7.1 Suggested default values for the parameters of the nonlinearity detection algorithm

Description	Value
Data ensemble size, Q	≤ 500
Samples per feature, S	$7 \leq S \leq 25$
Number of columns in embedded matrix, E	$E = floor(S)$
Prediction horizon, H	$H = E$
Number of features in the data set, C	$C \geq 12$
Number of near neighbours, k	$k = 8$
Number of surrogates, M	$M = 50$

7.3.2 Choice of Embedding Parameters E and H

The recommended default parameter values ensure that rows of the embedding matrix capture complete features in the time trend. If the times series is oscillatory, then a *feature* is a whole cycle of oscillation, as shown in Fig. 7.2 where the oscillation period is about 16–17 samples per cycle. This is within the recommended range of $7 \leq S \leq 25$ and the embedding parameter values are thus $E = H = 16$. The oscillation-detection methods described in Sect. 18.3.3 (in Chap. 18) of the book can be applied to calculate S and hence to give automated parameter selection for the embedding of a periodic time series.

Automated parameter selection for non-periodic data remains an area for further research; however, visual inspection of the time trend offers a judgement of what constitutes a significant feature. Figure 7.3 shows a non-periodic disturbance from an industrial process. In this case, each large deviation from the baseline is a

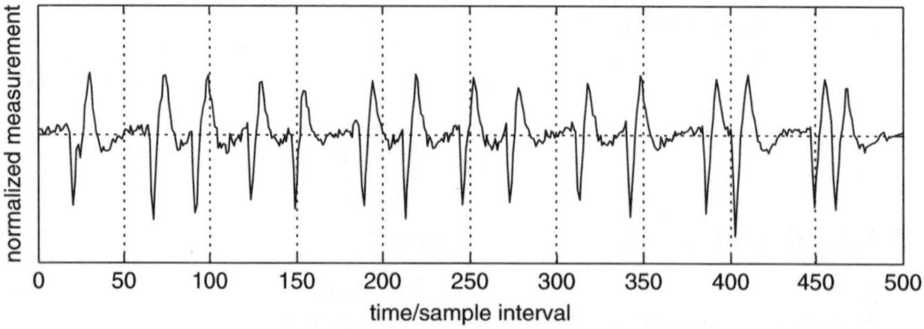

Fig. 7.3 Industrial data with non-periodic features

significant feature and it makes sense to determine S from the duration of each deviation. In Fig. 7.3, the average duration of a feature is about 12–13 samples and so E is chosen to be 12. The nonlinearity index of the time trend in Fig. 7.3 was $N = 3.6$.

7.3.3 Choice of Parameters C and k

Parameter C is the number of features in the data ensemble. The reason for the choice of $C \geq 12$ is that when the number of features becomes too few, the results start to become unreliable. On the basis of examples presented in Thornhill (2005), it seems necessary to use a data set with at least 12 cycles of oscillation or 12 features in the case of a non-periodic time trend.

A recommended value for the number of nearest neighbours, k, can be based on the number of cycles or features in the data set. From common sense reasoning, it is sensible to make sure that the number of near neighbours is smaller than the number of features because each near neighbour is a segment of the time trend equal in length to the duration of a feature if the $E = S$ recommendation is adopted. A conservative choice is $k = 8$ when the number of cycles is 12. The same conservative reasoning suggests that, for other cases, k should never be greater than $C - 4$, although in practice it has been found quite satisfactory to just set the value to $k = 8$.

The reason why the nonlinearity test gives less reliable results for values of k significantly larger than C is that the useful near neighbours run out. For instance, if $k = 20$ and if an oscillatory data set has 12 cycles, then any one cycle has eleven near neighbours that are closely matching cycles starting at the same position in the oscillation, like those in Fig. 7.2. The remaining near neighbours will have to be selected from other parts of the oscillation. They will not be such good matches and therefore less useful for prediction.

7.3.4 Default Data Ensemble Size, Q and Number of Samples Per Feature, S

It is practical to limit the size of the data ensemble (the total number of samples in the time trend) and the number of samples per feature, S, because the algorithm requires a distance measure to be ascertained between every pair of rows in the embedded matrix. The time taken for the computation increases roughly as Q^2, where $Q = C \times S$ is the data ensemble size. Therefore the number of samples per feature, S, and the number of features, C, cannot be increased arbitrarily. It is generally beneficial to keep the size of the data set below about 500 samples in an industrial implementation so that results are computed in a reasonable time. As discussed above, there is a lower bound of 12 for C and therefore the number of samples per feature has an upper bound.

There is extensive literature on the topic of how to choose the sampling interval correctly. What is needed is for adjacent columns of the Y matrix to be 'somewhat independent but not completely uncorrelated' (Lai and Lerner, 1998). The upper limit of 25 samples per feature was selected on the basis that 25 samples per cycle is enough to accurately capture the waveforms of oscillations observed in practical case studies. With $C = 12$ and $S = 25$, the data ensemble size would be 300. If additional data are available, then the data ensemble size can be increased to 500 by including more features. This has the effect of providing more near neighbours in the data set.

It is infeasible to operate with fewer than seven samples per cycle because high-frequency features would not be satisfactorily captured. For instance, in a periodic disturbance with $S = 7$, any third harmonic present is sampled at 2.33 samples per cycle, which just meets the Nyquist criterion of two samples per cycle. The reason for focussing the lower bound on the third harmonic is that it is the most prominent harmonic in symmetrical non-sinusoidal oscillations having square or triangular waveforms.

7.3.5 Choice of the Number of Surrogates, M

Sources of variability in the nonlinearity index are (i) the statistical variation associated with the random phase shuffling of the surrogates and (ii) variations due to the selection of the data ensemble. The number of surrogates has an effect because it is a statistical test, and there must be enough surrogates to properly define the reference distribution. There is, however, a penalty associated with the calculation of surrogates because increasing the number of surrogates also increases the computation time. Selection of the data has an effect because each data subset is a sample from the process that generates the data. Also, the nonlinearity present may vary between data subsets, for example, because of random external disturbances.

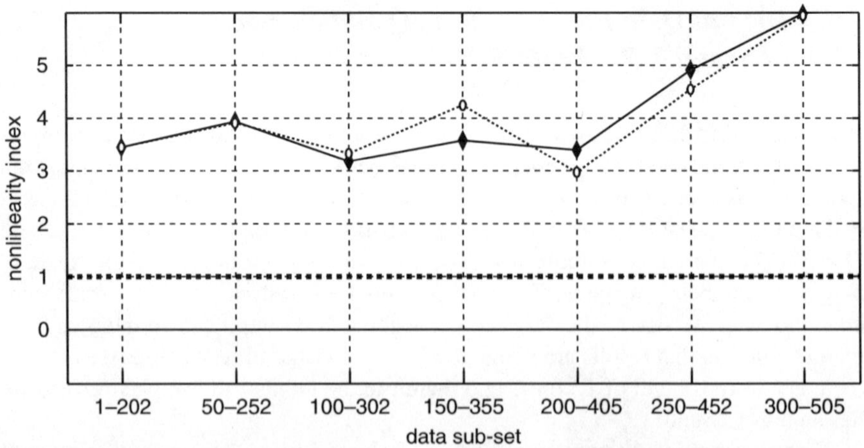

Fig. 7.4 The effects of data subsets on the nonlinearity index. *Black dots*: using 50 surrogates. *White circles*: using 250 surrogates

Figure 7.4 shows the variation in the nonlinearity index when different end-matched subsets of data from the same time trend are used. The second point from the left in this graph was calculated from the data set shown in Fig. 7.2 while other points are from other subsets of data from the same sensor.

The white dots in Fig. 7.4 show the effect of varying the number of surrogates. The difference in nonlinearity index between the results from 50 (black diamonds) and 250 surrogates (white circles) is about ± 0.1 overall and is less than the variability caused by the data subset. It is therefore concluded that 50 surrogates are enough.

7.4 Data-Preprocessing and End-Matching

7.4.1 False-Positive Results with Cyclic Data

False-positive results have been identified with the use of surrogate testing when the time series is strongly cyclic (Stam et al., 1998; Small and Tse, 2002). Nonlinearity testing based on strongly cyclic data can give rise to false detection of nonlinearity because artifacts in the DFT influence the surrogates when the time series is strongly cyclic. The surrogate is derived by taking the discrete Fourier transform (DFT) of the test data, randomization of the arguments, followed by an inverse DFT. Unless care is taken with end-matching, the test may give false-positive results and report a non-zero nonlinearity index for a linear time series.

The reason for the false-positive results is the phenomenon of spectral leakage in the discrete Fourier transform caused by the use of a finite length data sequence.

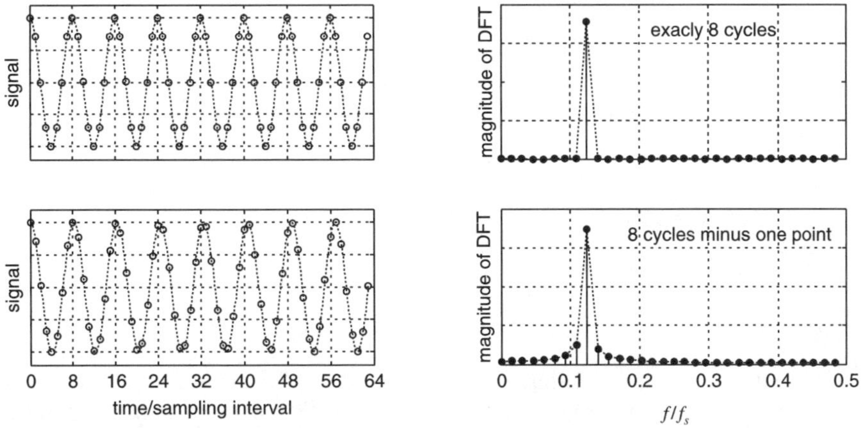

Fig. 7.5 Illustration of the importance of end-matching. For a strongly cyclic time trend, the data set should be an exact number of cycles of oscillation otherwise the Fourier transform will give spectral leakage into adjacent frequency channels

Figure 7.5 illustrates the effect of spectral leakage. The upper panel shows the DFT of a sine wave having eight samples per cycle when the total data length is an exact multiple of the period, in this case exactly eight cycles. The DFT is zero in all frequency channels except the one at $f/f_s = 0.125$ corresponding to the frequency of the oscillation. By contrast, the lower panel shows the DFT when the total data length is a complete number of cycles minus one sample. It has a non-zero magnitude in frequency channels adjacent to the channel containing the main spectral peak. A phase-randomized surrogate derived from the DFT in the lower panel therefore contains frequencies that were not present in the original signal and will thus be less predictable than the original sine wave giving a false indication of nonlinearity. The true surrogate of a sine wave is a phase shifted sine wave at the same frequency and is equally predictable.

It is therefore necessary to take special precautions when analysing cyclic time series. Stam et al. (1998) showed that the problem can be solved with an end-matching step that ensures the data length of the time series is an exact multiple of the period of the cycle to avoid false nonlinearity detection.

7.4.2 End-Matching

Surrogate data analysis requires a subset of the data such that the starting gradient and value match well to the final gradient and value. Hegger et al., (2000) recommend finding a subset of the data (denoted by samples x) with n samples starting at x_i and ending at x_{i+n-1}, which minimizes the sum of the normalized discontinuities $(d_0 + d_1)$ between the initial and end values and the initial and end gradients, where:

$$d_0 = \frac{(x_i - x_{i+n-1})^2}{\sum_{j=i}^{i+n-1} (x_j - \bar{x})^2}$$

and

$$d_1 = \frac{((x_{i+1} - x_i) - (x_{i+n-1} - x_{i+n-2}))^2}{\sum_{j=i}^{i+n-1} (x_j - \bar{x})^2}$$

where $\bar{x}$ is the mean of the sequence x_i, ..., x_{i+n-1}.

End-matching of an oscillating time trend as described above creates a time trend where the last value is the first sample of another cycle. An end-matched sequence that contains an exact number of cycles is x_i, ..., x_{i+n-2} and is derived from the x_i, ..., x_{i+n-1} sequence by omitting the last sample.

The end-matching procedure is also used for non-periodic time trends in order to reduce spurious frequencies in the DFT due to discontinuity between the values and gradients at the start and end of the trend. As with a periodic time trend, the aim is to select a subset of the data that minimizes the normalized discontinuities $(d_0 + d_1)$ subject to the constraint that the number of features in the data set remains at least 12 or more.

7.4.3 Summary of the Data-Preprocessing Steps

Step 1: The period of the plant-wide oscillation is determined;
Step 2: The number of samples per cycle, S, is adjusted to be no more than 25. The time trends are subsampled if necessary;
Step 3: The number of cycles of oscillation in the data set should be at least 12 for a reliable nonlinearity estimate;
Step 4: The selected data are end-matched to find a subset of the data containing an integer number of full cycles;
Step 5: The end-matched data are mean-centred and scaled to unit standard deviation.

7.4.4 Application to Oscillating Time Trends

Given the importance of end-matching, it is useful to examine the robustness of the nonlinearity index calculation to errors in end-matching that will inevitably be caused by noise and minor irregularities in the data.

Experimental data or simulations will give problems of the type outlined above when the experimental system is driven by a cyclic source such as a laboratory signal generator. This is termed in the literature as *strongly cyclic*. Industrial process data, however, even when cyclic, do not often suffer from the problems described above because the cyclic behaviour is not normally strong. Although they might be readily detectable, the cycles are generally not completely regular and the spectral power in the test signal is already spread across several frequency channels. Figure 7.6 shows an extreme example of oscillatory data from a separation column in an Australian refinery from Thornhill et al., (1996). The time trend has two shorter cycles at around samples 230–250 and two longer cycles between samples 100 and 150. The effects of spectral leakage in the DFT are less severe in this signal because a range of frequency channels is already occupied.

Table 7.2 shows nonlinearity calculations for the industrial data set of Fig. 7.6 and a pure sine wave of the same average period of oscillation. The correct result for the sine wave is $N < 1$, a result which is achieved when the subset of the data is an exact number of cycles of oscillation. If the subset is longer or shorter, even by one sample, there is a false nonlinearity detection because of spectral leakage contaminating the surrogates. By contrast, the industrial data set with its less regular cycles is not sensitive to minor variations in the end-match. No false nonlinearity was detected for the industrial data even when mismatched at the ends by one or two samples.

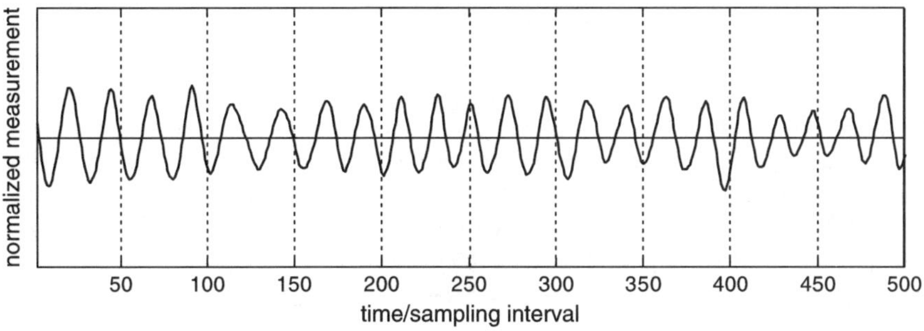

Fig. 7.6 Time trend from an industrial study. The trend is periodic but not strongly cyclic

Table 7.2 The effect of end-matching on false-positive results with synthesized and industrial data. The synthesized sine wave gives false-positives ($N > 1$) when the end-matching is not exact, however the industrial data set is robust to small end-mismatches

End matching	N for sine wave	N for industrial data
One point extra	**4.72**	0.89
End-matched	0.11	0.86
One point fewer	**5.07**	0.83
Two points fewer	**5.19**	0.84

7.5 Worked Examples

7.5.1 Identification of Nonlinear Root Causes

Empirically, both the bicoherence and surrogate nonlinearity measures do very well in the isolation of nonlinear root causes. The measurement point whose time trend is the most nonlinear is taken to be closest to the root cause. The reason why nonlinearity is strongest in the time trends of measurements nearest to the source of a disturbance is thought to be that the plant acts as a mechanical low-pass filter. As the disturbance propagates from its source to other variables such as levels, compositions and temperatures, the waveforms generally become more sinusoidal and more linear because plant dynamics destroy the phase coupling and remove the spectral harmonics. However, a full theoretical analysis is missing at present of why and how a nonlinearity changes as a disturbance propagates.

7.5.2 Application to the SE Asia Data Set

Nonlinearity diagnosis has been applied to the data set of the hydrogen reformer from the SE Asian refinery whose schematic is shown in Fig. 7.7. The plant was upset by a plant-wide oscillation with a period of about 16–17 min.

The root cause was known to be a valve fault in the recycle of fuel gas from the PSA unit to the reformer, and the process control engineer for the plant has confirmed that this, and not the pressure cycle swings of the PSA unit, was the cause. The aim of the analysis is to determine whether the nonlinearity test delivers the correct diagnosis by focussing attention on measurements closest to the root cause. The average oscillation period was 16–17 samples, the settings for the algorithm were therefore $E = H = 16$ and $k = 8$. A data set comprising 25 cycles of oscillation was used.

Figure 7.8 shows mean-centred normalized controller errors (1-min samples) with the Nonlinearity Index calculated using the surrogated method. The largest nonlinearity index is for tag 34. Therefore, the off-gas flow measured by tag 34 is identified as the measurement closest to the physical root cause.

7.5.3 The Mechanism of Propagation in the SE Asia Process

The above nonlinearity assessment is quantitative and is based on measurements from routine process operation. It indicates the location in the process where the disturbance originated and can distinguish measurements that have secondary-propagated disturbances from the primary cause. However, it does not by itself explain the propagation routes or the reasons why some other measurement points

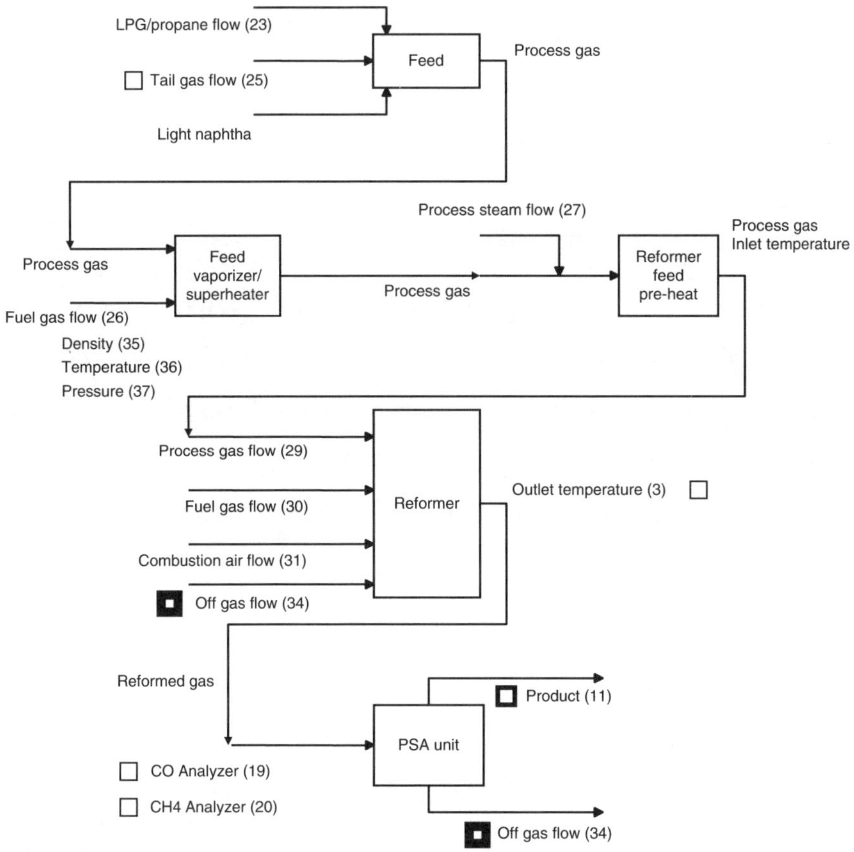

Fig. 7.7 Process schematic of SE Asia hydrogen plant showing the location of the oscillation (*faint □ symbols*) and highlighting nonlinearity (*heavy □ symbols*). Nonlinearity is strongest in tag 34

such as tag 25 are upset. To complete the task requires an examination of the process flows and connectivity by an engineer with knowledge and insights into the process.

The root cause, tag 34, is in the PSA unit, which is the last unit in the plant from where the final product leaves the plant. The mechanism for propagation to the reformer unit is the varying flow of off-gas from the PSA unit, which is recycled as fuel gas for the reformer.

Tag 25 is at the start of the process, yet it was still influenced by the oscillation. Inspection of the flow sheet alone might misleadingly suggest that tag 25 is the root cause because it is furthest upstream. Such a theory is hard to defend, however. It would require an explanation of how tag 25 can cause oscillations in the reformer and PSA unit without upsetting the feed vaporizer and preheat sections where no oscillations were detected. The reason why tag 25 is upset is that it is waste gas recycled from another downstream unit to which the oscillation has propagated. The nonlinearity test clearly shows that tag 25 is not the root cause because its time trend

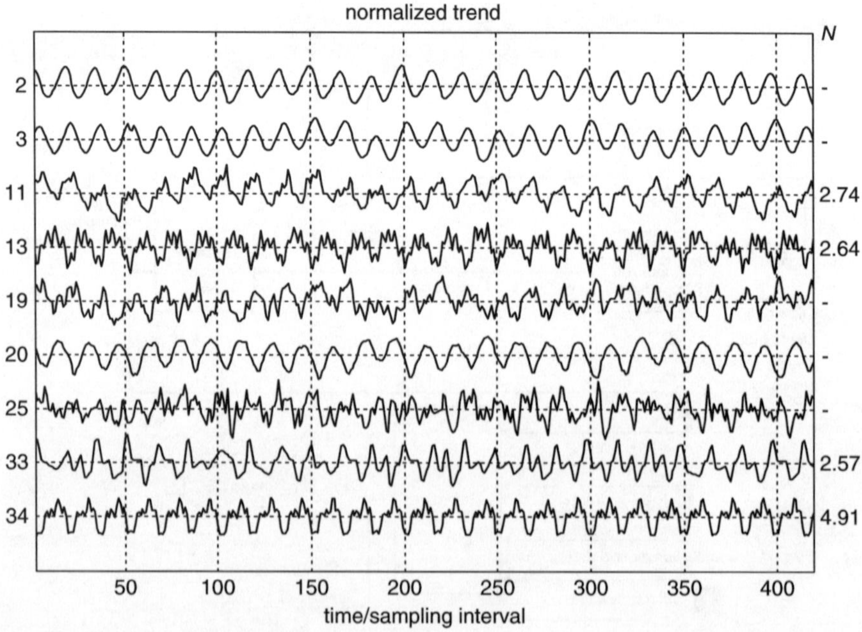

Fig. 7.8 Time trends of oscillating tags from SE Asia data set. The trends are for the controller errors. Nonlinearity indexes > 1 are shown on the right

has no nonlinearity. In fact, the lack of nonlinearity shows that tag 25 is one of the secondary oscillations furthest away from the root cause. Automation of the linkage of measurement-based analysis and process connectivity information for advanced diagnosis is an interesting research challenge at present, as highlighted in Thornhill and Horch (2007).

7.5.4 An Example with No Nonlinearity

Some plant-wide disturbances are not caused by nonlinearity. It is therefore desirable that nonlinearity testing gives negative results in such a case.

Table 7.3 and Fig. 7.9 show the schematic of a process with a recycle, and Fig.7.10 shows the time trends of the controller error and controller outputs (courtesy of BP Chemicals). Visually, it is possible to see that a plant-wide oscillation is present, since many of the time trends are moving in a coordinated way at the same frequency. Oscillation detection confirms the presence of a plant-wide oscillation with a period of 56 samples per cycle.

In this case, as reported in Bauer et al., (2004), the cause of the disturbance is the recycle loop that leads to a strong interaction between the reflux drum level

Table 7.3 Tag descriptions

Tag	Name	Description
1	LC1	Buffer tank level
2	LC2	Flash drum level
3	LC3	Reflux drum level
4	LC4	Reactor level
5	PC1	Reactor pressure
6	TC1	Reactor temperature
7	LC1.op	Buffer tank-level controller output
8	LC2.op	Flash drum-level controller output
9	LC3.op	Reflux drum-level controller output
10	LC4.op	Reactor-level controller output
11	PC1.op	Reactor pressure controller output
12	TC1.op	Reactor temperature controller output

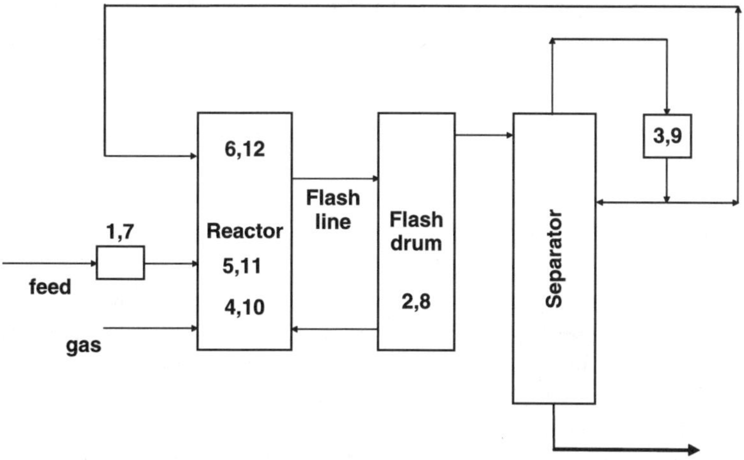

Fig. 7.9 Schematic of a process with recycle

controller (L3) and the variables associated with the reactor (LC4, PC1, TC1). This is therefore a linear dynamic disturbance. Testing using surrogates analysis showed no evidence of a nonlinear root cause, with the nonlinearity index of every time trend in the data set being below unity. These negative results show that the nonlinearity test can correctly indicate when nonlinearity is not a contributing factor to a plant-wide disturbance.

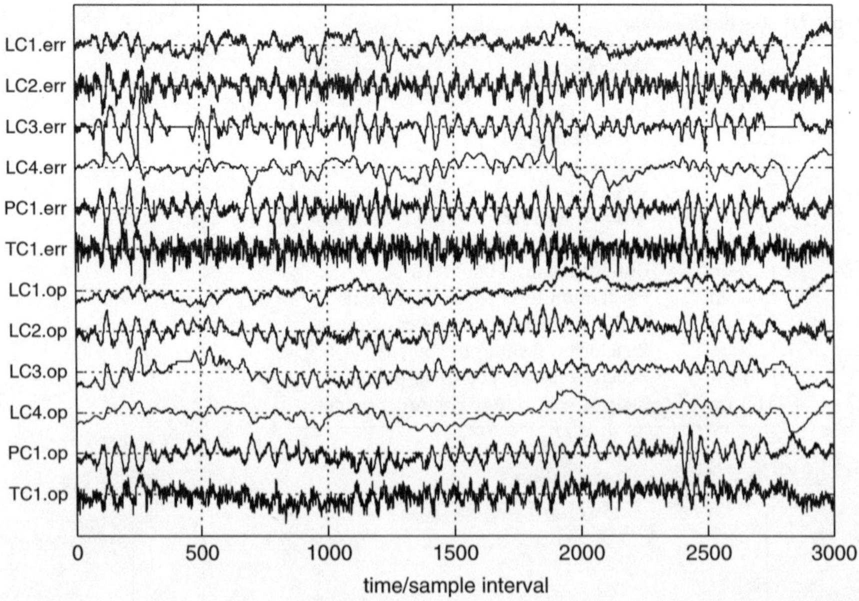

Fig. 7.10 Time trends of controller errors and controller outputs of the process with recycle

7.6 Summary

This chapter presented the methods of surrogate data analysis and nonlinear prediction for detection of the presence of nonlinearity in the time trends of measurements. The method relies on the matching of segments of the data set that are similar to each other, known as *near neighbours*. The basic concept is that prediction of the future time evolution of these near neighbours will be more similar to each other if the time trend is from a nonlinear source than in the general case where there is no phase coherence in the signal. A step-by-step algorithm was presented together with suggested default values for the parameters of the algorithm.

The surrogate data test was demonstrated with two industrial examples having plant-wide disturbances. It correctly detected and isolated a nonlinear root cause in one data set. The second data set had a plant-wide oscillation due to controller interaction in a recycle loop, which is a linear, not a nonlinear, cause. The surrogate data test correctly indicated that no nonlinearity was present.

Chapter 8
Nonlinearities in Control Loops

Nonlinearities in control loops may appear due to any one or combination of the following:

1. The process may be nonlinear in nature.
2. The control valve may have a nonlinear characteristic.
3. The control valve may develop nonlinear faults, e.g. stiction, deadband and hysteresis.
4. A nonlinear disturbance may enter the loop.

Usually, the process nonlinearities and nonlinear valve characteristics are called *soft nonlinearities* because of their nature of being locally linear in the presence of small excitations. Nonlinear faults in control valves, such as stiction, deadband, backlash and hysteresis, are termed as *hard nonlinearities* because of their dominant nonlinear characteristics. A nonlinear disturbance may fall either in the category of hard nonlinearity or soft nonlinearity depending on its severity.

In order to detect and diagnose any problem related to control-loop nonlinearities, all of the above should be investigated carefully. This chapter discusses the impact of these nonlinearities.

8.1 Process Nonlinearity

In linear control theory, it is often assumed that the process operates in a locally linear fashion. This section investigates this assumption using results obtained from the application of higher-order statistical tests developed in Chap. 6. Analysis of process nonlinearity using HOS will be illustrated using two examples.

8.1.1 Nonlinearity of a Spherical Tank

In linear systems theory, a nonlinear process is assumed locally linear if the principle of superposition holds for small deviations in the process variables around a nominal

M. A. A. S. Choudhury et al., *Diagnosis of Process Nonlinearities and Valve Stiction*,

operating point. This section examines the validity of this assumption using the total nonlinearity index (TNLI) through an amplitude- and frequency-dependent nonlinearity analysis. This analysis is performed using a simulation of the level control of a spherical tank (Agrawal and Lakshminarayanan, 2003). The example considers the control of water level in a spherical tank by manipulating the inlet volumetric flow rate, F_i (see Fig. 8.1).

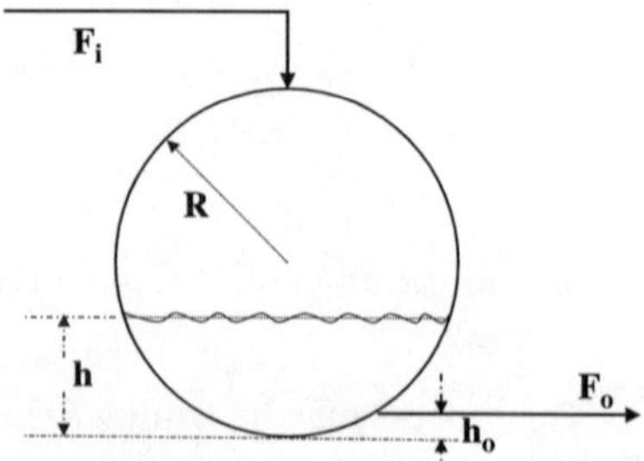

Fig. 8.1 A spherical tank system

The dynamics of the system can be modelled as:

$$\pi R^2 \left[1 - \frac{(R-h)^2}{R^2}\right] \frac{dh}{dt} = F_i(t-d) - F_o(t) \tag{8.1}$$

where $F_o(t)$ is the outlet flow rate at time t, h the height of the water level from the bottom of the tank, R the radius of the spherical tank and d corresponds to the delay in the input flow rate. The outlet flow rate, F_o, can be expressed as $F_o(t) = \sqrt{2g(h-h_o)}$, where g is the gravitational constant and h_o the height of the outlet pipe from the base of the vessel. For both open- and closed-loop simulations of the system, we use $R = 0.5$ m, $h_o = 0.01$ m, and the nominal operating point is selected to be, $h_{ss} = 0.40$ m. The steady-state operating point for the water height of the tank is chosen at 0.4 m instead of 0.5 m because the process is more nonlinear in the vicinity of the former operating point. The curvature of the surface contributes significantly to the nonlinearity.

8.1.1.1 Dependency of Nonlinearity on the Amplitude and Frequency of Input Signal

A linear system exhibits *sinusoidal fidelity*. Sinusoidal fidelity states that a sinusoidal input to a linear system will produce a sinusoidal output of identical frequency. The amplitude and phase of the output sinusoid may differ from those of the input sinusoid. But this is not the case for a nonlinear system. Generally, in response to a sinusoidal excitation, a nonlinear system will produce new frequencies in addition to the original frequency of the input sinusoid or even new frequencies instead of the original input frequency. This property suggests the use of a sinusoid as an excitation signal for testing the nonlinearity of a system. In this example, a sinusoidal excitation signal (added as a deviation to the steady state F_i) with varying frequencies and magnitudes is used. The amplitudes vary from 0.01 to 0.2 and the frequencies vary from 0.015 to 0.1 cycles/sample, i.e. periods of 67 samples/cycle to 10 samples/cycle. The sampling interval was 0.1 s. A total of 4,096 data points were used for each nonlinear analysis. The results are shown in Fig. 8.2. This figure shows that for small magnitude input signals, the process can be assumed locally linear. As the amplitude of excitation signal increases, so does the nonlinearity. However, the nonlinearity of this particular system is not very sensitive to changes in frequency. Therefore, it can be concluded that nonlinearity of the process strongly depends on the size of the excitation or input signal.

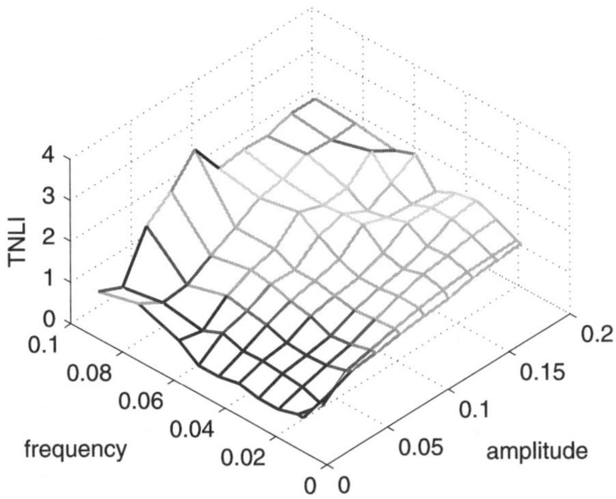

Fig. 8.2 Dependency of nonlinearity on the amplitude and frequency of the input signal in open loop operation

8.1.1.2 Operating Point Dependency of Nonlinearity

It is well known that the nonlinearity of a system depends on the operating point of the system. In order to investigate this, the spherical tank system was excited with a sinusoid of amplitude 0.1 m and 0.05 cycle/sample or 20 samples/cycle at various operating points. The results are shown in Fig. 8.3. It is clear from the figure that nonlinearity increases as the operating point moves away from the centre of the tank, i.e. 0.5 m. This is also expected because the curvature of the inner surface of the tank increases as it moves away from the centre. It is noteworthy that the result in Fig. 8.3 is not symmetric around 0.5. The reason for this finding is because the nonlinearity increases drastically at a point where the total surge volume of the tank drops below a certain level, for example, $h_{ss} = 0.35$ m. With the decrease in liquid surge volume, the capacity of the tank to attenuate disturbances decreases, which contributes to the increase of nonlinearity.

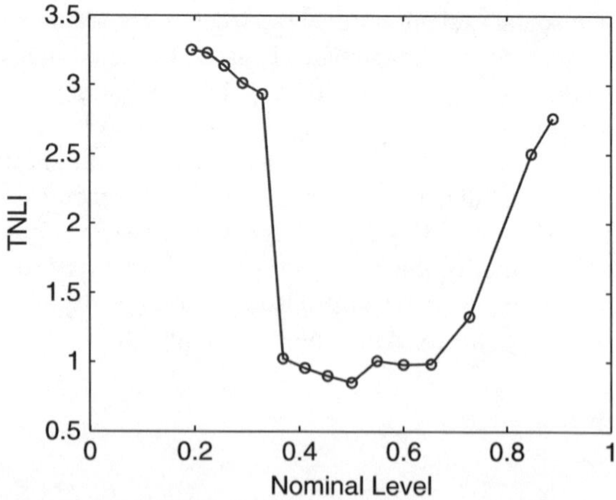

Fig. 8.3 Dependency of nonlinearity on the operating region of the system

8.1.1.3 Closed-Loop Operation of the Spherical Tank

The level of the spherical tank was controlled by manipulating the input flow-rate to the tank. The PI controller settings used for this simulation are given by:

$$C(z^{-1}) = \frac{0.3 - 0.1z^{-1}}{1 - z^{-1}} \tag{8.2}$$

Simulation was carried out as described in the open-loop case except the operating point was chosen at 0.3 m height of the tank in order to incorporate more nonlinearity. For all cases, 4,096 data points of the control error signal were used for nonlinear

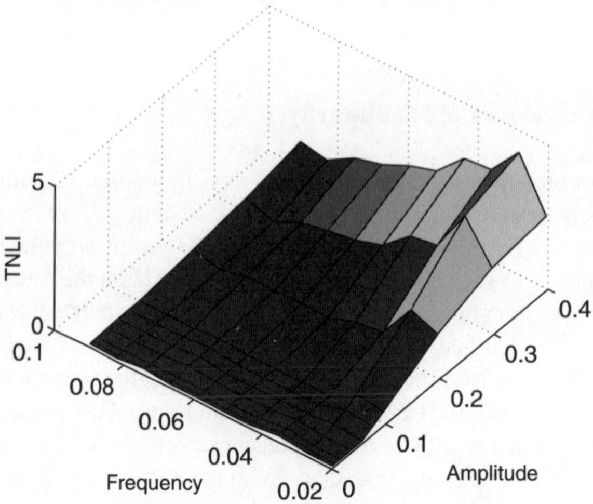

Fig. 8.4 Dependency of nonlinearity on the amplitude and frequency of the input signal in closed loop operation

analysis. The nonlinearity test results are shown in Fig. 8.4. The figure clearly shows that for the same amplitude and frequency, the TNLI is smaller in the closed-loop feedback configuration than in open loop. Also, at high frequency, the nonlinearity does not increase with the increase of amplitude because the tank essentially operates as a low-pass filter. The finding demonstrates the well-known linearizing effect of feedback control.

Remark 1. The assumption of linearity of a process is case-dependent. Generally, most control loops in chemical process industries operating under regulatory feedback control in order to reject disturbances will likely undergo small perturbations. Therefore, many of them can be assumed safely to be operating in a locally linear fashion.

8.1.2 Nonlinearities of a Continuous Stirred Tank Reactor (CSTR)

In this section, the nonlinearity of a non-isothermal continuously stirred tank reactor (CSTR) is measured around the operating point used for Case 1 described in Marlin (1995). The schematic of the CSTR process is shown in Fig. 8.5.

A simple first-order irreversible exothermic reaction of the form $A \longrightarrow B$ is considered here. Assuming perfect mixing in the reactor and the jacket, the CSTR model can be described by the following differential equations:

$$\frac{dc_A}{dt} = \frac{F}{V}(c_{A0} - c_A) - k_0 e^{-E/RT} C_A$$

$$\frac{dT}{dt} = \frac{F}{V}(T_0 - T) - \frac{aF_c^{b+1}}{V\rho C_p \left(F_c + \frac{aF_c^b}{2\rho_c C_{pc}}\right)}(T - T_{cin}) + \frac{1}{\rho C_p}(-\Delta H_{rxn}) k_0 e^{-E/RT} C_A \tag{8.3}$$

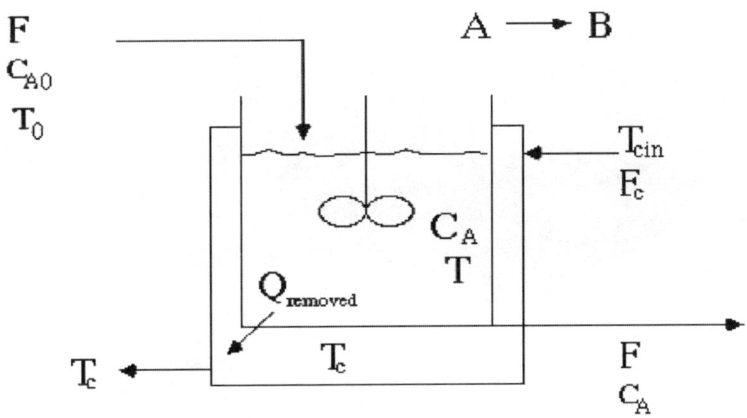

Fig. 8.5 Schematic of the CSTR

Table 8.1 CSTR parameters for the first-order reaction

Symbol	Description	Magnitude	Unit
F	Flow rate of reactant	1	m^3/min
C_{A0}	Initial concentration of A	2	kmole A/m^3
V	Volume of the reactor	1	m^3
C_p	Specific heat of outlet stream	1	cal/g°C
ρ	Density of outlet stream	10^6	g/m^3
T_o	Feed temperature	323	K
C_{pc}	Specific heat of coolant	1	cal/g°C
F_c	Flow rate of the coolant	15	m^3/min
T_{cin}	Inlet temperature of the coolant	365	K
ρ_c	Density of coolant	10^6	g/m^3
$-\Delta H_{r\times n}$	Heat of reaction	130×10^6	cal/kmole A
k_0	Constant	1.0×10^{10}	min^{-1}
E/R	Constant	8330.1	K^{-1}
b	Constant	0.5	K^{-1}
a	Constant	1.678×10^6	K^{-1}

The notation and the parameters for simulation are given in Table 8.1. The simulation of the CSTR with the above-mentioned parameters results in a steady-state value of 0.2645 $kmole/m^3$ for C_A and 393.9 K for the reactor temperature, T. Sinusoids with varying amplitudes and frequencies were added to the feed concentration, C_{A0} to excite the process, and the subsequent nonlinearity was quantified for each case. The results are shown in Fig. 8.6. The figure clearly shows that total

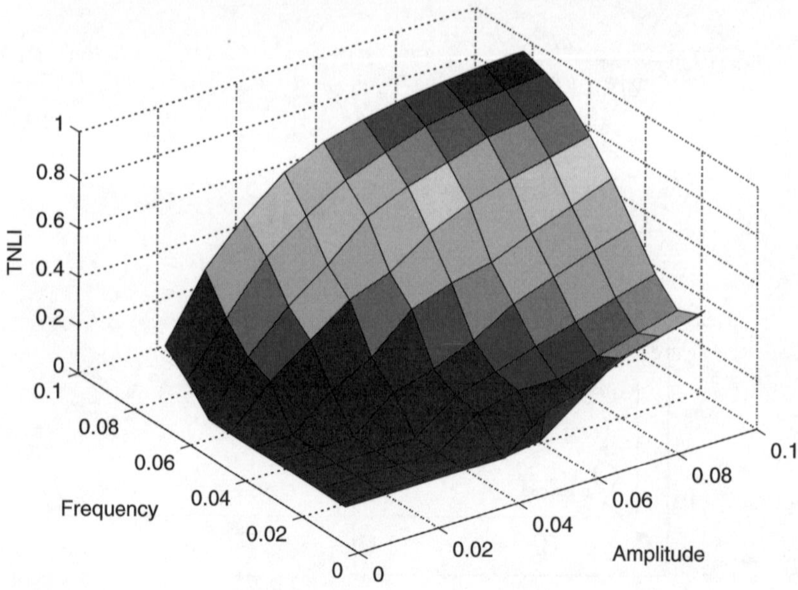

Fig. 8.6 Dependency of the CSTR nonlinearity on the amplitude and frequency of the input signal

nonlinearity increases with the increase of the amplitude of the excitation signal. Also, an increase in frequency of the input excitation causes an increase in nonlinearity for this particular system.

8.2 Nonlinear Valve Characteristic

The valve is the most commonly used actuator or final control element in a control loop. The flow through a valve is often described by the following relationship:

$$Q = KC_v(x)\sqrt{\frac{\Delta P_v}{s.g.}} \tag{8.4}$$

where Q is the volumetric flow rate through the valve, K is the constant that depends on the units used in the equation, x is the stem position expressed in the range 0 to 1, C_v is the valve coefficient that depends on the inherent valve characteristics and the stem position, ΔP_v is the pressure drop across the valve and $s.g.$ is the specific gravity of the fluid. There are three common valve characteristics:

- Linear characteristics: $C_v(x) = x$
- Equal percentage valve: $C_v(x) = \kappa^{(x-1)}$
- Quick opening or square-root valve: $C_v(x) = \sqrt{x}$

where κ is a constant, commonly taken to be 50.

These valve characteristics are illustrated in Fig. 8.7. For quick-opening valves, the sensitivity of the flow to the fraction opening or stem position is high at a low

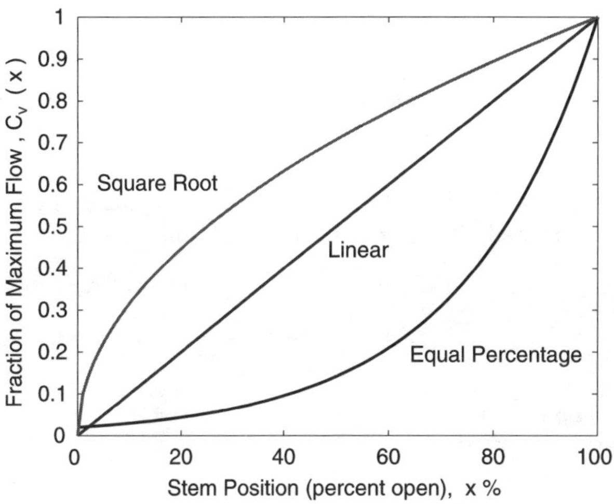

Fig. 8.7 Inherent characteristics for linear, equal percentage and square-root valve

flow rate and low at a high flow rate. The opposite is true for equal percentage valves. These relationships have been defined for constant pressure drop across the valve carrying an incompressible fluid. Thus, these relationships are valid only under such conditions.

8.2.1 Linear Valves

As the name suggests, a linear valve has a linear characteristic. There is no inherent nonlinearity in this type of valve. In the following text, the other two types of valves will be investigated for their nonlinear characteristics.

8.2.2 Equal Percentage Valves

Equal percentage valves cause a constant percentage increase of flow rate from the previous flow rate for an equal incremental increase in valve travel. For example, if a value of $\kappa = 50$ is used, for every 10% increase in stem position or valve travel x, there will be an increase of 47.88% in the flow rate from the previous flow rate. The effect of nonlinearity in the characteristic equation of this type of valve has been investigated using the following simulation example.

A simple single-input, single-output (SISO) system in a feedback control configuration (Fig. 8.8) was used to generate simulated data. The first-order process with time delay is given by the following transfer function:

$$G(z^{-1}) = \frac{z^{-3}(1.45 - z^{-1})}{1 - 0.8z^{-1}} \tag{8.5}$$

The process is regulated by a PI controller whose parameters are $K_c = 0.15$ and $I = K_c/\tau_i = 0.15\ \mathrm{s}^{-1}$. The simulation was performed at a steady state valve setting of 12 mA corresponding to the valve position at 50%. Random noise was used to drive the process. The variance of the random noise was increased to drive the valve from 50% nominal state to over a wide range of its travel span. The simulation was performed for 6,000 sampling intervals. To remove the effect of transients, the first

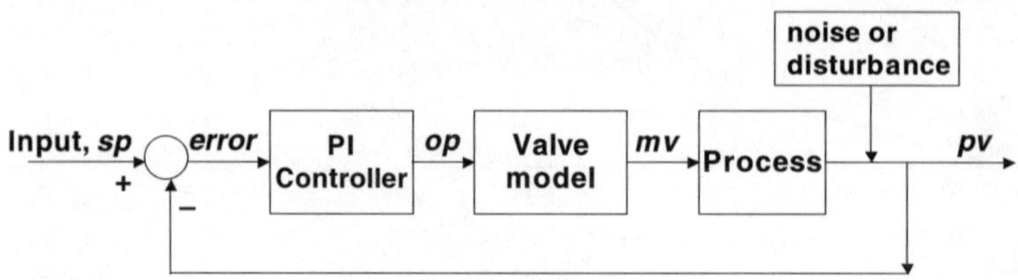

Fig. 8.8 Block diagram of a simple SISO process

several hundred data points were discarded. Transients or sharp changes in a signal have a serious contamination effect on the estimated bicoherence. The notations used here are *sp* for the set-point, *pv* for the controlled variable, *op* for the controller output and *mv* for the valve position or valve output. The last 4,096 points of the error signal to the controller (*sp*–*pv*) were analysed to detect the nonlinearity present in the system. Figure 8.9 shows the effect of increasing noise or valve travel range around the 50% nominal operating point. The abscissa of the figure represents the total travel range of the valve from its nominal operating point. For example, if the valve closes up to 10% and opens up to 90%, then the total travel range will be 80%. It is clear from the figure that even if the valve travel range is up to 40%, the equal percentage valve does not add any significant nonlinearity. It may be assumed to operate in a locally linear fashion.

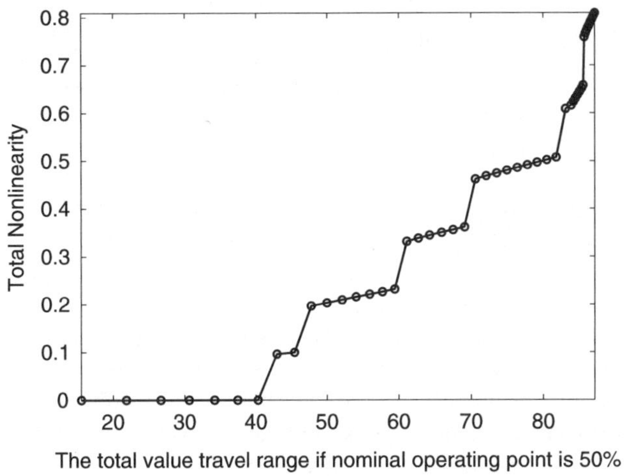

Fig. 8.9 Total nonlinearity index (TNLI) vs. the valve output range when the equal percentage valve is operating at 50% nominal point

8.2.3 Square-Root Valve

The quick opening or square-root valve provides large changes in flow for small changes in valve travel or stem position, especially in the lower region of the valve travel. Therefore, it has a high valve gain unsuitable for use in regulatory control. Its use is limited to on–off service. To investigate the nonlinear behaviour of this valve, a simulation study was performed using the procedure described in the case of the equal percentage valve, but with the use of a square-root valve instead of an equal percentage one. Since the characteristic curve (refer to Fig. 8.7) for square-root type valve is mainly nonlinear in the lower region of valve travel, the nominal operating point for this simulation was set at 25%. The valve was forced to travel in a wide range around this steady state (25%) by increasing the noise variance.

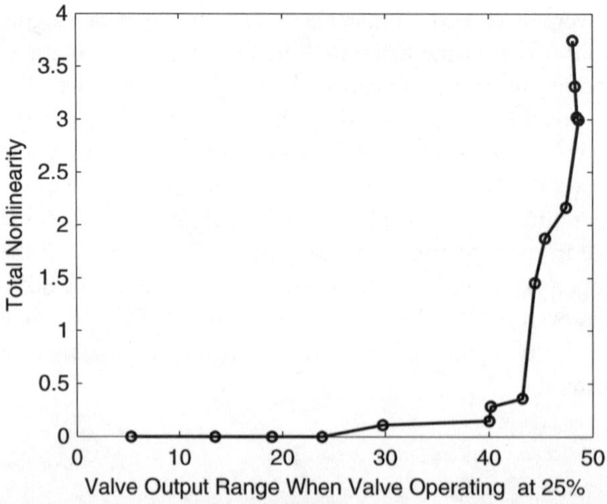

Fig. 8.10 Total nonlinearity index (TNLI) vs. the valve output range when the square-root valve is operating at 25% nominal point

As in the previous case, to remove the effect of transients, the first several hundred data points were discarded. The last 4,096 points of the error signal to the controller (*sp–pv*) were analysed to detect and quantify the nonlinearity present in the system. Figure 8.10 shows that with the increase of valve travel span, the nonlinearity increases. However, if the valve travel range is within 25% of full range, the square-root valve can be assumed linear even in the lower operating region. The nonlinearity increases very quickly after 40% of the valve output range from its 25% operating point because, in such cases, the valve is saturating at one end because in such cases the valve is almost closed at one end since $25 \pm 20\%$ makes the valve travel from 5% to 45% of the full range.

8.2.4 Remarks on Nonlinear Valve Characteristic

As demonstrated, if the valve travels within a small range of the whole travel span (0–100%), say within ±20%, then the valve characteristic can be assumed linear and does not add any nonlinearity to the loop. This observation can also be realized by investigating the valve characteristic curves in Fig. 8.7. For any $\Delta x = 20\%$, the characteristic curves can be assumed to be locally linear. These characteristic curves are called *inherent characteristics* of the valve. They are only valid for constant pressure drop across the valve and for an incompressible fluid flow. However, in real life, the pressure drop across the valve does not remain constant. Moreover, the valve is connected to other process equipment. Therefore, the characteristic curves of the valve should be re-evaluated after installing the valve in a real process. These characteristics are termed *installed characteristics*.

In reality, the flow through the valve also depends on the pressure drop across the valve. Manufacturers test valves in a rig, where the pressure drop is kept constant. Thus the performance they observe is the inherent characteristics of the valve. In a real plant, pressure drop varies as the flow changes. Therefore, the characteristic relationship seen between valve travel and flow will not be the same as that seen in the test rig. This *installed characteristic* is what really matters to a process engineer. That is why during the selection and design of a valve, considerable effort is made to ensure that the installed characteristics of the valve is as linear as possible (Fisher-Rosemount, 1999; Fitzgerald, 1995; Baumann, 1994; Riggs, 1999). As described in Fisher-Rosemount (1999), it is a good operating practice to keep valve swings below 5% so that the loop gain does not change much and the stability of the loop is ensured. Based on the experience of the authors in industrial data analysis, it has been found that the controller output in most cases swings below $\pm 20\%$. Therefore, the control valve characteristics can be assumed to be locally linear in most cases of real-world control-loop data analysis.

8.3 Nonlinear Disturbances

Disturbances entering a control loop may be either linear or nonlinear. If a disturbance is measurable, the nonlinearity test developed in Sect. 6.3 can be applied to assess its nonlinearity. If the disturbance is unmeasurable, it is difficult to infer anything about the linearity of the loop.

8.4 Summary

This chapter has examined the validity of an important assumption often made in linear control theory – *the process operates in a locally linear fashion*. Two simulation examples – a spherical tank and CSTR – have been used to investigate the operating point dependency of nonlinearity. It can be concluded that a process or system can be assumed to operate in a locally linear fashion if the excitation signal or input to the process is small in magnitude. This chapter has also shown that actuators or final-control elements even with nonlinear valve characteristics such as equal percentage or square root can safely be assumed to be locally linear for small deviations or excursions in their operating points.

Chapter 9
Diagnosis of Poor Control Performance

Poor performance of a control loop is usually caused by poor controller tuning, presence of disturbances, control-loop interactions and/or loop nonlinearities. The presence of nonlinearities in control loops is one of the main reasons for poor performance of a linear controller designed based on linear control theory. In a control loop, nonlinearities may appear either in the control valve or in the process and instrumentation. Among the control-valve nonlinearities, stiction, deadband, deadzone, hysteresis and saturation are most common. A nonlinear system often produces a non-Gaussian and nonlinear time series. The test of Gaussianity or nonlinearity of a control loop variable serves as a useful diagnostic aid towards diagnosing the causes of poor performance of a control loop. In this chapter two indices, the non-Gaussianity index (NGI) and the nonlinearity index (NLI), developed in Chap. 6 are used to detect the possible presence of nonlinearity in the loop. These indices together with specific patterns in the process output (*pv*) vs. the controller output (*op*) plot can be conveniently used to diagnose the causes of poor control-loop performance. The method has been successfully applied to many industrial data sets. A few of them are presented in this chapter. In all cases, the results of the analysis were confirmed and mitigated during routine maintenance by plant engineers.

9.1 Introduction

In the process industry, control and automation are increasingly the route to achieve various objectives such as maintaining world-class quality of the product, reducing operating and maintenance cost, enhancing operators' safety, meeting environmental and occupational health regulations, optimizing resource management and increasing profitability. One manifestation of this increased automation is an increasing number of controllers and control loops in process industries. A process control plant may have anywhere from a few control loops to several thousands, depending on the complexity of the plant from the perspective of control. The performance assessment and monitoring of the performance of these control loops are crucial to the achievement of the desired objectives. Performance demographics of 26,000 PID controllers collected over a period of 2 years and across a large range

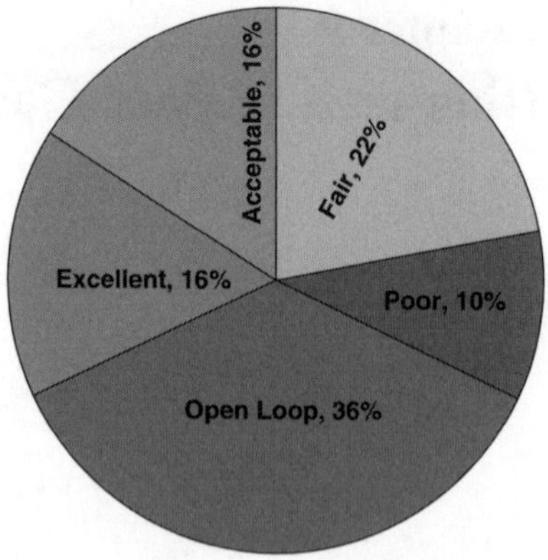

Fig. 9.1 Global multi-industry performance demographics (from Desborough and Miller, 2002)

of continuous process industries have been discussed by Desborough and Miller (2002). The results are reproduced in Fig. 9.1.

Each type of control loop (flow, pressure, level, temperature, etc.) was classified into one of the five categories – excellent, acceptable, fair, poor and open loop – based on a combined algorithm of minimum variance benchmark and an oscillation metric. The classifications were further refined through extensive validation and feedback from industry to reflect controller performance relative to practical expectations for each measurement type. It is evident from Fig. 9.1 that only a third of the loops are performing well or in an acceptable fashion. The other two thirds have significant opportunity for improvement. The key to improving their performance is to diagnose the causes behind their poor performance.

9.2 Problem Description

Figure 9.2 shows a typical control loop under feedback configuration. The objective of this control loop may be either set-point tracking or disturbance rejection. In practice, data for only three measurements for each control loop are available. They are set-point (*sp*), controlled variable (*pv*) and controller output (*op* or sometimes also termed *co*) signals. The measurement of control valve stem position (*mv*) is seldom available. Many authors (Desborough and Harris, 1992; Harris, 1989; Stanfelj et al., 1993; Huang and Shah, 1999) have discussed methods, for example, minimum variance benchmark and settling time benchmark, to estimate the controller or loop performance from routine operating data. In many cases, this information may also be available from plant engineers or operators, who are dissatisfied with the poorly performing loops. The challenge here is to identify the root cause of a poorly performing loop from routine operating (*sp*, *pv* and *op*) data.

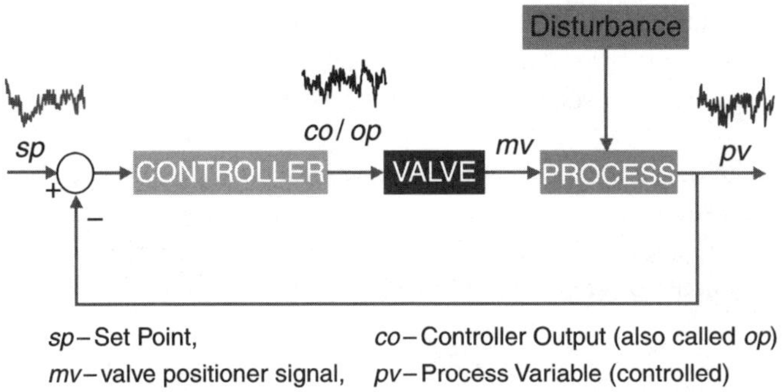

Fig. 9.2 A typical control loop under feedback control

9.3 Usual Causes of Poor Performance

Jelali (2006) has provided a comprehensive overview of control performance assessment technology and its limitations. Poor performance of a control loop is usually due to the following:

1. **Poor controller tuning**. Often times the controllers are left with the default tuning settings from commissioning of the plant, even though there are many changes in the process operation or process modification over the years, which demand retuning of the controller parameters.
2. **Equipment failures**. Failure or malfunction of sensors or actuators may result in poor control performance. In most process industries, process control engineers have limited time to devote to individual control loops and maintain them properly.
3. **Poor process design**. There should be a proper coordination between process design engineers and control engineers during the design phase of a chemical plant in order to ensure better controllability. It is hard to achieve good control performance on an improperly designed process. For example, the current practice in process design tries to minimize the liquid hold-up at the bottom of a distillation column; however, such a design offers little surge capacity to filter out or reject disturbances. A case study on this has been documented in Luyben and Hendershot (2004). The use of reduced inventory and small buffer capacity, recycle streams and tight energy integration contributes to strong loop interactions, plant-wide oscillation propagations and thereby leads to poor controller performance.
4. **Loop interactions**. Inadequate compensation for interacting loops, improper loop pairing and competing controllers may result in poor control performance.
5. **Not using the derivative component in the PID controller**. In most process applications, the derivative components of PID controllers are left out because of the inability or lack of expertise in designing a good filter required to use the D part of the controller. Birgitta (2003) and Adeleye (2006) have shown that with the use of either first- or second-order filters, a PID controller can

significantly improve the control performance over a PI controller. Also, for processes having significant time delays, substantial improvement in closed-loop performance can be achieved if PID controllers are used instead of PI controllers.

6. **Presence of nonlinearities**. Loop nonlinearities can cause poor performance because controllers are usually designed based on linear control theory assuming that everything in the loop is locally linear. The nonlinearities in a loop may arise due to the presence of actuator or valve nonlinearities and/or nonlinear nature of the process itself.

9.4 Diagnosis of Poor Control Performance

Poor control performance can be attributed to specific causes related to either linearity or nonlinearity. The linearity-related causes include tightly tuned controllers, linear oscillatory disturbances and loop interactions. The nonlinearity-related causes include valve nonlinearities or process nonlinearities. The following example demonstrates the application of HOS-based techniques in distinguishing linear and nonlinear causes of poor control performance. Similar results are obtained using Surrogate Data methods. The method is summarized via a signal flow diagram in Fig. 9.3.

First, the control error signal is examined by the HOS-based nonlinearity test. If a nonlinearity is not detected, the focus of the diagnosis should be on controller tuning or on the possible presence of an external linear oscillatory disturbance. If the method detects nonlinearity, then the nonlinearity must be isolated or localized. Is it in the valve or in the process? This study assumes that the process is locally linear.

A simple single-input, single-output (SISO) system in a feedback control configuration (Fig. 9.4) was used to generate simulated data. The first-order process with time delay is given by the following transfer function:

$$G(z^{-1}) = \frac{z^{-3}(1.45 - z^{-1})}{1 - 0.8z^{-1}} \tag{9.1}$$

The process is regulated by a PI controller. The structure of the PI controller is as follows:

$$C(s) = K_c + \frac{K_c}{\tau_i s} = K_c + \frac{I}{s} \tag{9.2}$$

An integrated white noise sequence generated by integrating random noise with a variance of 0.05 was added to the process output. The simulation was performed for 6,000 sampling intervals. To remove the effect of transients, the first several hundred data points were discarded. It is important not to use the data that contain transients. Hence only the last 4,096 points of the error signal to the controller (*sp*–*pv*) were analysed to detect the nonlinearity present in the system.

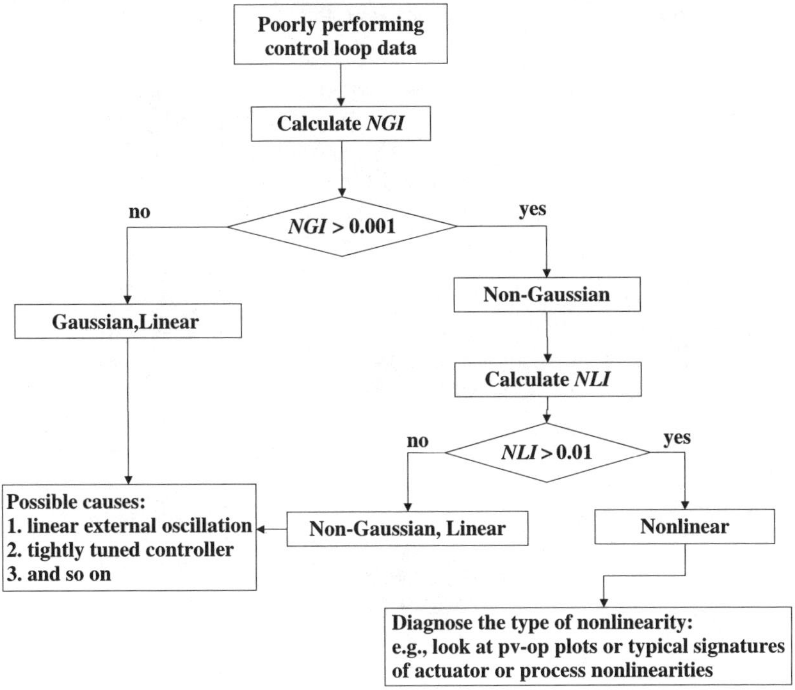

Fig. 9.3 Rule-based decision flow diagram

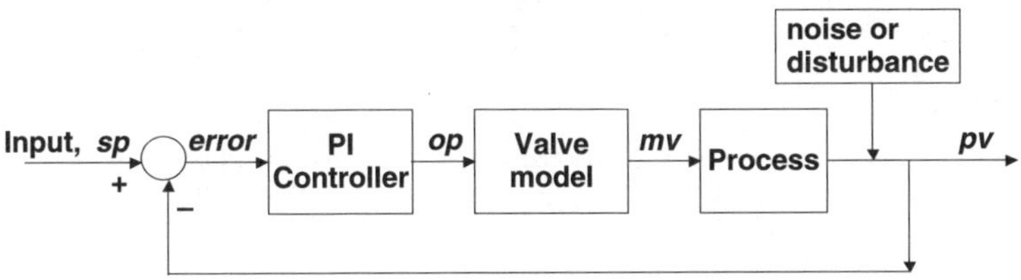

Fig. 9.4 Block diagram of a simple SISO process with stiction nonlinearity in the valve

9.4.1 Well Tuned Controller

The PI controller parameters for this case were $K_c = 0.15$ and $I = K_c/\tau_i = 0.15/1 = 0.15\,\text{s}^{-1}$. The valve model is linear in this simulation block diagram. The top row of Fig. 9.5 shows the results for this case. The proposed test shows that NGI $= -0.0008$. This indicates that the error signal is Gaussian and linear. The corresponding bicoherence plot is flat.

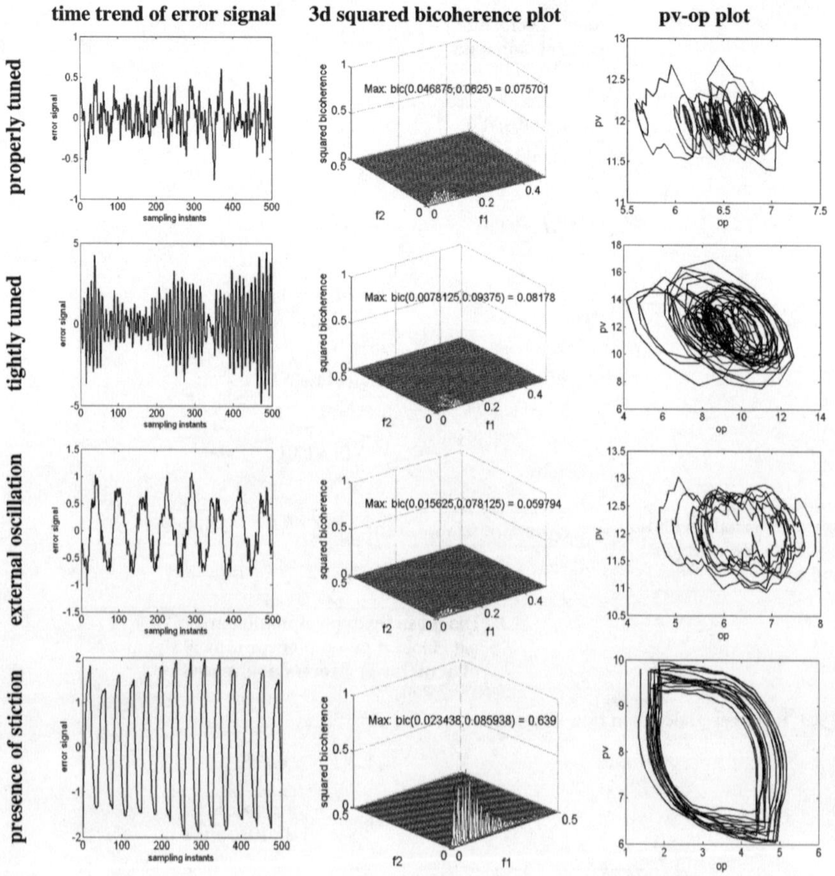

Fig. 9.5 Results of the simulation example. Bicoherence correctly detects the first three cases as linear and the last case as nonlinear. The *pv–op* plot for the stiction case shows elliptical patterns

9.4.2 Tightly Tuned Controller or Excessive Integral Action

For this case, the controller parameters were set at $K_c = 0.15$ and $I = K_c/\tau_i = (0.15/0.4) = 0.375\,\mathrm{s}^{-1}$. The second row of Fig. 9.5 shows the results. The presence of relatively large integral action produces large oscillations in the process variables. An NGI value of -0.0007 indicates that the poor performance is not due to nonlinearity. Also, in the absence of external oscillatory disturbances, the probable cause may be inferred to be due to a tightly tuned controller.

9.4.3 Presence of an External Oscillatory Disturbance

A sinusoid with amplitude 2 and frequency 0.016 or approximately 62 samples/cycle was added to the process output in order to feed an external oscillatory disturbance to the process. The results for this case are shown in the third row of Fig. 9.5. The proposed test gives an NGI value of −0.0003. This results shows that the reason for the oscillation is not due to presence of nonlinearity in the system.

9.4.4 Presence of Stiction

The two-parameter valve-stiction model described in Chap. 13, which also appeared in Choudhury et al. (2004a and 2005a), was used to perform this simulation. The model consists of two parameters: (1) deadband plus stickband, S, and (2) slip jump, J. Figure 13.3 in Chap. 13 summarizes the model algorithm.

To perform the simulation for this particular case, $S = 3$ and $J = 3$ were used. In order to initiate limit cycles or oscillations in a simple first-order time-delay process in the presence of valve stiction, a set-point change at the beginning of the simulation is required. Thereafter, the process is allowed to operate under regulatory control. The last row of Fig. 9.5 shows the time trend of the control error signal, the bicoherence plot and the *pv–op* plot. The presence of stiction produces oscillations in the process. The values of NGI and NLI are 0.28 and 0.59, respectively, which indicate the presence of nonlinearity in the process signal. After detecting the nonlinearity, the process variable versus controller output plot, i.e. *pv–op* plot can be used to diagnose the type of nonlinearity. Usually, the presence of distinctively shaped elliptical trajectories in the *pv–op* is an indication of the presence of stiction in the valve. If the valve position (*mv*) data are available, the *mv–op* plot can be used to identify more accurately the type of nonlinearity in the valve.

Remark 2. The *pv–op* plot alone should *not* be used for detecting stiction. First, the nonlinearity must be detected by HOS-based method or any other method and only then should the *pv–op* plot be used to diagnose the type of nonlinearity. The sole use of *pv–op* plot without a nonlinearity test may result in false-positives in the detection of stiction or other valve problems. There is more discussion of this important topic in Chap. 15.

9.5 Industrial Case Studies

The proposed method has been successfully applied to the detection and isolation of actuator or valve faults for many industrial control loops. There are many types of valve faults such as stiction, saturation, oversized valve and corroded valve seat. Examples of several types of valve faults will be discussed here.

9.5.1 Stiction in a Furnace Dryer Temperature Control Valve

This section describes a temperature control loop on a furnace feed dryer system at the Teck Cominco Limited mine in Trail, British Columbia, Canada. The temperature of the dryer combustion chamber is controlled by manipulating the flow rate of natural gas to the combustion chamber.

Figure 9.6a depicts time trends of the controlled variable (*pv*), the controller output and the set-point. It shows oscillations in both the controlled variable and the controller output. Figure 9.6b shows the bicoherence plot. The NGI and NLI indices obtained for this loop were 0.24 and 0.38, respectively, indicating the presence of nonlinearity in the loop. The presence of distinct elliptical pattern in the characteristic *pv*–*op* plot (Fig. 9.6c) together with the pattern obtained in Fig. 9.6d characterizes the presence of stiction in the control valve. Thus, this analysis was able to confirm the cause of poor loop performance due to the presence of valve stiction.

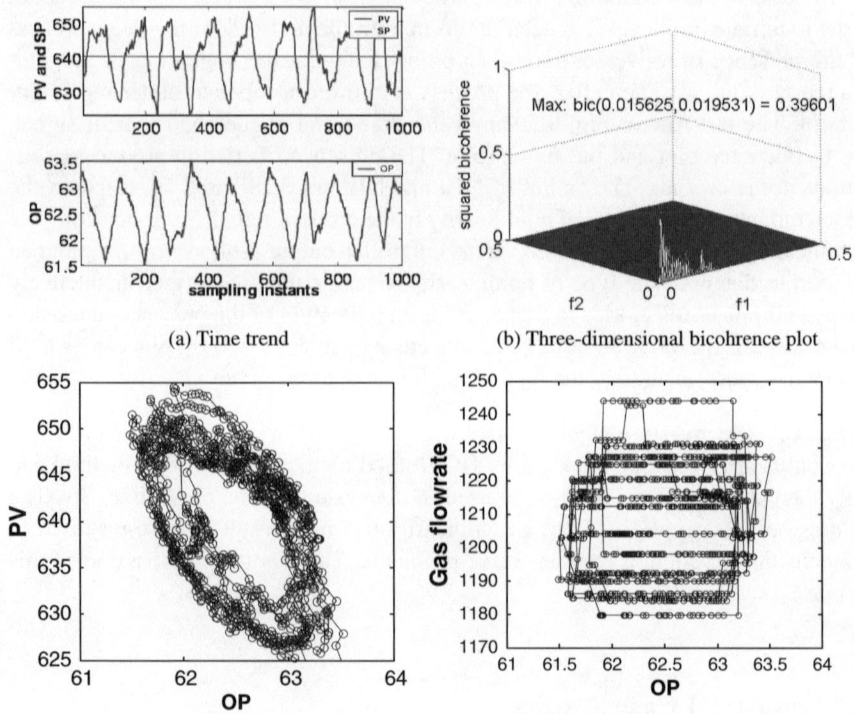

(a) Time trend (b) Three-dimensional bicohrence plot

(c) pv – op characteristic plot (d) mv (flow rate) – op plot

Fig. 9.6 Analysis results of an industrial temperature control-loop data

9.5.2 Valve Saturation

Figure 9.7 illustrates an example of an industrial control loop where the valve suffers from saturation. The time trends of the *pv*, *op* and *sp* are depicted in Fig. 9.7a showing the presence of oscillations. The bicoherence test gives NGI $= 0.16$ and NLI $= 0.26$, indicating a nonlinear loop. The corresponding bicoherence plot also shows a number of peaks indicating significant nonlinearity.

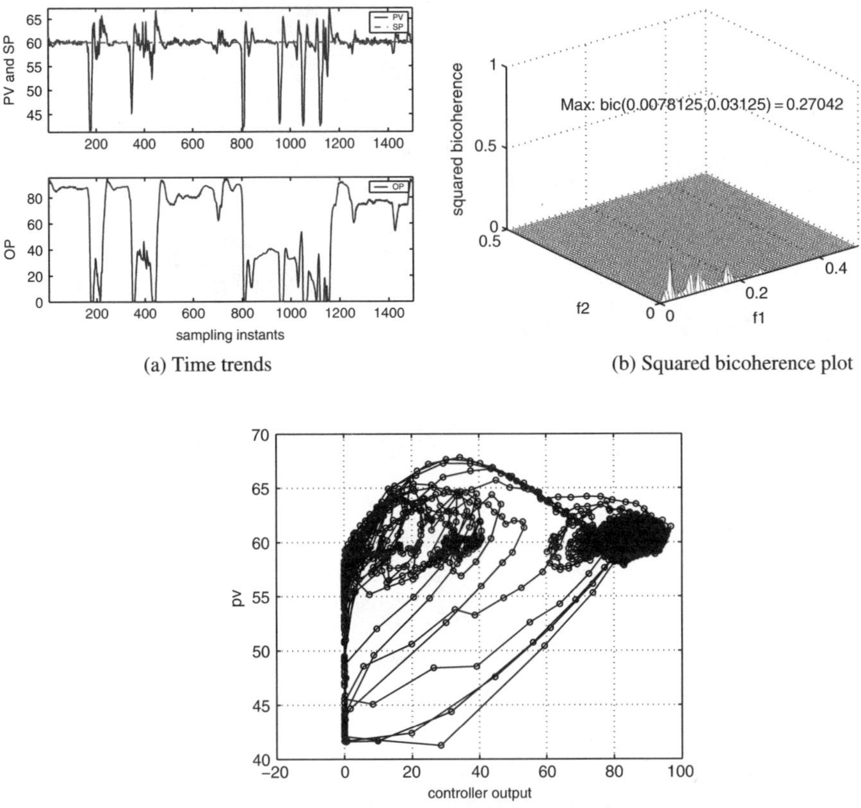

(a) Time trends

(b) Squared bicoherence plot

(c) Pv–op plot

Fig. 9.7 Results of the analysis of an industrial loop with valve saturation

To diagnose the type of nonlinearity, specific patterns in the *pv–op* plot are found to be useful. The *pv–op* mapping for this loop is shown in Fig. 9.7c. A vertical straight line with some random cycles in the *pv–op* plot is a signature of valve saturation. If the valve saturates at both ends (that is, for both full close and full open conditions), there will be two vertical straight lines in the *pv–op* plot. The explanation of this pattern lies in the use of anti-wind up algorithm in the integral action of a PI(D) controller. Because of anti-wind up action, the controller output

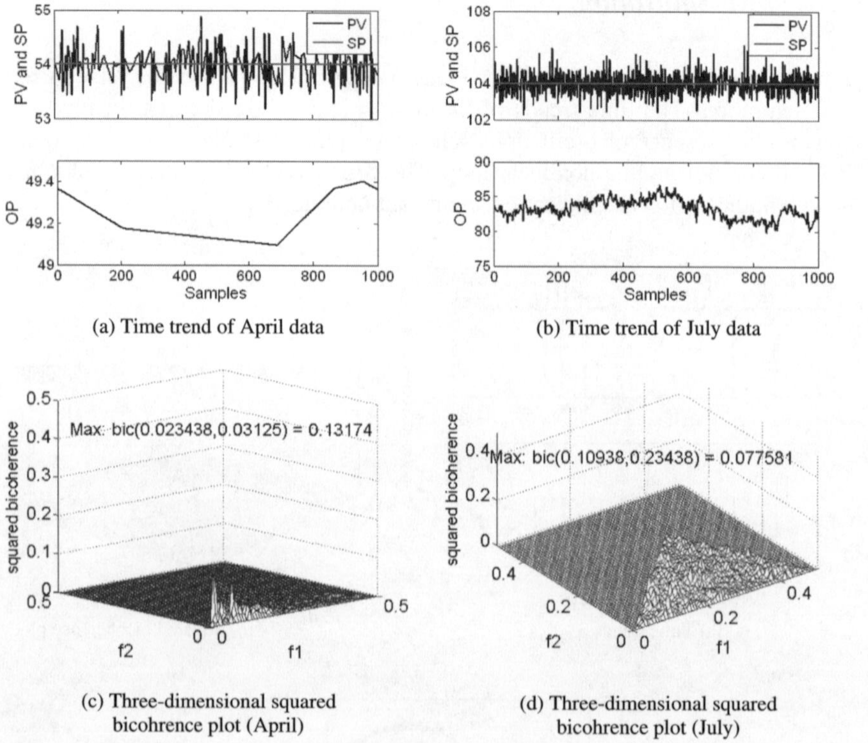

(a) Time trend of April data

(b) Time trend of July data

(c) Three-dimensional squared bicohrence plot (April)

(d) Three-dimensional squared bicohrence plot (July)

Fig. 9.8 Analysis of flow loop 1 data before (April – *left*) and after the (July – *right*) plant maintenance shutdown period

is kept constant while the process output may change. This ends up as a vertical straight line in the *pv–op* plot.

9.5.3 Valve Problems in Some Flow Control Loops

This analysis is for two flow control loops at Celanese Canada Ltd., a chemical complex located in Edmonton, Canada. Data were collected with a sampling interval of 1 min over two periods of time: 10–17 April 2001 and 1–15 July 2001, the latter following an annual maintenance shutdown of the plant. Results for both these loops are discussed below.

9.5.3.1 Oversized Valve

The first example is a recycle flow control loop. Detailed diagnostic plots are shown in Fig. 9.8. Time series of the data collected in April and July are shown in Figs. 9.8a,b. The *op* trend in Fig. 9.8a shows that the valve movement was very slow and insignificant compared to the change in the error signal, (*pv*–*sp*). The values of NGI and NLI are 0.11 and 0.12, respectively. An NGI value of 0.11 shows that the signal is non-Gaussian. The NLI value of 0.12 indicates the presence of nonlinearity in the error signal. The *op* time trend in Fig. 9.8a shows that a small change in *op* caused a large change in *pv* value (note the range of *y*-axis for *op*, 49.4–50). Therefore, it was suggested that the nonlinearity in this loop was most likely due to an oversized valve. The 6-in. diameter valve was replaced by a 3-in. diameter valve during the annual maintenance shutdown of the plant in May 2001. In order to confirm the result of the analysis, additional data were collected in the month of July of the same year. The results of the post-maintenance data analysis are shown in the right panel of the Fig. 9.8. For the new data set, NGI = 0, indicating a Gaussian linear system characteristics. Thus the nonlinearity was eliminated by replacing the oversized valve with a suitably sized smaller valve.

9.5.3.2 Corroded Valve Seat

The example in this section is also a flow control loop at the outlet of a pump located at the bottom of a distillation column. Analysis of the April 2001 data for this loop revealed that the loop had nonlinearity problems with NGI = 0.28 and NLI = 0.63. The diagnostic plots are shown in the top panel of Fig. 9.9. The *pv*–*op* characteristic

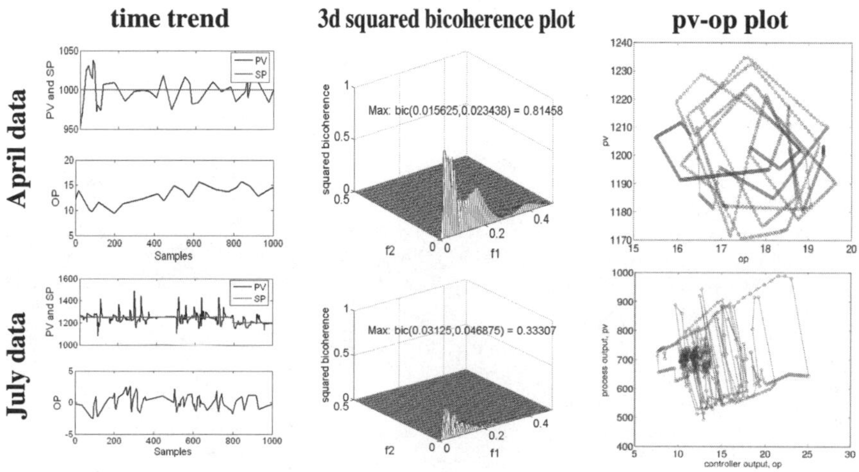

Fig. 9.9 Analysis of flow loop 2 data before (April – *top*) and after (July – *bottom*) the plant maintenance shutdown period

plot indicated a type of nonlinear characteristic in the process or the valve that had not been observed before. During the annual maintenance, plant instrument personnel noticed that the valve seat and the plug were severely corroded. Consequently, the valve seat and the plug were replaced during this maintenance period. The results of the post-maintenance analysis are shown in the lower panel of Fig. 9.9. For the post-maintenance data set, NGI and NLI are 0.18 and 0.28, respectively, indicating yet again the presence of a nonlinearity but in a substantially reduced form. The *pv–op* plot still shows an unfamiliar pattern for unknown sources of nonlinearities or some other problems. However, the overall controller performance of this loop had improved significantly to the point where additional analysis was deemed unnecessary.

9.6 Summary

This chapter has shown the potential of HOS-based indices developed in Chap. 6 in diagnosing the poor performance of control loops. The NGI and NLI indices together with the specific patterns in the process output (*pv*) vs. the controller output (*op*) mapping can be conveniently used to diagnose the causes of poor control-loop performance. The method has been successfully applied to many industrial data sets. Several representative examples of such studies have been presented here. In all cases, the results of the analysis were confirmed by plant engineers.

Part IV
Control Valve Stiction – Definition, Modelling, Detection and Quantification

Chapter 10
Different Types of Faults in Control Valves

This chapter introduces the physical makings of a control valve, describes its various components and discusses the typical faults that may afflict a valve during the course of normal operation.

10.1 What Is a Control Valve

Control valves are the most commonly used actuators or final control elements. They are also the main or the only mechanically moving part of a control loop. A cross-sectional diagram of a typical pneumatic control valve is shown in Fig. 10.1. This figure presented a cross-sectional diagram of a typical pneumatic control valve. The purpose of the valve is to restrict the flow of process fluid through the pipe that can be seen at the very bottom of the figure. The valve plug is rigidly attached to a stem that is attached to a diaphragm in an air pressure chamber in the actuator section at the top of the valve. When compressed air is applied, the diaphragm moves up and the valve opens. The spring is compressed at the same time. The valve illustrated in Fig. 10.1 is a fail-closed type of valve because when the air pressure is reduced the spring forces the valve to close.

A control valve has three basic components:

1. **Actuator:** Most of the actuators are pneumatic. Usually, the actuator works with the help of a diaphragm and instrument air. This is the device which positions the throttling element, i.e. the valve plug inside the valve body.
2. **Valve body subassembly:** This is the part where the valve plug, valve seats and the valve casing are located.
3. **Accessories:** These include positioners, I/P (current to pressure) transducers and position sensors.

Baumann (1994) proposed a control valve model consisting of three components:

1. **Gain:** Every control valve has its gain (K_v), which is determined by the installed flow characteristics of the valve. This is quite different from the gain listed in the vendors' catalogues. K_v determines the flow through the valve for per unit change of controller output signal.

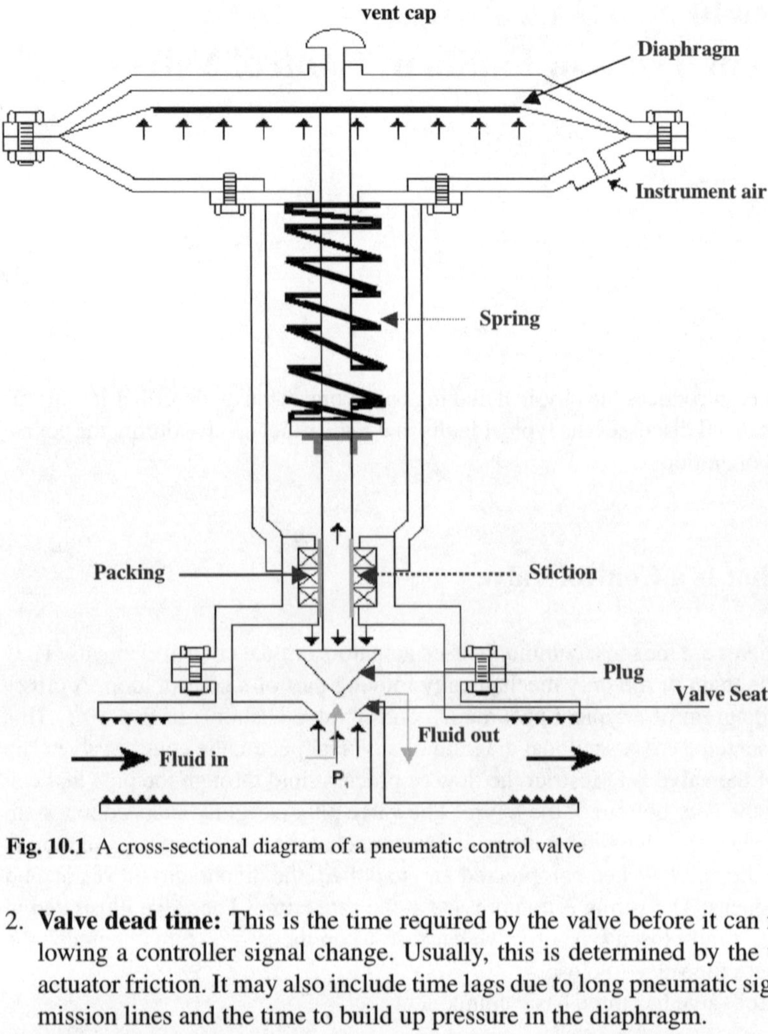

Fig. 10.1 A cross-sectional diagram of a pneumatic control valve

2. **Valve dead time:** This is the time required by the valve before it can move following a controller signal change. Usually, this is determined by the valve and actuator friction. It may also include time lags due to long pneumatic signal transmission lines and the time to build up pressure in the diaphragm.
3. **Time constant of a valve:** Time constant of a valve is related to the stroking speed of the actuator, or the actuator and positioner combination. It determines how fast the valve responds to an upset in the controlled variable.

An ideal valve should have a constant gain throughout the valve travel span, i.e. a linear installed flow characteristic, no dead time with properly adjusted packing and a small time constant.

10.2 Faults in Control Valve

Control valves are mechanical devices subject to wear and tear with respect to time. They require regular maintenance. With time, they may develop various problems such as large deadband, excessive static friction or stiction, saturation, backlash, seat corrosion, diaphragm wear, etc. These problems reduce the performance of the control loop. Detection and diagnosis of such problems are an important objective of the methods proposed in this book.

10.2.1 Oversized Valve

A study by Baumann (1994) reported that 99 out of 100 valves were oversized. An oversized valve is slow to respond to the change in controller output signal, i.e. it increases the time constant of the valve. Oversizing causes valve cycling and hunting and reduces the rangeability of the valve. Interestingly, it was noticed that if a valve is excessively oversized, it may introduce some nonlinearity in the control loop in addition to the poor performance of such loops.

10.2.2 Undersized Valve

As mentioned in the previous section, most of the control valves are oversized. However, in practice one may encounter a few cases of undersized valves. Undersized control valves are not capable of handling the required flow rate through the valves. Therefore, they cannot offer a good control performance and, in general, will lead to saturation-type nonlinearities.

10.2.3 Corroded Valve Seat

The possibility of corrosion or erosion of the valve seat should be taken into account when performing a control-loop performance diagnosis. The wear of plug and seat material causes leaky valves, alters flow characteristics and may create a new flow path in cases of severe erosion. Thus it may contribute to the poor control performance.

10.2.4 Faulty Diaphragm

The diaphragm is a flexible pressure-responsive element that transmits force to the actuator stem. It utilizes the fluid pressure to develop a force to move the actuator stem. Since control valves are constantly throttling and a typical control valve may stroke several thousand times a day, it is not uncommon that the diaphragm gradually loses its efficiency and develops faults. The diaphragm may eventually rupture or develop leaks.

10.2.5 Packing Leakage

A loose or worn stem packing causes external leakage of the process fluids, which may violate EPA (Environmental Protection Agency) regulations. On the other hand, a tight packing may cause excessive friction that can make the loop performance unsatisfactory.

10.2.6 Valve Hysteresis

Valve hysteresis effects are believed to cause less harm than the valve deadband or backlash or stiction. The definition of hysteresis can be found in Sect. 11.2.1 of Chap. 11.

10.2.7 Valve Stiction

Stiction is a portmanteau word formed from the two words *static friction*. Valve stiction is a topic of major interest in this book and is discussed in detail in the following chapters.

10.2.8 Large Deadband

A valve deadband keeps the valve from responding instantly when the signal changes its direction. Therefore it adds dead time to the control loop. An ideal valve should not have a deadband of more than 1% with good stem packing.

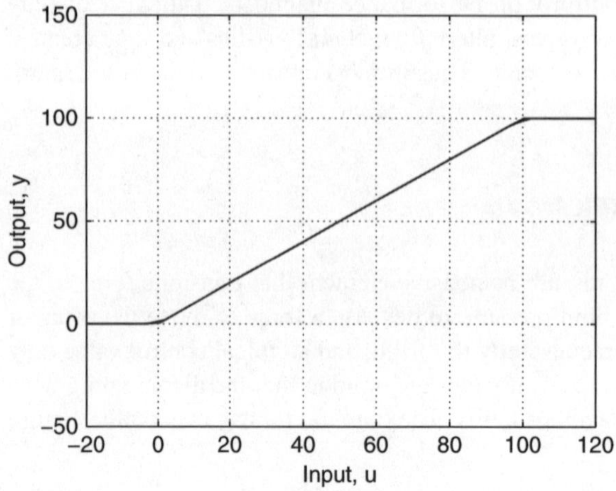

Fig. 10.2 Input–output characteristics of valve saturation

10.2.9 Valve Saturation

All equipment has physical limits. A valve cannot open more than 100% or cannot close more than 0%. If the controller output demands a valve to be open more than 100% or close more than 0%, the valve saturates. The input–output behaviour of saturation is shown in Fig. 10.2. If there is persistent saturation in a control valve, control performance deteriorates and oscillations may arise in the process variable. Therefore, techniques for diagnosing poor performance should also be capable of detecting valve saturation. In some cases, saturation may be deliberately applied. For example, an MPC may deliberately drive a control valve to saturation to achieve maximum throughput.

10.3 Summary

This chapter has introduced the physical makings of a control valve, described its various components and discussed the typical faults that may afflict a valve during the course of normal operation. It is important to have correctly sized control valves in the process in order to ensure good control performance. Also, control valves should not have excessive hysteresis, stiction or deadband.

Chapter 11
Stiction: Definition and Discussions

The presence of nonlinearities such as stiction, deadband and backlash in a control valve limits the control-loop performance. Stiction is the most common problem in spring-diaphragm-type valves, which are widely used in the process industry. There have been many attempts to understand, define and model the stiction phenomenon, and several of these will be reviewed in this chapter. The chapter also presents an approach to stiction modelling that can be related directly to the practical situation as observed in real valves in the process industry. It focusses on the definition of stiction and understanding of the mechanism that causes stiction.

11.1 Introduction

The moving part in a process control loop is a control valve. If the control valve develops nonlinearities such as stiction, backlash and deadband, the valve output may be oscillatory, which in turn can cause oscillations in the process output. Among the many types of nonlinearities in control valves, stiction is the most common and one of the long-standing problems in the process industry. It hinders the achievement of good performance of control valves as well as control loops. Many studies (Horch, 2000; McMillan, 1995; Sharif and Grosvenor, 1998; Horch and Isaksson, 1998; Horch et al., 2000; Aubrun et al., 1995; Wallén, 1997; Taha et al., 1996; Gerry and Ruel, 2001; Ruel, 2000; Armstrong-Hélouvry et al., 1994) have been conducted to define and detect static friction or stiction. However, a unique definition and description of the mechanism of stiction have not been available previously.

11.2 What Is Stiction?

Terms such as deadband, backlash and hysteresis are often misused and wrongly used in describing valve stiction. For example, a deadband in a valve is commonly referred to as *backlash* or *hysteresis*. Therefore, before proceeding to the definition of stiction, these terms are defined for a better understanding of the stiction mechanism and a formal definition of stiction.

M. A. A. S. Choudhury et al., *Diagnosis of Process Nonlinearities and Valve Stiction*,

11.2.1 Definition of Terms Relating to Valve Nonlinearity

This section reviews the American National Standard Institution's (ANSI) formal definition of terms related to stiction. The aim is to differentiate clearly between the key concepts that underlie the ensuing discussion of friction in control valves. These definitions can also be found in EnTech, (1998) and Fisher-Rosemount, (1999), which also refer to ANSI. ANSI (ISA-S51.1-1979, Process Instrumentation Terminology) defines the above terms as follows:

- Backlash: *"In process instrumentation, it is a relative movement between interacting mechanical parts, resulting from looseness, when the motion is reversed"*.
- Hysteresis: *"Hysteresis is that property of the element evidenced by the dependence of the value of the output, for a given excursion of the input, upon the history of prior excursions and the direction of the current traverse."*
  - *"It is usually determined by subtracting the value of deadband from the maximum measured separation between upscale going and downscale going indications of the measured variable (during a full range traverse, unless otherwise specified) after transients have decayed."* Figure 11.1(a) and (c) illustrate the concept.
  - *"Some reversal of output may be expected for any small reversal of input. This distinguishes hysteresis from deadband."*
- Deadband: *"In process instrumentation, it is the range through which an input signal may be varied, upon reversal of direction, without initiating an observable change in output signal."*
  - *"There are separate and distinct input-output relationships for increasing and decreasing signals (See Fig. 11.1(b))."*
  - *"Deadband produces phase lag between input and output."*
  - *"Deadband is usually expressed in percent of span."*

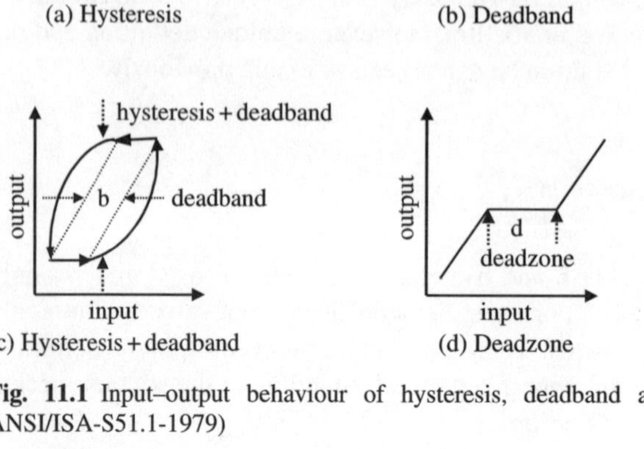

Fig. 11.1 Input–output behaviour of hysteresis, deadband and dead zone (redrawn from ANSI/ISA-S51.1-1979)

Deadband and hysteresis may be present simultaneously. In such an event, the characteristics in the lower left panel of Fig. 11.1 would be observed.

- Dead zone: *"It is a predetermined range of input through which the output remains unchanged, irrespective of the direction of change of the input signal."* (*There is another definition of deadzone for multi-position controller. It is a zone of input for which no value of output exists. It is usually intentional and adjustable*).
  - *"There is but one input-output relationship (See Fig. 11.1(d))."*
  - *"Dead zone produces no phase lag between input and output."*

The above definitions show that the term 'backlash' applies specifically to the slack or looseness of the mechanical part when the motion changes direction. Therefore, in control valves, it may add deadband effects only if there is some slack in rack-and-pinion-type actuators (Fisher-Rosemount, 1999) or loose connections in a rotary valve shaft. ANSI (ISA-S51.1-1979) definitions and Fig. 11.1 show that hysteresis and deadband are distinct effects. Deadband is quantified in terms of input signal span (i.e. on the horizontal axis), while hysteresis refers to a separation in the measured (output) response (i.e. on the vertical axis).

11.2.2 Discussion of the Term 'Stiction'

Various studies and organizations have defined stiction in various ways. Several definitions are reproduced below:

- According to the Instrument Society of America (ISA) (ISA Subcommittee SP75.05, 1979), "*stiction is the resistance to the start of motion, usually measured as the difference between the driving values required to overcome static friction upscale and downscale*". The definition was first proposed in 1963 in American National Standard C85.1-1963,"Terminology for Automatic Control" and has not been updated. This definition was adopted in ISA 1979 Handbook (ISA Subcommittee SP75.05, 1979) and has remained the same in the revised 2001 edition.
- According to Entech (1998), "*stiction is a tendency to stick-slip due to high static friction. The phenomenon causes a limited resolution of the resulting control valve motion. ISA terminology has not settled on a suitable term yet. Stick-slip is the tendency of a control valve to stick while at rest, and to suddenly slip after force has been applied*".
- According to Horch (2000), "*The control valve is stuck in a certain position due to high static friction. The (integrating) controller then increases the set point to the valve until the static friction can be overcome. Then the valve breaks off and moves to a new position (slip phase) where it sticks again. The new position is usually on the other side of the desired set point such that the process starts in the opposite direction again*". This is the extreme case of stiction. On the contrary, once the valve overcomes stiction, it might travel smoothly for some time and then stick again when the velocity of the valve is close to zero.

- Ruel (2000) reported: "*stiction as a combination of the words stick and friction, created to emphasize the difference between static and dynamic friction. Stiction exists when the static (starting) friction exceeds the dynamic (moving) friction inside the valve. Stiction describes the valve's stem (or shaft) sticking when small changes are attempted. Friction of a moving object is less than when it is stationary. Stiction can keep the stem from moving for small control input changes, and then the stem moves when there is enough force to free it. The result of stiction is that the force required to get the stem to move is more than the required force to go to the desired stem position. In presence of stiction, the movement is jumpy*". This definition resembles stiction as measured online in process industries – putting the control loop in manual and then increasing the valve input in small increments until there is a noticeable change in the process variable.
- In Olsson (1996), stiction is defined as "*short for static friction as opposed to dynamic friction. It describes the friction force at rest. Static friction counteracts external forces below a certain level and thus keeps an object from moving*".

The above discussion reveals the lack of a formal and general definition of stiction and the mechanism(s) that cause(s) it. All of the above definitions agree that stiction is the static friction that prevents an object from moving and when the external force overcomes the static friction the object starts moving. However, these definitions disagree in the way stiction is measured and how it can be modelled. Also, there is no clear description of what happens at the moment when the valve just overcomes the static friction. Several modelling approaches described this phenomenon using a Stribeck effect model (Olsson, 1996).

11.2.3 A Formal Definition of Stiction

The motivation for a new definition of stiction is to capture the descriptions cited earlier within a definition that explains the behaviour of a valve with stiction in terms of its input–output behaviour, as is carried out in the ANSI definitions for backlash, hysteresis and deadband (see Fig. 11.2). The new definition of stiction as proposed by the authors is based on careful investigation of real process data.

A typical phase plot of the input–output behaviour of a valve suffering from stiction is described in Fig. 11.2. It consists of four components: deadband, stickband, slip-jump and the moving phase. When the valve comes to rest or changes direction at point A in Fig. 11.2, the valve sticks. After the controller output overcomes the deadband (AB) and the stickband (BC) of the valve, the valve jumps to a new position (point D) and continues to move. Due to very low or zero velocity, the valve may stick again between points D and E in Fig. 11.2 while travelling in the same direction (EnTech, 1998). In such a case, the magnitude of deadband is zero and only the stickband is present. An example of a sticking valve in industry that matches the phase plot shown in Fig. 11.2 is shown in Fig. 11.3. It is uncommon to see features such as the sticking between D and E in industrial data because the valve generally

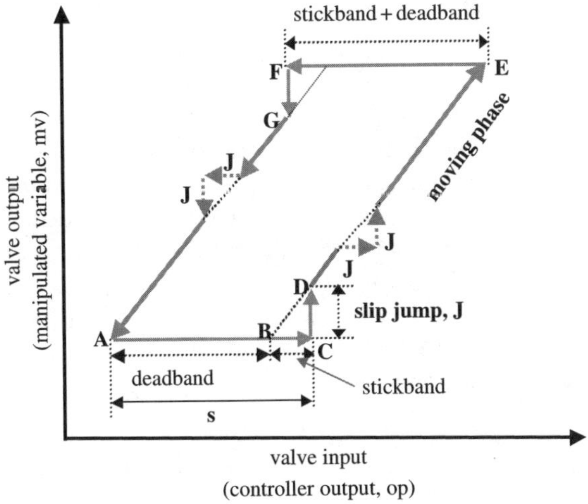

Fig. 11.2 Typical input–output behaviour of a sticky valve

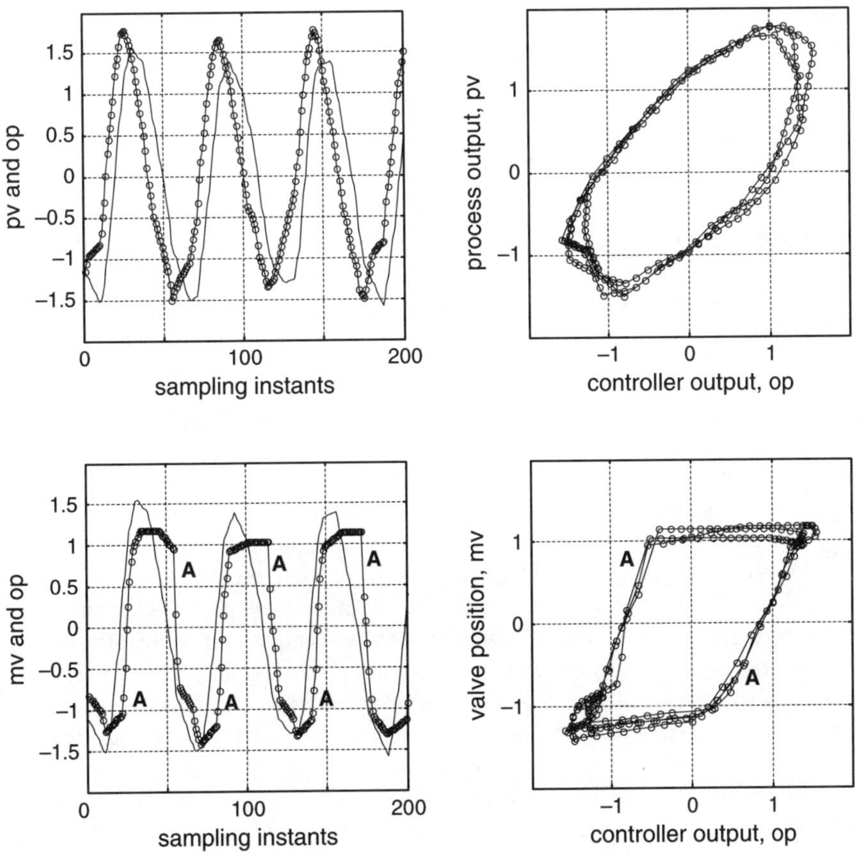

Fig. 11.3 Flow control cascaded to level control in an industrial setting, the line with circles is *pv* and *mv*, and the thin line is *op*

keeps moving while the input is changing. The valve movement stops only when the input signal to the valve stops changing or reverses its direction.

The deadband and stickband represent the behaviour of the valve when it is not moving, even though the input to the valve keeps changing. Slip–jump represents the abrupt release and conversion to kinetic energy of potential energy that was stored in the actuator chamber due to the high static friction. The magnitude of the slip–jump is critical in determining the limit-cyclic behaviour introduced by stiction (McMillan, 1995; Pipponen, 1996). Once the valve slips, it continues to move until it sticks again (point E in Fig. 11.2). In this moving phase, dynamic friction may be much lower than the static friction. As depicted in Fig. 11.2, this section has proposed a rigorous description of the effects of friction in a control valve. Therefore, we propose the following new definition of stiction:

The presence of stiction impairs proper valve movement, i.e. the valve stem may not move in response to the output signal from the controller or the valve positioner. The smooth movement of the valve in response to a varying input from the controller or the valve positioner is preceded by a stickband and an abrupt jump termed as slip-jump. Its origin in a mechanical system is static friction, which exceeds the dynamic friction during smooth movement of the valve.

11.3 Practical Examples of Valve Stiction

The objective of this section is to explore the effects of stiction through the investigation of data from industrial control loops. The observed effects reinforce the need for a rigorous definition of stiction. This section analyses four data sets. The first set is from a power plant, the next two are from a petroleum refinery and the fourth is from a furnace dryer system. To preserve the confidentiality of the plants, all data are scaled and reported as mean-centred with unit variance.

- Loop 1 is a level control loop that controls the level of condensate in the outlet of a turbine by manipulating the flow rate of the liquid condensate. In total, 8,640 samples for each tag were collected at a sampling rate of 5 s. Figure 11.3 shows a portion of the time domain data. The left panel shows time trends for level (*pv*), the controller output (*op*), which is also the valve demand, and valve position (*mv*), which can be taken to be the same as the condensate flow rate. The plots in the right panel show the characteristics *pv–op* and *mv–op* plots. The bottom plot clearly indicates both the stickband plus deadband and the slip–jump effects. The slip–jump is large and visible from the bottom plot, especially when the valve is moving in a downward direction. It is marked as 'A' in the figure. It is evident from this figure that the valve output (*mv*) can never reach the valve demand (*op*). This kind of stiction is termed as 'undershoot case' of valve stiction. The *pv–op* plot does not show the jump behaviour clearly. The slip–jump is very difficult to observe in the *pv–op* plot because the process dynamics (i.e. the transfer function between *mv* and *pv*) destroy the pattern. This loop shows one possible type of stiction phenomenon clearly. The stiction model developed later in Chap. 13 based on the control signal (*op*) is able to simulate this kind of behaviour.

- Loop 2 is a liquid flow slave loop of a cascade control loop. The data were collected at a sampling rate of 10 s and the data length for each tag was 1,000 samples. The left plot of Fig. 11.4 shows the time trend of *pv* and *op*. A closer examination of this figure reveals that the *pv* (flow rate) is constant for a given period of time, although the *op* changes over that period. This is the period during which the valve was stuck. Once the valve overcomes deadband plus stickband, the *pv* changes very quickly (denoted as 'A' in the figure) and moves to a new position where the valve sticks again. It is also evident that sometimes the *pv* overshoots the *op*, and sometimes it undershoots. The *pv–op* plot has two distinct parts – the lower part and the upper part extended to the right. The lower part corresponds to the overshoot case of stiction, i.e. it represents an extremely sticky valve. The upper part corresponds to the undershoot case of stiction. These two cases have been separately modelled in the data-driven stiction model. This example serves to illustrate a mixture of undershoot and overshoot cases of stiction. The terminologies regarding different cases of stiction will be clarified in Sect. 13.2 in Chap. 13
- Loop 3 is a slave flow loop cascaded to a master-level control loop. A sampling rate of 6 s was used for the collection of data and a total of 1,000 samples for each tag were collected. The top panel of Fig. 11.5 shows the presence of stiction with a clear indication of stickband plus deadband and the slip–jump phase. The slip–jump appears as the control valve just overcomes stiction (denoted as point 'A' in Fig. 11.5). This slip–jump is not very clear in the *pv–op* plot of the closed-loop data (top-right plot) because both *pv* and *op* jump together due to the probable presence of a proportional only controller. However, it shows the presence of deadband plus stickband clearly. Sometimes, it is best to look at the *pv–sp* plot when examining a cascaded loop with the slave loop operating under proportional control only. The bottom panel of Fig. 11.5 shows the time trend

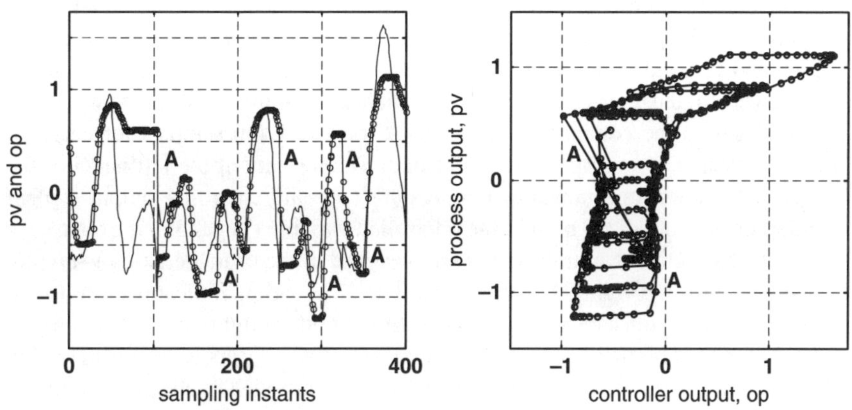

Fig. 11.4 Data from a flow loop in a refinery, time trend of pv and op (*left*) – the line with circles is pv and the thin line is op, and the pv–op plot (*right*)

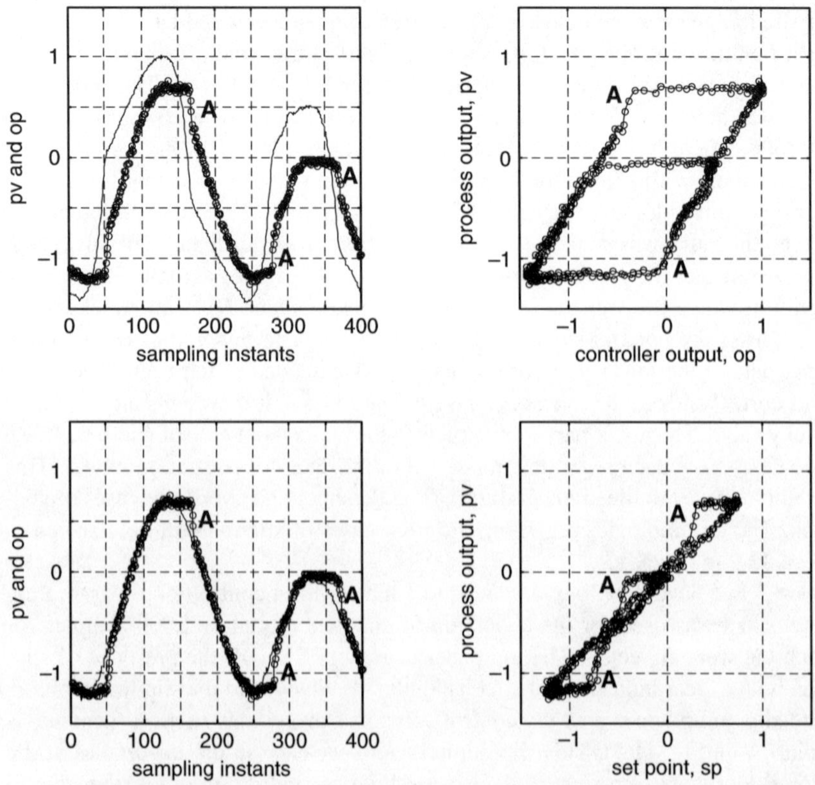

Fig. 11.5 Data from a flow loop in a refinery, time trend of pv and op (*top left*) – the line with circles is pv and the thin line is op, the pv–op plot (*top right*), time trend of pv and sp (*bottom left*), line with circles is pv and thin line is sp, and the pv–sp plot (*bottom right*)

and phase plot of *sp* and *pv*, where the slip–jump behaviour is clearly visible. This example represents a case of pure stick-slip or stiction with no offset.

- Loop 4 is a temperature control loop on a furnace feed dryer system at the Teck Cominco Limited mine in Trail, British Columbia, Canada. The temperature of the dryer combustion chamber is controlled by manipulating the flow rate of natural gas to the combustion chamber. A total of 1,440 samples for each tag were collected at a sampling rate of 1 min. The top plot of the left panel of the Fig. 11.6 shows time trends of temperature (*pv*) and controller output (*op*). It shows clear oscillations both in the controlled variable (*pv*) and in the controller output. Distinctive oval trajectories are observed in the characteristic *pv–op* plot (see Fig. 11.6 top right). For this loop, there was a flow indicator close to the valve and indicator data were available. In the bottom-right panel of this figure the flow rate is plotted versus *op*. The flow-rate data appears to be quantized but the presence of stiction in this control valve was confirmed by the plant engineer. The two bottom plots clearly show the stickband and the slip–jump of the valve.

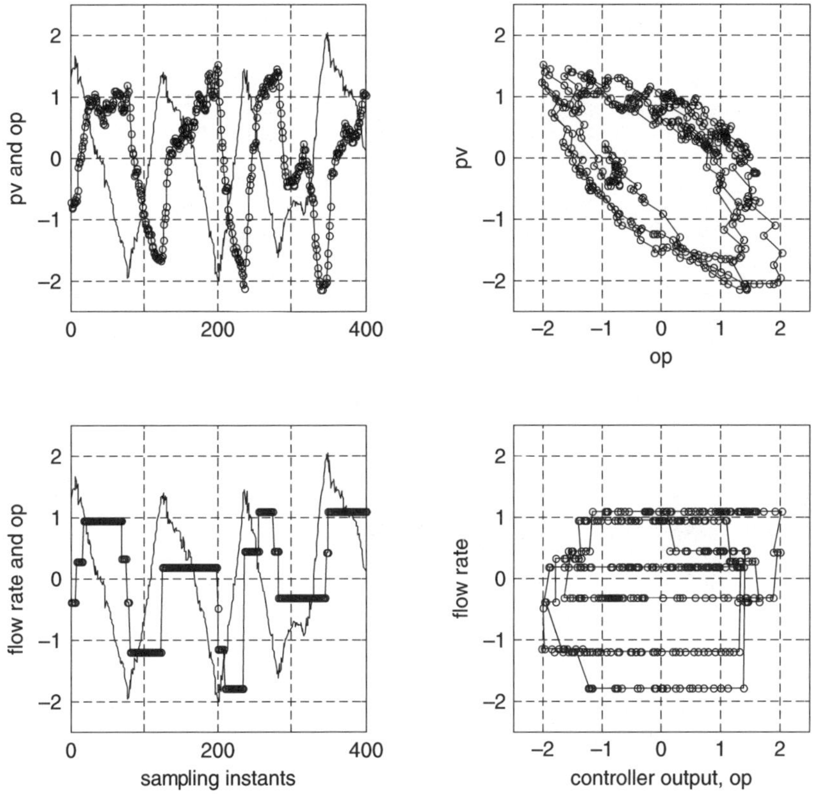

Fig. 11.6 Industrial dryer temperature control-loop data, lines with circles are pv and flow rate, the thin line is op (*the bottom-left panel*)

The moving phase of the valve is almost absent in this example. After the valve overcomes stiction, it jumps to the new position and sticks again.

11.4 Summary

This chapter has presented a discussion of the terms related to valve nonlinearity, followed by the proposal of a new formal definition of stiction. Some practical examples of valve stiction from real-world data are provided to gain insights into the mechanism of stiction. Practical examples of valve stiction are given to justify the newly proposed definition of stiction.

Chapter 12
Physics-Based Model of Control Valve Stiction

In this book, stiction in control valves is modelled using two approaches. One is physics based and the other is data driven. A physics-based model is outlined in this chapter, while Chap. 13 will present a data-based model that has the same input–output behaviour. The physics-model discussed in this chapter was published in Choudhury et al. (2005a).

12.1 Introduction

Friction in the valve arises principally in the packing (See Fig. 10.1). It is the packing that stops process fluid from leaking out of the valve but the valve stem nevertheless has to move freely relative to the packing. There is a trade-off because too tight packing reduces emissions and leaks from the valve but at the same time increases the friction. Loose packing reduces friction but there is a potential for process fluids to leak. Other effects that cause excessive friction are corrosion of the valve stem, which makes it rough or nonsmooth, and deposits on the valve seat, which can make the valve plug stick in the seat.

Sticking control valves have deadband and stick–slip behaviour (stiction) caused by excessive static friction. Friction effects have been thoroughly studied in the literature, for instance by Karnopp (1985), Dewit et al. (1995), Olsson (1996), Kayihan and Doyle III (2000) and Choudhury et al. (2005a, 2005b).

12.2 Physical Modelling of Valve Friction

12.2.1 Physics of a Control Valve

The purpose of this section is to understand the physics of valve friction and reproduce the behaviour seen in real plant data.

M. A. A. S. Choudhury et al., *Diagnosis of Process Nonlinearities and Valve Stiction*,

For a pneumatic sliding stem valve, the force balance equation based on Newton's second law can be written as:

$$M\frac{d^2x}{dt^2} = \sum Forces = F_a + F_r + F_f + F_p + F_i, \tag{12.1}$$

where M is the mass of the moving parts; x is the relative stem position; $F_a = Au$ is the force applied by pneumatic actuator, where A is the area of the diaphragm and u is the actuator air pressure or the valve input signal; $F_r = -kx$ is the spring force where k is the spring constant; $F_p = -A_p\Delta P$ is the force due to fluid pressure drop, where A_p is the plug unbalance area and ΔP is the fluid pressure drop across the valve; F_i is the extra force required to force the valve to be into the seat and F_f is the friction force (Fitzgerald, 1995; Kayihan and Doyle III, 2000; Whalen, 1983). Following Kayihan and Doyel III, F_i and F_p will be assumed to be zero because of their negligible contribution to the model.

12.2.2 Friction Model

The friction model is adopted from Karnopp (1985) and Olsson (1996). It includes static and moving friction. The expression for the moving friction is in the first line of Eq. (12.2) and comprises a velocity-independent term F_c known as Coulomb friction and a viscous friction term vF_v that depends linearly upon the velocity. Both act in opposition to the velocity, as shown by the negative signs.

$$F_f = \begin{cases} -F_c\, sgn(v) - vF_v & \text{if } v \neq 0 \\ -(F_a + F_r) & \text{if } v = 0 \text{ and } |F_a + F_r| \leq F_s \\ -F_s\, sgn(F_a + F_r) & \text{if } v = 0 \text{ and } |F_a + F_r| > F_s \end{cases}. \tag{12.2}$$

The second line in Eq. (12.2) is the case when the valve is stuck. F_s is the maximum static friction. The velocity of the stuck valve is zero, therefore the acceleration is zero also. Thus, the right-hand side of Newton's law is zero, such that $F_f = -(F_a + F_r)$. The third line of the model represents the situation at the instant of breakaway. At that instant the sum of forces is $(F_a + F_r) - F_s\, sgn(F_a + F_r)$, which is not zero if $|F_a + F_r| > F_s$. Therefore, the acceleration becomes non-zero and the valve starts to move.

Many refinements exist for this friction model, including the Stribeck effect, which gives a detailed treatment of the static friction spike between F_c and F_s, and the variable break away forces such that the value of F_s depends on how long the valve has been stuck. The classic friction model presented here is sufficient to capture the main features and can model deadband and stick–slip effects.

12.2.3 Model Parameters

The physical model of a control valve requires several parameters (M, F_s, F_c, F_v) to be known. The mass M and typical friction forces (F_s, F_c, F_v) depend upon the design of the valve. Kayihan and Doyle III (2000) used manufacturer's values suggested by Fitzgerald (1995) and similar values have been chosen here apart from a slightly increased value of F_s and a smaller value for F_c in order to make the demonstration of the slip-jump more obvious (see Table 12.1). Figure 12.1 shows the friction force characteristic in which the magnitude of the moving friction is smaller than that of the static friction. The friction force opposes velocity (see Eq. (12.2)), thus the force is negative when the velocity is positive.

The calibration factor of Table 12.1 is introduced because the required stem position x_r is the input to the simulation. In the absence of stiction effects, the valve

Table 12.1 Nominal values used for physical valve simulation

Parameters	Kayihan and Doyle III, 2000	Nominal case
M	3 lb (1.36 kg)	1.36 kg
F_s	384 lbf (1708 N)	1750 N
F_c	320 lbf (1423 N)	1250 N
F_v	3.5 lbf.s/in (612 N.s/m)	612 N.s/m
Spring constant, k	300 lbf/in (52500 N/m)	52500 N/m
Diaphragm area, A	100 in^2 (0.0645 m^2)	0.0645 m^2
Calibration factor, k/A	–	807692 Pa/m
Air pressure	10 psi (68950 Pa)	68950 Pa

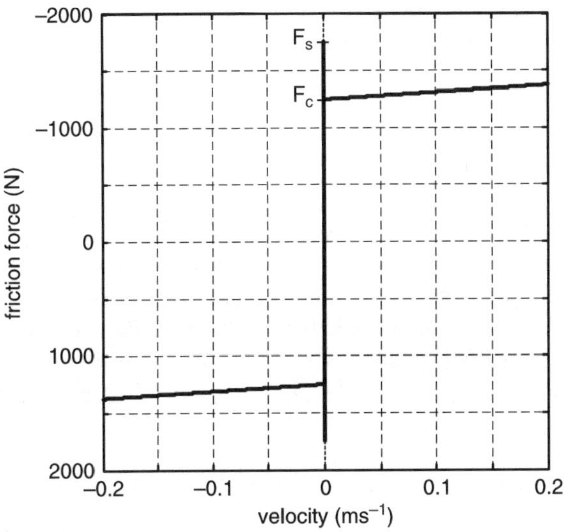

Fig. 12.1 Friction characteristic plot

moving parts come to rest when the force due to air pressure on the diaphragm is balanced by the spring force. Thus $Au = kx$ and the calibration factor relating the air pressure u to x_r is k/A. One consequence of miscalibration is that the valve can overshoot, as discussed later.

12.2.4 Detection of Zero Velocity

A numerical simulation of the valve requires a condition for the velocity to be zero in order for the second and third lines of the friction model to be activated. The velocity in metres per second is taken to be zero if it falls in the range $-0.0005 \leq v \leq 0.0005$.

12.2.5 Model of the Pressure Chamber

It was pointed out by Kano et al. (2004) that numerical simulations of valve models do not always respond correctly to a stochastic input in which the air pressure changes randomly at each simulation step. The reason is that the zero velocity condition is not met unless the amplitude of the stochastic input is very small.

A physical valve would not experience a stochastic air pressure signal because the air pressure chamber has a smoothing effect. A solution to the numerical problem identified by Kano can make use of this physical insight by introducing a first-order lag transfer function model to represent the dynamics of the pressure chamber in place of the algebraic model $F_u = Au$. The adjusted model is:

$$F_a = \frac{A}{\tau s + 1} u, \tag{12.3}$$

where τ is a time constant that can be set to some realistic value such as 1 s or any value that is significantly faster than the time constant of the process. If the stochastic input u has a large enough amplitude, then the valve does not stick even when the pressure chamber dynamics are taken into account. This effect is also observed with a real valve and such an input is called a dither signal. Adding a dither signal to the valve air pressure is well known as a means of reducing the impact of a control valve deadband.

12.3 Valve Simulation

The purpose of simulation of the valve was to determine the influence of the three friction terms in the model. The nonlinearity in the model is able to induce limit-cycle oscillations in a feedback control loop. The aim is to understand the contribution of each friction term to the character and shape of the limit cycles.

12.3.1 Open-Loop Response

Simulations using the model have been reported in Choudhury et al., (2005a) and are reproduced here. Figure 12.2 shows the valve position when the valve model is driven by a sinusoidal variation in *op* in the absence of the controller. The left-hand column shows the time trends while the right-hand panels are plots of valve demand (*op*) versus valve position (*mv*). Several cases are simulated using the parameters shown in Table 12.2. The linear values are those suggested by Kayihan and Doyle III for the best case of a smart valve with Teflon packing.

In the first row of Fig. 12.2, the Coulomb friction F_c and the static friction F_s are small and linear viscous friction dominates. The input and output are almost in phase in the first row of Fig. 12.2 because the sinusoidal input is of low frequency compared to the bandwidth of the valve model and is on the part of the frequency response function where input and output are in phase.

Valve deadband is due to the presence of Coulomb friction F_c, a constant friction that acts in the opposite direction to the velocity. In the deadband simulation case, the static friction is the same as the Coulomb friction, $F_s = F_c$. The deadband arises

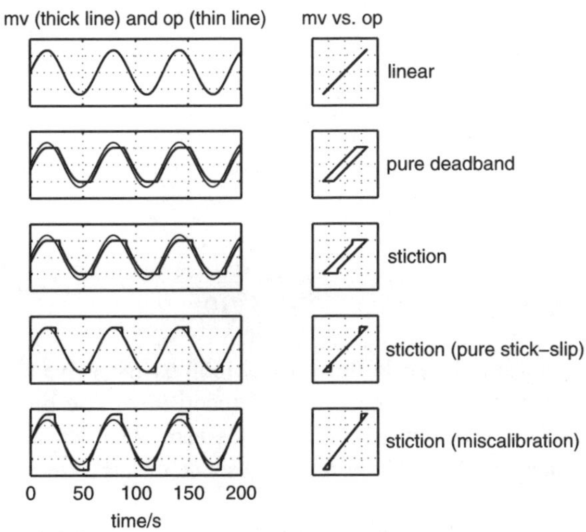

Fig. 12.2 Open-loop responses of the physics-based model of a valve

Table 12.2 Friction values used in simulation of physical valve model

Parameters	Linear	Pure deadband	Stiction (undershoot) (open loop)	Stiction (undershoot) (closed loop)	Stiction (no offset)
F_s (N)	45	1250	1750	1000	1750
F_c (N)	45	1250	1250	400	0
F_v (N.s/m)	612	612	612	612	612

because, on changing the direction, the valve remains stationary until the net applied force is large enough to overcome F_c. The deadband becomes larger if F_c is larger.

A valve with high initial static friction such that $F_s > F_c$ exhibits a jumping behaviour that is different from a deadband, although both behaviours may be present simultaneously. When the valve starts to move, the friction force decreases abruptly from F_s to F_c. There is therefore a discontinuity in the model on the right-hand side of Newton's second law and a large increase in acceleration of the valve moving parts. The initial velocity is therefore higher than in the $F_s = F_c$ case, leading to the jump behaviour observed in the third row of Fig. 12.2. If the Coulomb friction F_c is absent, then the deadband is absent and the slip-jump allows the *mv* to catch up with the *op* (fourth row).

If the valve is miscalibrated, then swings in the valve position (*mv*) are larger than swings in the demanded position (*op*). In that case the gradient of the *op*–*mv* plot is greater than unity during the moving phase. The bottom row of Fig. 12.2 shows the case when the calibration factor is larger by 25%. A slip-jump was also used in this simulation.

12.3.2 Closed-Loop Response

For assessment of closed-loop behaviour, the valve output drives a first-order plus time-delay (FOPTD) process $G(s)$ and receives its *op* reference input from a PI controller $C(s)$, where:

$$G(s) = \frac{3e^{-10s}}{10s+1} \qquad C(s) = 0.2\left(\frac{10s+1}{10s}\right) \qquad (12.4)$$

Figure 12.3 shows the limit cycles induced in this control loop by the valve together with the plots of valve position (*mv*) versus valve demand (*op*). The limit cycles were present even though the set point to the loop was zero. That is, they were internally generated and sustained by the loop in the absence of any external set-point excitation.

There was no limit cycle for the FOPTD process in the linear case dominated by viscous friction or in the case with deadband only when $F_s = F_c$. It is known that

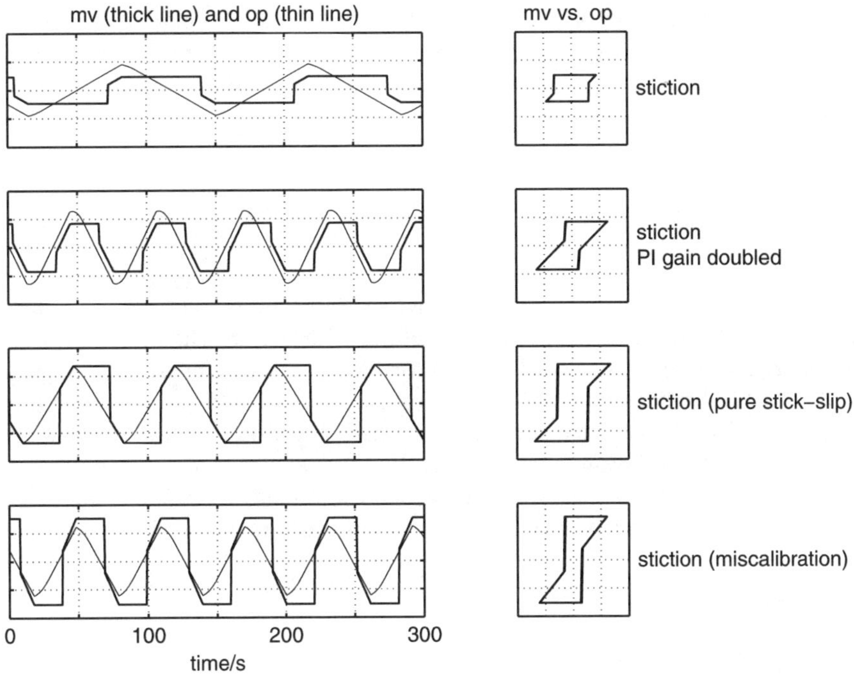

Fig. 12.3 Close loop responses of the physics-based model of a valve.

deadband alone cannot induce a limit cycle unless the process $G(s)$ has integrating dynamics, as will be discussed further in Chap. 14.

The presence of stiction ($F_s > F_c$) induces a limit cycle with a characteristic triangular shape in the controller output. Cycling occurs because an offset exists between the set point and the output of the control loop while the valve is stuck, which is integrated by the PI controller to form a ramp. By the time the valve finally moves in response to the controller *op* signal, the actuator force will have grown quite large and the valve will then move quickly to a new position where it will stick again. Thus, a self-limiting cycle is set up in the control loop.

If stiction and deadband are both present, then the period of the limit cycle oscillation can become very long. The combination $F_s = 1750\,N$ and $F_c = 1250\,N$ gave a period of 300 s while the combination $F_s = 1000\,N$ and $F_c = 400\,N$ had a period of about 140 s (top row, Fig. 12.3). In both cases, the period is much longer than the time constant of the controlled process. The period of oscillation can also be influenced by altering the controller gain. If the gain is increased, the linear ramps of the controller output signal are steeper, the actuator force moves through the deadband more quickly and the period of the limit cycle becomes shorter (second row, Fig. 12.3). The technique of changing the controller gain is used by industrial control engineers to test the hypothesis of a limit cycle induced by valve nonlinearity while the plant is still running in closed loop.

In the pure stick–slip case shown in the third row of Fig. 12.3, the Coulomb friction is negligible and the oscillation period is shorter because there is no deadband. The bottom row in Fig. 12.3 shows that miscalibration causes an overshoot under closed loop.

12.4 Summary

This chapter gave an analysis of the behaviour of feedback control loops with a sticking valve through physics-based modelling of the valve. The model successfully reproduced a range of closed-loop limit cycles observed in industrial processes.

The physical model has fulfilled its purpose of giving in-depth understanding of the origin of deadband and stiction effects in a control valve. However, it is not useful for routine use because the values of the physical parameters such as friction forces and spring constant are not generally known. A data-driven empirical model will be discussed in the next chapter as a means of overcoming this disadvantage.

Chapter 13
Data-Driven Model of Valve Stiction

In order to use the physical model of valve stiction, several parameters such as valve friction force, mass and spring constant are required to be known. Explicit values of these parameters depend on the size of the valve and its manufacturer and are difficult to obtain in practice. Therefore, physical models of valve friction are not easily available for routine use. On the other hand, a data-driven stiction model is simple to use and does not require knowledge of the parameters of the physical model. This chapter first discusses a one-parameter valve-stiction model and its limitations, followed by a description of the two-parameter data-driven model of valve stiction that first appeared in Choudhury et al. (2005a, 2005b). It also validates the two-parameter stiction model with the physical model described in Chap. 12 by comparing simulation results with experimental results from process instrumentation.

13.1 One-Parameter Data-Driven Stiction Model

Stenman et al. (2003) with reference to a private communication with Hägglund reported a one-parameter data-driven stiction model. The model is described as follows:

$$y(k) = \begin{cases} u(k-1) & \text{if } \mid u(k) - y(k-1) \mid < d \\ u(k) & \text{otherwise} \end{cases} \tag{13.1}$$

where u and y are the valve input and output, respectively, and d is the valve stiction band. This model has also been used by Srinivasan et al. (2005a).

Figure 13.1 shows the simulation results when the one-parameter model was excited with a sinusoidal input. Clearly, the input–output phase plot (the right plot) does not match with the plot that is observed in the case of a sticky valve (see Fig. 13.2). This model has some serious drawbacks:

1. The model compares the difference of the current input and the previous output with the stickband (see Eq. (13.1)). However, it is the deadband that should be compared with the cumulative increment of the input signal from the moment when the valve became stuck.

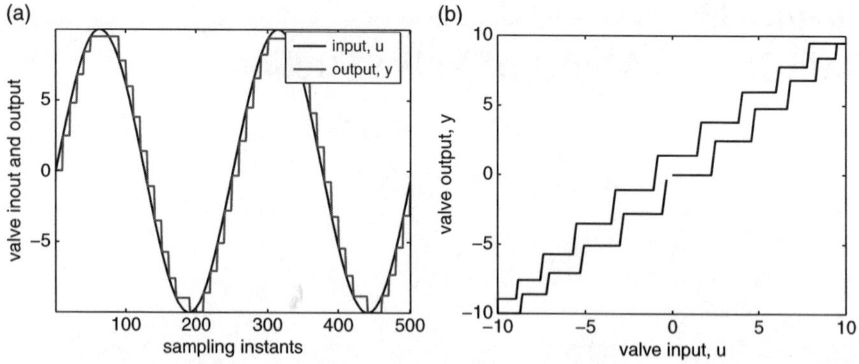

(a) Time trends (b) Valve input–output plot

Fig. 13.1 Simulation results for a one-parameter valve-stiction model under sinusoidal excitation (d = 2 was used in the simulation)

2. A real valve can stick anywhere whenever the input reverses direction. The problem with using the model in Eq. (13.1) as a dynamic model is that the valve sticks while it is moving whereas it should only stick when the velocity goes to zero.
3. The phase plot (valve output vs. input), as shown on the right-hand plot in Fig. 13.1, does not match the typical phase plot of a sticky valve. An example of a typical phase plot of a sticky valve is presented in Fig. 11.3. The phase plot shown in Fig. 13.1 looks similar to the one when a slow, increment stroke test is performed on the valve.

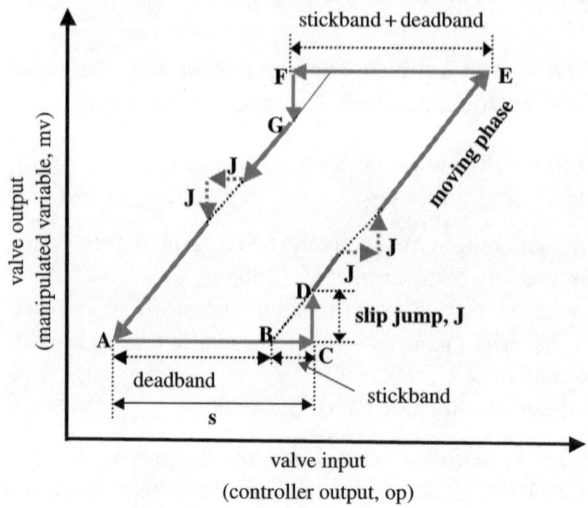

Fig. 13.2 Typical input–output behaviour of a sticky valve

Therefore, the one-parameter model is not adequate as a dynamic data-driven model of valve stiction. For a dynamic stiction model, the challenges are (i) to model the tendency of the valve to stay moving once it has started until the input changes direction or the velocity goes to zero, and (ii) to include the effects of deadband and the slip–jump. The following section describes a two-parameter stiction model that can meet these challenges.

13.2 Two-Parameter Data-Driven Model of Valve Stiction

Parameters of the two-parameter data-driven valve stiction model can be related directly to plant data, and furthermore such a model produces the same open- and closed-loop behaviour as the physical model. The model only requires the controller output signal and the specification of deadband plus stickband (*S*) and slip–jump (*J*). It overcomes the main disadvantages of physical modelling of a control valve, namely it does not require the knowledge of the mass of the moving parts of the actuator, spring constant and the friction forces. The effect of the change of the parameters of a physical model cannot easily be determined analytically because the relationship between the values of the parameters and the observation of the deadband/stickband as a percentage of valve travel is not straightforward. On the other hand, in a data-driven model, the parameters are easy to choose and the effects of the parameter changes are simple to realize.

13.2.1 Model Formulation

The valve sticks only when it is at rest or changing direction. When the valve changes its direction it comes to rest momentarily. Also, when the velocity of the valve is very small, the valve may stick. Once the valve overcomes stiction, it starts moving and may keep moving for sometime, depending on the valve input and the amount of stiction present in the valve. In the moving phase (see Fig. 13.2), the valve suffers only dynamic friction, which is usually smaller than static friction. It continues to move until its velocity is again very close to zero or it changes its direction.

In the process industry, stiction is generally measured as a percentage of the valve travel or the span of the control signal (Gerry and Ruel, 2001). For example, 2% stiction means that when the valve gets stuck, it will start moving only after the cumulative change of its control signal is greater than or equal to 2%. If the range of the control signal is from 4 to 20 mA, then a 2% stiction means that a change of the control signal less than 0.32 mA in magnitude will not be able to move the valve.

In the modelling approach described herein, it is assumed that the controller has a filter to remove the high-frequency noise from the signal (see Sect. 13.2.2 for more discussion on this topic). Then, the control signal is translated to the percentage of valve travel with the help of a linear look-up table. The model consists of two parameters – namely the size of deadband plus stickband *S* (specified in the input axis) and slip–jump *J* (specified on the output axis). The term '*S*' contains both

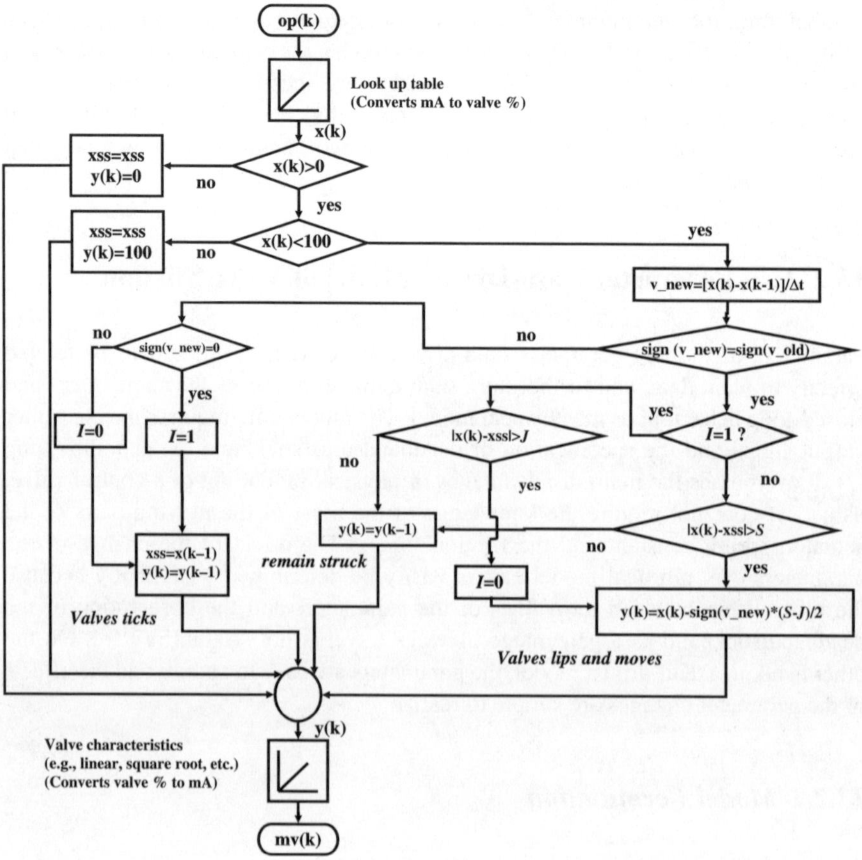

Fig. 13.3 Signal and logic flow chart of the two-parameter data-driven stiction model

the deadband and stickband. Figure 13.3 summarizes the model algorithm, which consists of the following steps:

- First, the controller output (mA) is converted to valve travel percentage using a look-up table.
- If the transformed controller output signal (%) is less then 0 or more than 100, the valve is considered to be saturated (i.e., fully close or fully open).
- If the signal is within the 0–100% range, the algorithm calculates the slope of the controller output signal.
- Next, the change of the direction of the slope of the input signal is taken into consideration. If the *sign* of the slope changes or remains zero for two consecutive instants, the valve is assumed to be stuck and does not move.

  The *sign* function of the slope gives the following outputs:

  – If the slope of input signal is positive, the *sign*(*slope*) returns +1.
  – If the slope of input signal is negative, the *sign*(*slope*) returns −1.
  – If the slope of input signal is zero, the *sign*(*slope*) returns 0.

Therefore, when *sign*(*slope*) changes from +1 to −1 or vice versa, this means that the direction of the input signal has changed and the valve is at the beginning of its stick position (points A and E in Fig. 13.2). The algorithm detects the stick position of the valve at this point. Now, the valve may stick again while travelling in the same direction (opening or closing direction) only if the input signal to the valve does not change or remains constant for two consecutive instants, which is uncommon in practice. For this situation, the *sign*(*slope*) changes to 0 from +1 or −1 and vice versa. The algorithm again detects the stick position of the valve in the moving phase and this stuck condition is denoted with the indicator variable $I = 1$. The value of the input signal when the valve gets stuck is denoted as *xss*. This value of *xss* is kept constant and does not change until the valve gets stuck again. The cumulative change of input signal to the model is calculated from the deviation of the input signal from *xss*. It is noteworthy that this can also be performed by using a velocity threshold below which the valve will come to a rest or stop.

- For the case where the input signal changes direction (i.e. the *sign*(*slope*) changes from +1 to −1 or vice versa), if the cumulative change of the input signal is more than the amount of the deadband plus stickband (S), then the valve slips and starts moving.
- For the case when the input signal does not change direction (i.e. the *sign*(*slope*) changes from +1 or −1 to 0, or vice versa), if the cumulative change of the input signal is more than the amount of the stickband (J), then the valve slips and starts moving. This takes care of the case when valve sticks again while travelling in the same direction (Entech, 1998).
- The output is calculated using the equation:

$$output = input - sign(slope) \times (S - J)/2 \tag{13.2}$$

  and depends on the type of stiction present in the valve. It can be described as follows:

  – Deadband: If $J = 0$, then the valve has a pure deadband without any slip-jump.
  – Stiction (undershoot): If $J < S$, then the valve output can never reach the valve input. There is always some offset or deviation between the valve input and the valve output. This represents the undershoot case of stiction.
  – Stiction (no offset): If $J = S$, the algorithm produces pure stick–slip behaviour. There is no offset between the valve input and the valve output. Once the valve overcomes stiction, the valve output tracks or reaches the valve input exactly. This is the well-known 'stick-slip case'. For this case, the value of deadband is zero.
  – Stiction (overshoot): If $J > S$, the valve output overshoots the valve input due to excessive stiction. This is termed as overshoot case of stiction. This may happen due to miscalibration of the valve as discussed in the previous chapter.

- The parameter J is an output quantity measured on the vertical axis. It signifies the slip-jump start of the control valve immediately after it overcomes the

deadband plus stickband. It accounts for the offset or deviation between the valve input and output signals.
- Finally, the output is converted back to a mA signal using a look-up table based on the characteristics of the valve such as linear, equal percentage or square root, and the new valve position is reported.

13.2.2 Dealing with Stochastic or Noisy Control Signals

The two-parameter stiction model uses the *sign* function to detect the change of valve direction. In the case of a noisy signal, this may cause some numerical problems in the simulation. In real life, a noisy control signal causes unnecessary valve movements resulting in rapid wear and tear. In order to handle noisy or stochastic control signal, either of the following two approaches can be used:

1. A time domain filter, e.g. an exponentially weighted moving average filter (EWMA), can be used before or after the controller to filter the noise. The work presented in this chapter uses the following filter:

$$G_f(z) = \frac{\lambda z}{z - (1 - \lambda)} \tag{13.3}$$

   In this work, $\lambda = 0.1$ was used for all simulations. The magnitude of λ will depend on the extent of noise used in the simulation.
2. A transfer function similar to Eq. (12.3) in Chap. 12 can be used to filter out the high-frequency noise.

13.2.3 Open-Loop Response of the Model Under a Sinusoidal Input

Figure 13.4 shows the open-loop behaviour of the two-parameter data-driven stiction model in the presence of deadband and varying amounts of slip–jump. Plots in the left panel show the time trend of the valve input *op* (thin solid line) and the output *mv* (thick solid line). The right panel shows the input–output behaviour of the valve on a X–Y plot.

- The first row shows the case of a linear valve without stiction.
- The second row corresponds to pure deadband without any slip–jump, i.e. $J = 0$. For this case, the magnitude of stickband is zero and deadband itself equals 'S'.
- The third row shows the undershoot case of a sticky valve where $J < S$. This case is also illustrated in the first and second examples of industrial control loops in Figures 11.3 and 11.4. In this case, the valve output can never reach the valve input. There is always some offset. Both deadband and stickband are active for this case.

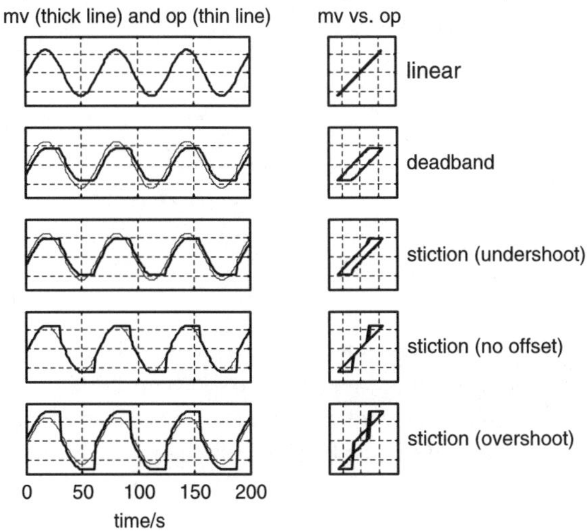

Fig. 13.4 Open-loop simulation results of the data-driven stiction model

- The fourth row represents pure stick–slip behaviour. There is no offset between the input and output. Once the valve overcomes stiction, the valve output tracks the valve input accurately. For this case, the deadband is zero.
- In the fifth row, the valve output overshoots the desired set position or the valve input due to excessive stiction or miscalibration. This is termed 'overshoot case of stiction'.

13.2.4 Stiction in Reality

In reality, a combination of these stiction phenomena may be observed. Although the two-parameter data-driven stiction model is not directly based on the dynamics of the valve, its strength is that it is simple to use for the purpose of simulation and can quantify stiction as a percentage of valve travel or span of input signal. Also, the parameters used in this model are simple to understand, realize and relate to real stiction behaviour.

13.2.5 Closed-Loop Behaviour of the Model

Closed-loop behaviour of the stiction model has been studied for two different cases, namely a concentration loop and a level loop. The concentration loop has a PI controller and a process model having slow dynamics and a large dead time. The level

Table 13.1 Process models, controllers and stiction model parameters

Loop	Process	Controller	Deadband		Stiction					
					Undershoot		No Offset		Overshoot	
			S	J	S	J	S	J	S	J
Concn.	$\frac{3e^{-10s}}{10s+1}$	$0.2(\frac{10s+1}{10s})$	5	0	5	2	5	5	5	7
Level	$\frac{1}{s}$	$0.4(\frac{2s+1}{2s})$	3	0	3	1.5	3	3	3	4.5

loop has a PI controller and the process is represented by an integrator. The transfer functions, controllers and parameters used in the simulation are shown in Table 13.1. The magnitudes of S and J are specified as a percentage (%) of valve input span and output span, respectively. The results for each of the loops are discussed in separate sections below.

13.2.5.1 Concentration Loop

The transfer function model for the concentration loop was obtained from Horch and Isaksson, (1998). This transfer function with a PI controller in a feedback closed-loop configuration is used here for simulation. Steady-state results of the simulation for different stiction cases are presented in Fig. 13.5a and 13.5b. In both figures, the thin lines are the controller output. If an integrator is not present in the process, the triangular shape of the time trend of controller output is one of the characteristics of stiction (Horch, 2000). If it is an integrating process with a PI controller, the valve signal is integrated twice and appears as a series of parabolic segments. Figure 13.7 shows such an example.

Figure 13.5a shows the controller output (*op*) and valve position (*mv*). Mapping of *mv* vs. *op* clearly shows the stiction phenomena in the valve. It is a common practice to use a mapping of *pv* vs. *op* for valve diagnosis (see Fig. 13.5b). However, in this case, such a mapping only shows elliptical trajectories with sharp turn-around points. The reason is that the *pv*–*op* map captures not only the nonlinear valve characteristic but also the dynamics of the process, $G(s)$, which in this case is a first-order lag plus dead time. Therefore, if the valve position data are available, one should plot the valve position (*mv*) against the controller output (*op*). Apart from the case of liquid flow loops where the flow through the valve (*pv*) can be taken to be proportional to valve opening (*mv*), the *pv*–*op* maps should be used with caution.

13.2.5.2 A Level Control Loop

The closed-loop simulation of the stiction model using only an integrator as the process is performed to investigate the behaviour of a typical level loop in the presence of valve stiction. Results are shown in Figs. 13.6 and 13.7. The second row of the figure shows that the deadband can produce oscillations. Again, it is observed that if there is an integrator in the process dynamics, then even a pure deadband can

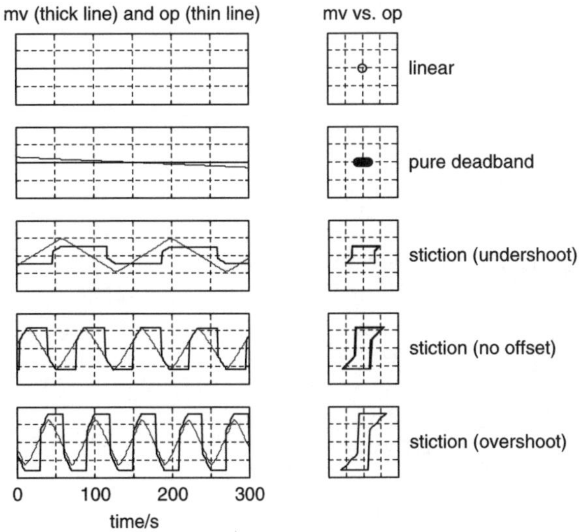

(a) Left: Time trends of *mv* and *op*; Right: *mv* – *op* mappings

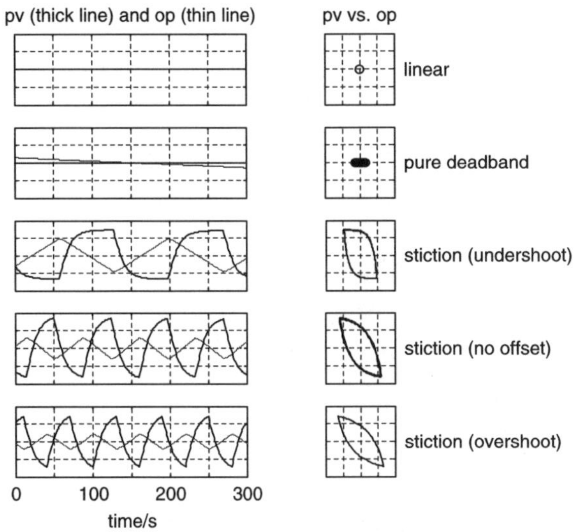

(b) Left: Time trends of *pv* and *op*; Right: *pv* – *op* mappings

Fig. 13.5 Closed-loop simulation results of a concentration loop using the data-driven stiction model

produce limit cycles. The *mv–op* mappings depict various cases of valve stiction. The *pv–op* plots show elliptical loops with sharp turn-around. Therefore, as was noted in an earlier example, the *pv–op* map is not a very reliable diagnostic tool for valve faults in a level loop. A diagnostic technique discussed in Chap. 15 as developed by Choudhury et al. (2004b) based on higher-order statistical analysis of data is able to detect and diagnose the presence of stiction in control loops.

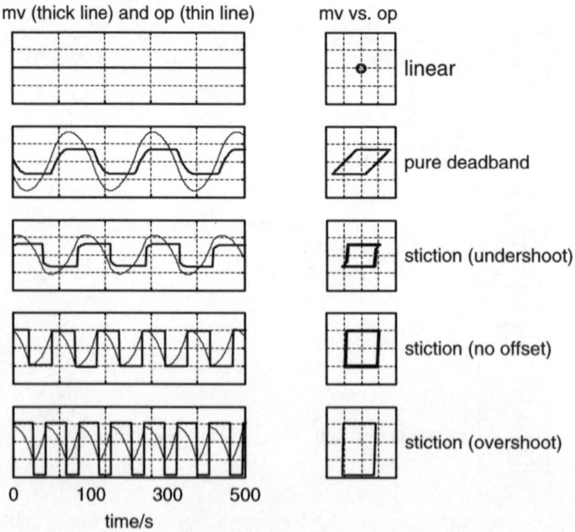

Fig. 13.6 Closed-loop simulation results of a level loop using the data-driven stiction model (*mv* and *op*)

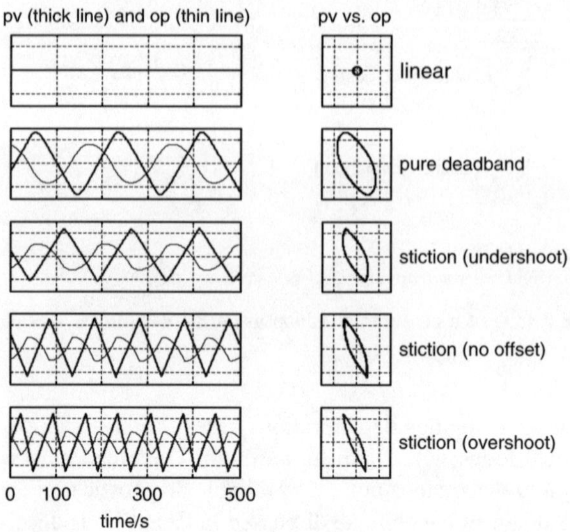

Fig. 13.7 Closed-loop simulation results of a level loop using the data-driven stiction model (*pv* and *op*)

13.3 Comparison of Physics-Based Model and Data-Driven Model

Although the two-parameter data-driven model is an empirical model and not based on physics, it is observed that this model can correctly reproduce the behaviour of the physics-based stiction model described in Chap. 12. The data for Figs. 13.5a and 12.3 are obtained from the simulation of the same process and same controller, but with different stiction models. By comparing Fig. 13.5 with Fig. 12.3, the following features can be observed:

- For a first-order plus time-delay model, both stiction models show no limit cycle for the case of pure deadband. This confirms the findings of Pipponen (1996), and McMillan (1995), who stated that the presence of pure deadband only adds dead time to the process and the presence of deadband together with an integrator produces a limit cycle (discussed further in Sect. 14.4).
- Both models show that a certain amount of slip-jump is required in order to initiate limit cycles if the process does not contain an integrator and the controller is a PI controller.
- If the process contains an integrator, both models show that limit cycles are generated even in the case of pure deadband.
- Both models produce identical results for other cases of stiction.

The open-loop simulation results for both models look qualitatively very similar in Fig. 13.4 and 12.2. A one-to-one comparison of these figures cannot be made because there is no direct one-to-one relation among the parameters of the empirical data-driven model and that of the physics-based model.

13.4 Summary

This chapter first discussed the one-parameter stiction model and its limitations, followed by the detailed formulation of a two-parameter data-driven stiction model. The two-parameter stiction model has been validated using the simulation results of the physics-based valve model. This chapter showed that the two-parameter stiction model can adequately simulate stiction phenomena. This model can be easily used to perform simulations in order to develop methodologies for stiction compensation, oscillation diagnosis and loop-interaction diagnosis.

Chapter 14
Describing Function Analysis

Under some circumstances, a control valve suffering from stiction induces a limit-cycle oscillation in the control loop. This chapter gives an analysis to gain insights into the conditions under which such limit cycles might arise and also provides some understanding of how the magnitude and frequency of the limit cycle are influenced by tuning of the controller. Describing function analysis uses a quasilinear approach in which the nonlinearity is approximated by a signal-dependent linear gain. The key assumption, which is met in the systems considered here, is that the behaviour of the control loop can be understood by considering just the fundamental sinusoidal component of the limit cycle. Limit-cycle oscillations caused by valve stiction can be studied using describing function analysis, and valuable insights about the signal-generating system can be obtained. The analysis of this chapter focusses on the describing function of the two-parameter stiction model developed in Chap. 13.

14.1 Introduction

A nonlinear actuator with a stiction characteristic may cause limit cycling in a control loop. Further insights into the behaviour of such systems may be achieved through a describing function analysis (Cook, 1986). The nonlinearity is modelled by a nonlinear gain N. The assumptions inherent in the approximation are that (1) there are periodic signals present in the system and (2) the controlled system is low pass and responds principally to the fundamental Fourier component. The conditions for oscillation in a negative feedback loop arise when the loop gain is -1:

$$G_o(i\omega) = -\frac{1}{N(X_m)} \tag{14.1}$$

where $G_o(i\omega)$ is the open-loop frequency response that includes the controlled system and the proportional plus integral controller, and $N(X_m)$ is the describing function that depends on the magnitude of the controller output X_m. When the condition $G_o(i\omega) = -1/N(X_m)$ is met, the system will oscillate spontaneously with a limit cycle. The variation of the quantity $-1/N(X_m)$ with signal amplitude means that signals initially present in the loop as noise can grow until they are large enough to satisfy the equality and hence provide a self-starting oscillation. The solution to the complex equation $G_o(i\omega) = -1/N(X_m)$, if one exists, may be found graphically by superposing plots of $G_o(i\omega)$ and $-1/N$ on the same set of axes.

14.2 Describing Function Analysis for Two-Parameter Stiction Model

The describing function of a nonlinearity is:

$$N = \frac{Y_f}{X} \tag{14.2}$$

where X is a harmonic input to the nonlinearity of angular frequency ω_o and Y_f is the fundamental Fourier component at angular frequency ω_o of the output from the nonlinearity. Thus, a Fourier analysis is needed on the output signals shown as bold lines in Fig. 14.1a. The quantity N depends upon the magnitude of the input X_m. N is complex for the stiction nonlinearity because the output waveform has a phase lag compared to the input.

14.2.1 Derivation of the Describing Function

The input–output behaviour of stiction nonlinearity is shown in Fig. 14.1. The output from stiction nonlinearity (i.e. the solid line in Fig. 14.1a) is not analytic. The term d in this figure is used in place of (S–J), which is equal to the deadband. It is useful to consider a sine wave input (dotted line in Fig. 14.1a) with angular frequency of $1\,\text{rad}\cdot\text{s}^{-1}$ and period 2π. The output (the solid line in Fig. 14.1a) is then:

$$y(t) = \begin{cases} k\left(X_m \sin(t) - \frac{S-J}{2}\right) & 0 \leqslant t \leqslant \frac{\pi}{2} \\ k\left(X_m - \frac{S-J}{2}\right) & \frac{\pi}{2} \leqslant t \leqslant \pi - \phi \\ k\left(X_m \sin(t) + \frac{S-J}{2}\right) & \pi - \phi \leqslant t \leqslant \frac{3\pi}{2} \\ k\left(-X_m + \frac{S-J}{2}\right) & \frac{3\pi}{2} \leqslant t \leqslant 2\pi - \phi \\ k\left(X_m \sin(t) - \frac{S-J}{2}\right) & 2\pi - \phi \leqslant t \leqslant 2\pi \end{cases}$$

where X_m is the amplitude of the input sine wave, S the deadband plus stickband, J the slip–jump, $\phi = \sin^{-1}\left(\frac{X_m - S}{X_m}\right)$ and k the slope of the input–output characteristic in the moving phase. k is assumed to be 1 for a valve with linear characteristics.

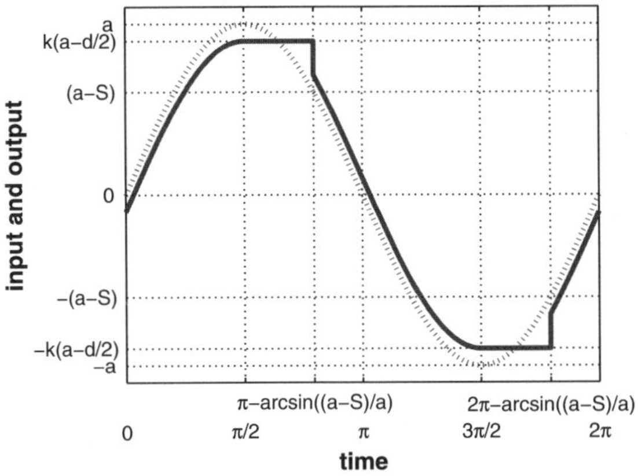

(a) Input (dotted line) and output (solid line) trends for stiction

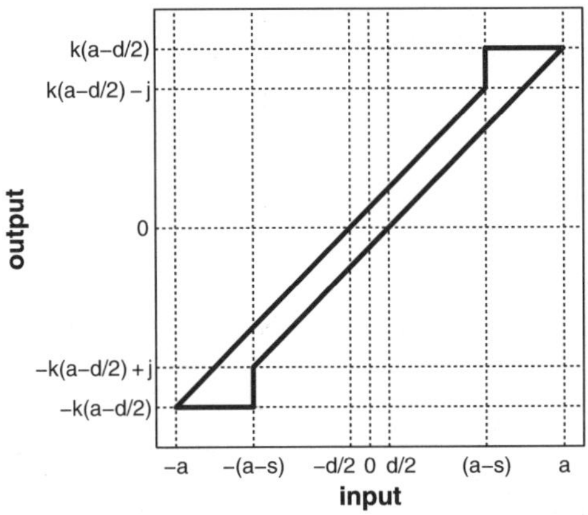

(b) Input–output phase plot for stiction

Fig. 14.1 Describing function analysis for stiction nonlinearity

14.2.1.1 Evaluation of the Fundamental Fourier Component

The fundamental component of the complex Fourier series is:

$$\frac{1}{2\pi}\int_{t=0}^{2\pi} y(t)\, e^{-it} dt$$

which after substitution of $\sin(t) = \frac{1}{2i}\left(e^{it} - e^{-it}\right)$ gives:

$$\int_{t=0}^{2\pi} y(t)\, e^{-it} dt = \int_{t=0}^{\pi/2} k\left(\frac{X_m}{2i}\left(e^{it} - e^{-it}\right) - \frac{S-J}{2}\right) e^{-it} dt + \int_{t=\pi/2}^{\pi-\phi} k\left(X_m - \frac{S-J}{2}\right) e^{-it} dt$$

$$+\int_{t=\pi-\phi}^{3\pi/2} k\left(\frac{X_m}{2i}\left(e^{it}-e^{-it}\right)+\frac{S-J}{2}\right)e^{-it}dt \;+\int_{t=3\pi/2}^{2\pi-\phi} k\left(-X_m+\frac{S-J}{2}\right)e^{-it}dt$$

$$+\int_{t=2\pi-\phi}^{2\pi} k\left(\frac{X_m}{2i}\left(e^{it}-e^{-it}\right)-\frac{S-J}{2}\right)e^{-it}dt \qquad (14.3)$$

Stating it compactly:

$$\int_{t=0}^{2\pi} y(t)\,e^{-it}dt = T_1+T_2+T_3+T_4+T_5$$

where $T_1 = \int_{t=0}^{\pi/2} k\left(\frac{X_m}{2i}\left(e^{it}-e^{-it}\right)-\frac{S-J}{2}\right)e^{-it}dt$, and so on. Evaluation term-by-term gives:

$$T_1 = \frac{k}{2}\left(X_m-S+J\right)+ik\left(\frac{S-J}{2}-\frac{X_m\pi}{4}\right)$$

$$T_2 = -k\left(X_m-\frac{S-J}{2}\right)(1-\sin\phi)-ik\left(X_m-\frac{S-J}{2}\right)\cos\phi$$

$$T_3 = k\left(\frac{X_m}{4}(1+\cos 2\phi)-\frac{S-J}{2}(1+\sin\phi)\right)$$
$$+ik\left(\frac{X_m}{4}\sin 2\phi-\frac{X_m}{2}\left(\frac{\pi}{2}+\phi\right)+\frac{S-J}{2}\cos\phi\right)$$

$$T_4 = -k\left(X_m-\frac{S-J}{2}\right)(1-\sin\phi)-ik\left(X_m-\frac{S-J}{2}\right)\cos\phi$$

$$T_5 = -k\left(\frac{X_m}{4}(1-\cos 2\phi)+\frac{S-J}{2}\sin\phi\right)$$
$$-ik\left(\frac{X_m\phi}{2}+\frac{S-J}{2}-\frac{X_m}{4}\sin 2\phi-\frac{S-J}{2}\cos\phi\right)$$

Collecting all the terms gives the required fundamental Fourier component of the output:

$$\frac{1}{2\pi}\int_{t=0}^{2\pi} y(t)\,e^{-it}dt = \frac{1}{2\pi}(B+iA)$$

where $A = \left(k\frac{X_m}{2}\sin 2\phi-2kX_m\cos\phi-kX_m\left(\frac{\pi}{2}+\phi\right)+2k(S-J)\cos\phi\right)$
and $B = -3k\frac{X_m}{2}+k\frac{X_m}{2}\cos 2\phi+2kX_m\sin\phi-2k(S-J)\sin\phi$

The fundamental component of the complex Fourier series of the input sine wave is $X_m/2i$. Therefore, the describing function is:

$$N = \frac{B+iA}{2\pi}\times\frac{2i}{X_m} = -\frac{1}{\pi X_m}(A-iB) \qquad (14.4)$$

14.3 Asymptotes of the Describing Function

There is no output from the nonlinearity when $X_m < S/2$. The limiting cases considered are therefore $X_m = S/2$ and $X_m \gg S$.

When $X_m \gg S$, $\phi = \sin^{-1}\left(\frac{X_m - S}{X_m}\right) = \frac{\pi}{2}$, $A = -k\pi X_m$, $B = 0$ and thus $N = k$. This result is to be expected because the influence of the stickband and jump are negligible when the input has a large amplitude and the output approximates a sine wave of magnitude kX_m. The slope of the moving phase for a valve with a deadband is $k = 1$ when the input and output to the nonlinearity are expressed as a percentage of full range. Therefore for a valve with stiction, $N = 1$, when $X_m \gg S$.

When $X_m = S/2$, the result depends upon the magnitude of the slip–jump, J. For the case with no deadband ($S = J$), $\phi = -\frac{\pi}{2}$, $A = 0$, $B = -4kX_m$ and $N = -ik\frac{4}{\pi} = k\frac{4}{\pi}e^{-i\pi/2}$. For a valve with $k = 1$, $N = \frac{4}{\pi}e^{-i\pi/2}$. This result describes the situation where the output is a square wave of amplitude X_m lagging the input sine wave by one quarter of a cycle, as shown in Fig. 14.2.

For intermediate cases, where both deadband and slip–jump are present such that $|S - J| > 0$, then the $X_m = S/2$ limit gives $\phi = -\frac{\pi}{2}$, $A = 0$, $B = -2kJ$ and $N = -ik\frac{2J}{\pi X_m} = k\frac{2J}{\pi X_m}e^{-i\pi/2}$. For instance, if $J = S/2$ and $k = 1$, then the $X_m = S/2$ limit gives $N = \frac{2}{\pi}e^{-i\pi/2}$ and the output is a square wave of amplitude $X_m/2$ lagging the input sine wave by one quarter of a cycle.

When the nonlinearity has a deadband only and no slip–jump ($J = 0$), the describing function has a limit given by $N = \varepsilon e^{-i\pi/2}$ where $\varepsilon \to 0$ as $X_m \to S/2$.

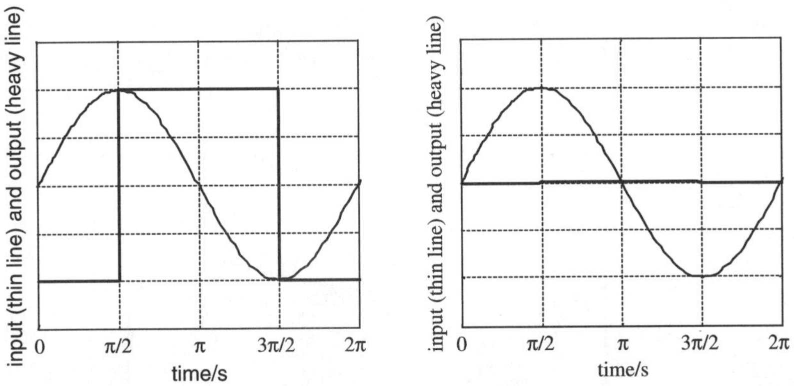

Fig. 14.2 Input (*thin line*) and output (*heavy line*) time trends for the limiting case as $X_m = S/2$. **Left panel**: slip–jump only with $S = J$. **Right panel**: deadband only with $J = 0$. The output in the right plot has been magnified for visualization; its amplitude becomes zero as X_m approaches $S/2$

14.4 Insights Gained from the Describing Function

Figure 14.3 shows graphical solutions to the limit-cycle equation $G_o(i\omega) = -1/N(X_m)$ for the composition control loop (left panel) and level control loop (right panel) presented earlier. The describing function is parameterized by X_m and the open-loop frequency-response function of the controller and controlled system is parameterized by ω. Both systems are closed loop stable and thus intersect the negative real axis between 0 and -1. The plots explain the behaviour observed in simulation.

It is clear from the left plot of Fig. 14.3 that there will be a limit cycle for the composition control loop if a slip–jump is present. The slip-jump forces the $-1/N$ curve onto the negative imaginary axis in the $X_m = S/2$ limit. Thus, the frequency-response curve of the FOPTD composition loop and its proportional plus integral controller is guaranteed to intersect with the describing function because the integral action means that open loop phase is always below $-\pi/2$ (i.e. it is in the third quadrant of the complex plane at low frequency).

Figure 14.3 also shows the $-1/N$ curve for the deadband limit cycle. In the $X_m = S/2$ limit, the curve becomes large, negative and imaginary. The composition loop does not have a limit cycle if the nonlinearity is a pure deadband, because the frequency-response curve does not intersect the $-1/N$ curve. The lack of a limit cycle in this case has been noted by other authors (Pipponen, 1996; McMillan, 1995).

The level loop with proportional plus integral control has a frequency response for which the phase becomes $-\pi$ at low frequency. The right-hand panel of Fig. 14.3 shows that it will intersect the $-1/N$ curve for the slip–jump cases and also for the pure deadband case. Therefore, a valve with a deadband and no slip–jump can cause a limit-cycle oscillation for an integrating process with a P + I controller.

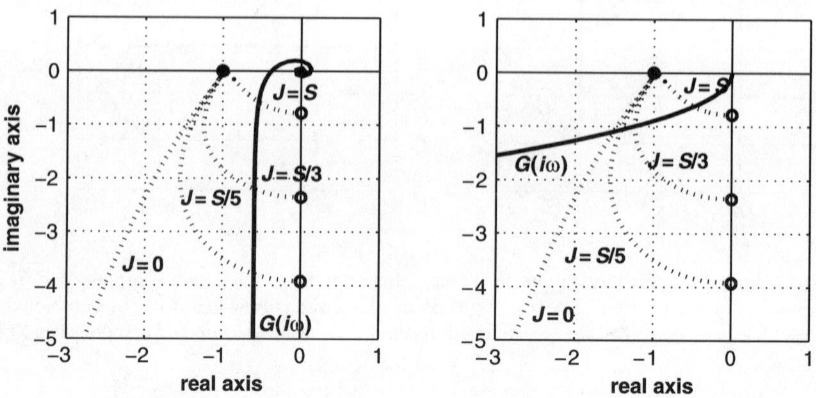

Fig. 14.3 Graphical solutions for limit-cycle oscillations. **Left panel**: composition control loop. **Right panel**: level control loop. Dotted lines are the $-1/N$ curves and the solid line is the frequency response function

The frequency of oscillation is higher and the period of oscillation is shorter when the slip–jump is present because the $-1/N$ curves with the slip–jump intersect the frequency-response curve at higher frequencies than the $-1/N$ curve for the deadband.

14.4.1 The Impact of the Controller on the Limit Cycle

It has been known for many years that control loops with sticking valves do not always have a limit cycle (McMillan, 1995; Pipponen, 1996; Olsson and Åström, 2001). The insights from the above section have explained many of these observations and are summarized in Table 14.1, which lists the behaviour depending on the process, controller and the presence or not of deadband and stick-slip.

Gerry and Ruel (2001) have reviewed methods for combating stiction online including conditional integration in the PI algorithm and the use of a dead zone for the error signal in which no controller action is taken. A short-term solution is to change the controller to P-only. The oscillation should disappear in a non-integrating process and while it may not disappear in an integrating process its amplitude will probably decrease. A further observation is that changing the controller gain changes the amplitude and period of the limit-cycle oscillation. In fact, observing such a change is a good test for a faulty control valve (Thornhill et al., 2003a; Rossi and Scali, 2005; Choudhury et al., 2007). The aim is to reduce the magnitude of the limit cycle in the short term until maintenance can be carried out. In practice, since the expected change in amplitude and period is complicated to anticipate in advance, one can try a 50% reduction in gain first or a similar increase in gain if the trend seems to be going the wrong way. More sophisticated algorithmic control solutions for friction compensation have also been proposed (Pipponen, 1996; Kayihan and Doyle III, 2000; Hägglund, 2002; Srinivasan and Rengaswamy, 2005).

Table 14.1 Limit cycles in control loops

Process and controller	Deadband only	Stick–slip
Integrating, PI	Limit cycle	Limit cycle
Integrating, P-only	No limit cycle	Limit cycle
Non-integrating, PI	No limit cycle	Limit cycle
Non-integrating, P-only	No limit cycle	No limit cycle

14.5 Summary

The physics-based valve model of Chap. 12 and the two-parameter data-driven model of Chap. 13 both captured the limit-cycling behaviour of control loops when the control valve has stiction.

This chapter has provided analytical insights into these limit cycles using describing function analysis. The analysis was successful in showing when a limit cycle will arise and when it will not, and has provided an explanation for observations reported in the literature that only integrating processes oscillate if the valve has a pure deadband. For non-integrating processes, the valve must also have stiction before a limit cycle can arise.

The analysis in this chapter specified the deadband and slip–jump, as defined in Fig. 11.2. These are not known in a practical situation; however, so a challenge remains to detect the presence of stiction and to quantify the amount of stiction for a faulty valve using measurements from closed-loop operation.

Chapter 15
Automatic Detection and Quantification of Valve Stiction

Earlier chapters have emphasized that stiction is a common problem in control valves. Following from the analysis in Chap. 14 of valve stiction, this chapter addresses the practical problem of detecting and quantifying stiction. Stiction is a major problem in spring diaphragm type valves, which are widely used in the process industry. Invasive methods such as the valve travel test can easily detect stiction, but are expensive and tedious to apply to hundreds of valves present in a typical plant. Thus, there is a clear need in the process industry for a non-invasive method that cannot only detect but also quantify stiction so that the valves that need repair or maintenance can be identified, isolated and repaired. This chapter describes a method for detecting and quantifying stiction in control valves using routine operating data obtained from the process. No additional excitation of the plant or experimentation on the plant is required. More than a dozen industrial case studies have demonstrated the wide applicability and practicality of this method as a useful diagnostic aid in control-loop performance monitoring.

15.1 Introduction

As discussed in Chap. 10, control valves frequently suffer from problems such as stiction, leaks, tight packing and hysteresis. Based on a large-scale industrial study, Bialkowski (1992) reported that about 30% of control loops are oscillatory due to control valve problems. Desborough and Miller (2002) reported that control valve problems account for about one third of the 32% of controllers classified as 'poor' or 'fair' in an industrial survey (Desborough et al., 2000). If the control valve has deadband and stiction, then it may case limit-cycle oscillations in the control loop, which in turn can cause oscillations in the process output. Stiction hinders the proper movement of the valve stem and consequently affects control-loop performance. Stiction can be detected easily using invasive methods such as the valve travel or bump test. However, to apply such invasive methods across an entire plant site is neither feasible nor cost-effective because they are cost and time intensive in nature. Desborough and Miller (2002) have highlighted the desirability of a non-invasive method, saying: *'a passive or non-invasive method that can reliably and automatically classify valve performance in closed loop is desperately needed in process industry'* (Desborough and Miller, 2002).

This chapter describes a method from Choudhury et.al (2006c), which combines the powerful features of nonlinearity detection with a shape analysis of the *pv–op* map for the quantification of the amount of stiction present in a control valve. An effective non-intrusive data-based monitoring method such as the one considered here can reduce the cost of maintenance by screening and short-listing only those control loops and/or valves that need maintenance.

15.2 Stiction Detection – A Literature Review

Invasive methods for the detection of valve stiction put the control loop in manual and then drive the valve over its full travel span. This is now termed the *valve travel test* in the standards of the Instrument Society of America (ISA) (ISA-75.13-1996; ANSI/ISA-75.05.01-2000). Using this type of test, stiction can be quantified as the amount of change required in the control signal to move the valve from the position where it was stuck.

Control valve diagnosis is also straightforward in closed loop using routine operating data if the controller output signal, *op*, and either the flow through the valve or the valve position, *mv*, are measured and made available. A *op–mv* plot is a straight line at 45° for a healthy linear valve, and any deviations such as deadband can be easily diagnosed by visual inspection or by automated methods sensitive to features such as the sections where the valve position does not change.

The flow through a control valve often is *not* measured, however. Similarly, the position, while it may be measured on a modern valve with a positioner, is not always available in the data historian. The challenge in the analysis of valve problems is to discover and quantify the types of fault present using *op* and *pv* data only. The *pv* is the measured or controlled variable of the control loop, for instance, the level in the case of a level control loop. The difficulty is that the process dynamics (integration in the case of a level loop) greatly influence the shape of the waveforms.

A number of investigators have suggested methods to determine the presence of a deadband directly from the shape of the waveform. Rengaswamy et al. (2001) proposed an automated qualitative shape analysis with the ability to detect square, triangular and other waveforms and to give a rule-based interpretation of the causes. The basic ideas were extended for use with closed-loop *pv* and *op* measurements by Srinivasan et al. (2005a), again using pattern-matching templates for square, triangular and saw-toothed features in the temporal domain with dynamic time warping to accommodate cycles of irregular length. Yamashita (2006a) determined several waveforms and matched patterns in a two-dimensional *op–pv* plot against a bank of templates, successfully detecting stiction in the valves of level and flow control loops. Singhal and Salsbury (2005) have observed that the *pv* in a non-integrating process with a sticking valve tends to have a left–right asymmetry, and their method is based on the calculation of areas before and after the peak of an oscillating signal. A benefit of the waveform analysis tools is that they do not require the waveform to be oscillating and can therefore achieve a diagnosis even when no limit cycle is present.

Other methods have used signal analysis. Horch (1999) and Horch et al. (2000) made the observation that for non-integrating processes, control loop oscillations due to a sticking valve tend to have an odd cross-correlation function between *op* and *pv*. Stiction in a loop with an integrating process has also been successfully detected by examination of the probability density function of the derivatives of the *pv* signal (Horch, 2002; Yamashita, 2006b), while Stenman et al., (2003) looked for evidence of jumps in the control-loop signals. Srinivasan et al. (2005b) have devised a model identification method for detecting the presence of deadband in a control valve on the basis of closed-loop *pv* and *op* measurements using the identification of a Hammerstein model, which uses the one-parameter stiction model described in Chap. 13 and then identifies the deadband as a parameter in the nonlinear part of the model.

A study by Horch (2007) in the book edited by Uduehi et al. compared several of the above methods on a benchmark data set. It is encouraging that several of them are able to utilize *op* and *pv* data successfully. Rossi and Scali (2005) have also made performance comparisons of nonlinearity detection by bicoherence, cross-correlation and a relay test based on waveform shapes. The Horch (2007) study also included nonlinearity detection by bicoherence and surrogate data methods (Chaps. 6 and 7 in this book), both of which performed very well although they do not give the specific diagnosis of valve stiction.

15.3 Detection of Stiction Using Nonlinearity Information and the *pv–op* Mapping

In a control loop, a nonlinearity may be present either in the process itself or in the control valve. The analysis presented here assumes that the process nonlinearity is negligible in the steady-state operating region during which the data have been collected. This is a reasonable assumption because the method works with routine operating data of a control loop under regulatory control. As discussed in Chap. 8, when processes are fairly well regulated at standard operating conditions, the plant can be assumed to behave in a locally linear manner, since a linear controller is capable of satisfactory regulation of the plant. The method discussed in this chapter first examines the presence of nonlinearity in a control loop in an SISO configuration. If a nonlinearity is detected, then the process variable (*pv*), set-point (*sp*) and controller output (*op*) signals are used to diagnose whether the nonlinearity is due to valve stiction.

15.3.1 Detection of Loop Nonlinearity

A control loop containing valve nonlinearities often produces non-Gaussian (i.e., a signal with asymmetric distribution) and nonlinear time series, namely process output (*pv*) and controller output (*op*) data. Bicoherence-based nonlinearity assessment can be used as a diagnostic tool for troubleshooting of hardware faults that may be present in the control loop (Choudhury et al., 2002; Choudhury et al., 2004b). As described in Chap. 6, the tests of Gaussianity and nonlinearity of the control error signal (*sp–pv*) are useful diagnostic aids for determining the poor performance of a control loop. The test described in Chap. 6 uses the normalized bispectrum or bicoherence to detect the presence of nonlinear interactions in the signal. A distinctive characteristic of a nonlinear time series is the presence of phase coupling such that the phase of one frequency component is determined by the phases of others. Phase coupling leads to higher-order spectral features that can be detected in the bicoherence. Bicoherence is defined as:

$$bic^2(f_1,f_2) \triangleq \frac{|B(f_1,f_2)|^2}{E[|X(f_1)X(f_2)|^2]E[|X(f_1+f_2)|^2]} \tag{15.1}$$

where $B(f_1,f_2)$ is the bispectrum at frequencies (f_1,f_2) and is given by

$$B(f_1,f_2) \triangleq E[X(f_1)X(f_2)X^*(f_1+f_2)], \tag{15.2}$$

$X(f_1)$ is the discrete Fourier transform of the time series $x(k)$ at the frequency f_1, $X^*(f_1)$ is the complex conjugate and E is the expectation operator. A key feature of the bispectrum is that it has a non-zero value if there is a significant phase coupling in the signal x between frequency components at f_1 and f_2. The bicoherence gives the same information but is normalized as a value between 0 and 1.

For checking the possible presence of a nonlinear element in control loop, the control error signal (*sp–pv*) is tested by using the non-Gaussianity index (NGI) and the nonlinearity index (NLI) described in Chap. 6. If both NGI and NLI are greater than zero, then the signal is described as non-Gaussian and nonlinear. The details of the procedure are summarized in Fig. 15.1. The test can be applied to any time series to check its non-Gaussianity and nonlinearity. For a control loop, this test is applied on the error signal (*sp–pv*) to the controller because the error signal is more stationary than *pv* or *op* signal. If the error signal is found to be non-Gaussian and nonlinear, it is inferred that the loop in question exhibits significant nonlinearity.

The nonlinearity can be attributed to the control valve under the following assumptions:

- The process is locally linear.
- No nonlinear disturbance(s) enter the loop.
- The installed characteristic of the control valve is reasonably linear in the current operating region.

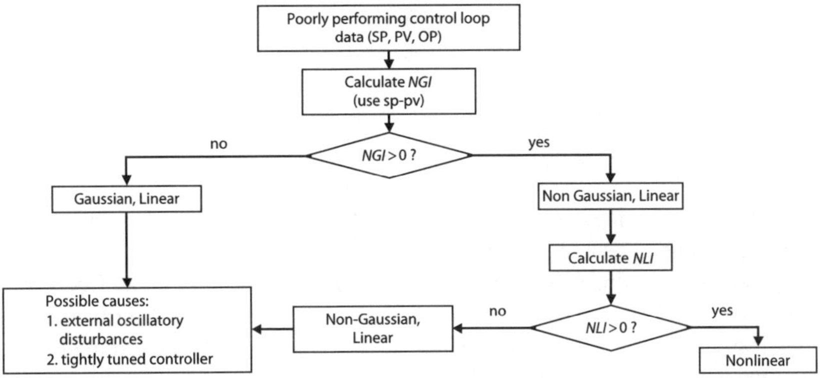

Fig. 15.1 Decision flow diagram of the method for the detection and diagnosis of loop nonlinearity

If the disturbance is measurable, then the test can be applied to check the linearity of the disturbance. One may also argue that the valve itself may have a nonlinear characteristic, e.g. a square-root or equal percentage characteristic, which is definitely not a fault. To clarify this situation, a simulation study was performed and the results have been presented in Chap. 8 for equal percentage and square-root valve characteristics in a simple feedback system. It was found that if the movement of the valve stem or the change in input signal to valve is within 30% of the full span (0–100%) of the valve travel, then a control loop exhibits linear behaviour under steady-state regulatory control. The same conclusion can also be realized by careful investigation of the valve characteristic curves.

Therefore, the higher-order statistics-based NGI and NLI indices can be calculated for each control loop in an entire plant site, and the loops that exhibit nonlinear behaviour can be isolated for further diagnosis. After a control loop is identified as nonlinear, the causes of nonlinearity should be diagnosed. With the assumptions listed above, it can be concluded that the valve is most likely responsible for the nonlinearity. The next problem is to diagnose whether this valve nonlinearity is due to stiction or some other problem. The *pv–op* plot is useful for solving this problem.

15.3.2 Use of *pv–op* Plot

The use of the *pv–op* plot for the detection of valve problems was discussed in Sect. 13.2.5, where its difficulties were highlighted. Except for liquid flow loops, the process dynamics generally mask features that clearly reveal the stiction signature. If the manipulated variable (*mv*) is not measured, however, then only the *pv–op* data are available, and therefore any method for stiction quantification must be able to accommodate the restriction to *pv* and *op* data only. In the method presented here, the *pv–op* plot will be used as a second step to diagnose the valve nonlinearity problem, not for the detection of the valve problems. The latter is carried out by using higher statistical-based NGI and NLI indices. If nonlinearity is detected, then the *pv–op* plot is used to diagnose the cause of this nonlinearity.

15.3.2.1 Data Filtering

Because of the contamination of real-world data with noise and disturbances, a *pv–op* plot is often unclear and ambiguous. It necessitates the use of a filter to clean the data. Since the nonlinearity detection is a frequency domain method, frequency domain-based filtering has been chosen here. Upon detection of nonlinearity, the frequencies responsible for significant nonlinear interactions can be determined from the significant peaks in the squared bicoherence plot. Then, a frequency domain Wiener filter is used to obtain those parts of the signal that contribute significantly to signal nonlinearity. Both *pv* and *op* signals are filtered. The Wiener filter sets the power in unwanted frequency channels to zero. The detailed filter design algorithm is given in Thornhill et al. (2003), which explains how to deal with aliased frequencies above the Nyquist frequency and constraints on the filter width, and Sect. 15.8.7 outlines an application of the filter. The frequency range for the filter is selected by inspecting the peaks in the bicoherence plot. It is preferable to use a large number of data points (e.g. 4,096 samples) for the nonlinearity-detection algorithm because a large data set ensures the consistency and unbiasedness of the bicoherence estimator (for details refer to Chap. 3). Filtering is also performed on the large data sets. However, the use of such a large number of data points in the pv_f–op_f mapping often produces a signature that is difficult to read and match with a known pattern of valve problems. Therefore, a segment of the data consisting of only several hundred data points should be chosen for the construction of the pv_f–op_f plot. In the above pv_f and op_f are the filtered *pv* and *op* signals.

15.3.2.2 Choosing an Appropriate Segment of the Data

The question that naturally arises is how to select a segment of the data for a useful pv_f–op_f plot. This problem can be resolved by choosing the segment that has regular oscillations because valve problems often manifest themselves as limit cycles. Thornhill et al. (2003) described a method for the assessment of the period and the regularity of oscillation of a time series. The zero-crossings of the autocovariance function of the time series are used to estimate the period of oscillation. An oscillation is considered to be regular if the standard deviation of the intervals between zero crossings (σ_{T_p}) is less than one third of the mean value of the intervals ($\bar{T}_p$). The statistic used is:

$$r = \frac{1}{3}\frac{\bar{T}_p}{\sigma_{T_p}} \qquad (15.3)$$

A value of r greater than 1 indicates a regular oscillation with a well-defined period. In this work, the filtered controller output signal (op_f) is divided into several segments of user-defined length that can be selected based on the period of oscillation. The segment of op_f corresponding to the highest value of r is used for the pv_f–op_f plot, where pv_f is the corresponding counterpart of op_f. The data segment corresponding to the highest value of r is chosen because valve nonlinearities, e.g. stiction and deadband, are measured as the maximum width of the cycles in the direction of valve input signal in a valve characteristic plot, i.e. the plot of the valve input vs. valve output signal.

15.4 Stiction Quantification

Some control valves may have an acceptable level of stiction, while others may have severe stiction that demands immediate maintenance. Therefore, it is important to be able to quantify stiction so that a list of sticky valves can be prepared in order of their maintenance priority.

Many of the modern control valves have positioners to measure actual stem positions (*mv*). However, the stem position data are not always available from a data historian or the DCS system. If the valve positioner data are available, a plot of valve output signal (*mv*) vs. valve input signal (*op*) can be used to quantify stiction. The challenge here is to estimate stiction from the available data – the controlled output (*pv*), the controller output (*op*) and the set point (*sp*).

It is well known that the presence of stiction in the control valve in a control loop produces limit cycles in the controlled variable (*pv*) and the controller output (*op*) (Hägglund, 1995; Horch, 1999; Ruel, 2000; Rengaswamy et al., 2001). For such a case, if *pv* is plotted against *op*, cyclic patterns are found in the resulting *pv*–*op* plot. A large number of such plots can be found in Choudhury et al. (2004a) and Choudhury et al. (2005a, 2005b), where stiction models were used in a closed-loop SISO system to produce data for these plots (see also Sect. 13.2.5, Chap. 13). The pv_f–op_f plot together with either of the following two methods can be used to quantify stiction in the unit of the *op* signal. The quantified stiction is termed as *apparent stiction* because the actual amount of stiction as obtained from the *mv*–*op* plot may differ from the estimated quantity owing to the effect of loop dynamics on the controlled variable, *pv*, in particular, the effect of the controller to compensate or fight stiction.

15.4.1 Clustering Techniques of Quantifying Stiction

Clustering methods are generally used for dividing scattered groups of data into several groups. Because the *pv*–*op* plot for a control loop with a sticky valve exhibits elliptic trajectories, the data corresponding to a narrow strip along the mean of *pv* and parallel to the *op* axis can be collected (see Fig. 15.2(c)) and used for quantifying stiction with the help of C-means or fuzzy C-means clustering techniques.

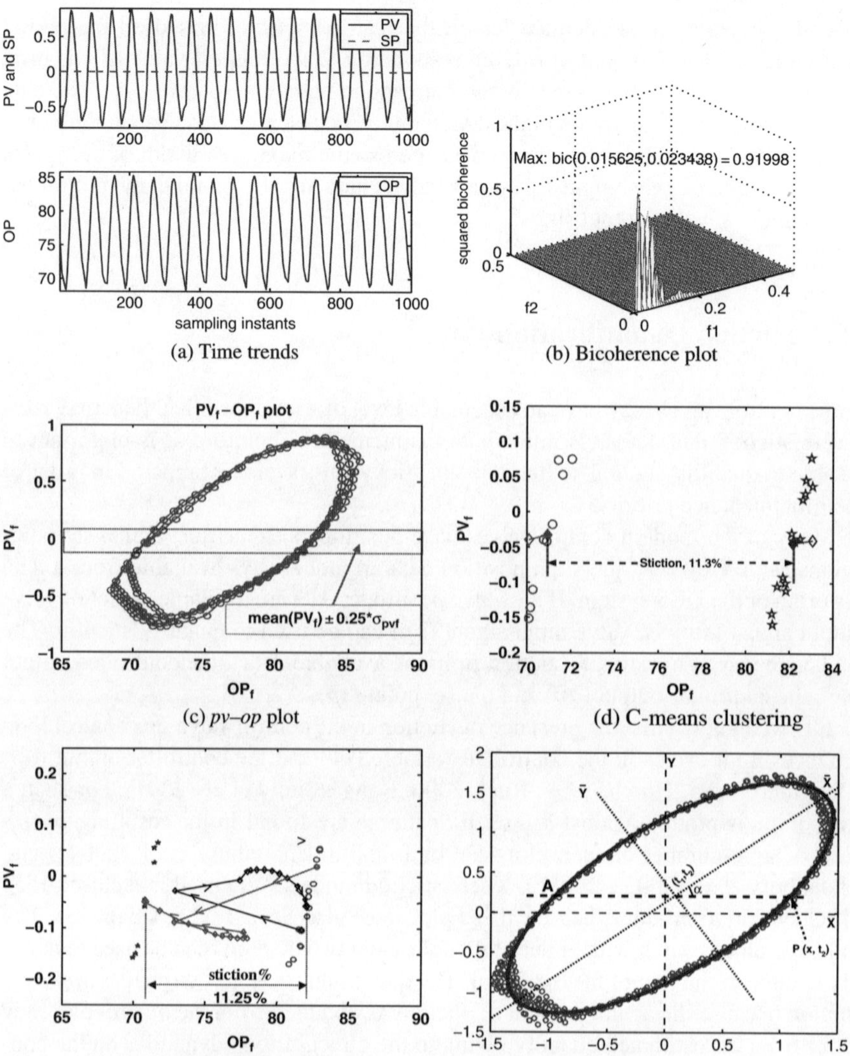

(a) Time trends (b) Bicoherence plot

(c) *pv–op* plot (d) C-means clustering

(e) Fuzzy C-means clustering (f) Fitted ellipse

Fig. 15.2 Analysis results for a level control loop. The biocherence plot is shown in subplot (**b**). Subplots (**c**), (**d**) and (**e**) demonstrate C-means and fuzzy C-means method of quantifying stiction while the ellipse fitting technique is shown in subplot (**f**)

15.4.1.1 C-means Clustering Technique

In this method, data are partitioned in to a pre-defined number of initial clusters. Then proceeding through all data points, each point is assigned to the nearest cluster (in terms of Euclidean distance). The centroids for the cluster receiving the new item and for the cluster losing the item are recalculated. This procedure is repeated until no more reassignments take place. Details of the method are given in Johnson and Wichern (1998). This method requires the initialization of the centres of the clusters. There are only two clusters, and the centres can be specified as [min(op_f), mean(pv_f)] and [max(op_f), mean(pv_f)] calculated from the data obtained along the strip shown in Fig. 15.2(c).

15.4.1.2 Fuzzy C-means Clustering Technique

The fuzzy C-means clustering method as described in Dulyakarn and Rangsanseri (2001) and Bezdek (1981) minimizes the following objective function.

$$J_m(U,V) = \sum_{j=1}^{n} \sum_{i=1}^{c} u_{ij}^m \| X_j - V_i \|^2, \quad 1 \le m < \infty \tag{15.4}$$

where U is a fuzzy partition of the data set, V is a set of K prototypes, m is any real number greater than or equal to 1, u_{ij} the degree of membership of X_j in the cluster i, X_j is the jth observation of the measured data, V_i is the centre of the cluster and $\| * \|$ is any norm expressing the similarity between a measured data point and the centre. Through an iterative optimization of Eq. (15.4), fuzzy partitioning is carried out with the update of membership of u_{ij} and the cluster centres V_i using the following equations.

$$u_{ij} = \frac{1}{\sum\limits_{k=i}^{c} \left(\frac{d_{ij}}{d_{ik}} \right)^{\frac{2}{m-1}}} \tag{15.5}$$

$$V_i = \frac{\sum\limits_{j=1}^{n} u_{ij}^m X_j}{\sum\limits_{j=1}^{n} u_{ij}^m}, \tag{15.6}$$

where d_{ij} is the Euclidean distance between the observation $X(j)$ and the centre of the cluster V_i. The criteria to end the iteration is when $\max(|u_{ij} - \hat{u}_{ij}|) < \varepsilon$, where ε is a number close to zero (e.g. 10^{-5}) and $\hat{u}_{ij}$ is the membership number at the previous iteration step. In this study, the number of clusters is two and 10^{-5} was used as the magnitude of ε.

The amount of stiction can be estimated from the absolute value of the difference between the x co-ordinates of the centres of the two clusters. If the co-ordinates of the final centres of the clusters are (op_1, pv_1) and (op_2, pv_2), then the amount of apparent stiction is determined using the following expression:

$$Apparent \;\; stiction = |op_1 - op_2|. \tag{15.7}$$

15.4.2 Fitted Ellipse Technique for Quantifying Stiction

An ellipse in the least-square sense can be fitted in the pv_f–op_f plot and then used to quantify stiction. The given data points are designated as **op** and **pv** where

$$\mathbf{op} = [op(1), op(2), \cdots, op(N)]^T \tag{15.8}$$
$$\mathbf{pv} = [pv(1), pv(2), \cdots, pv(N)]^T \tag{15.9}$$

Starting with the general equation of a conic, the equation of an ellipse is developed. The equation for a conic is given by the following (Gander et al., 1994):

$$a_1 x_1{}^2 + a_{12} x_1 x_2 + a_2 x_2{}^2 + b_1 x_1 + b_2 x_2 + c = 0 \tag{15.10}$$

or,

$$\Phi\Theta = 0 \tag{15.11}$$

where $\Phi = [x_1{}^2 \;\; x_1 x_2 \;\; x_2{}^2 \;\; x_1 \;\; x_2 \;\; 1]$, $\Theta = [a_1 \;\; a_{12} \;\; a_2 \;\; b_1 \;\; b_2 \;\; c]^T$, x_1 corresponds to data from the op signal and x_2 corresponds to data from the pv signal. Now, for a given data set, the above equation can be solved as constrained least-squares problem: minimize $||\Phi\Theta||$ w.r.t Θ subject to the constraint $||\Theta|| = 1$.

Often real-world data sets require a linearly shifted and rotated conic. Therefore, there is a need to fit a rotated and shifted conic in a transformed co-ordinates (see Fig. 15.2(f)). Equation (15.10) can be rewritten as:

$$\mathbf{x}^T \mathbf{A} \mathbf{x} + \mathbf{b}^T \mathbf{x} + c = 0 \tag{15.12}$$

with $\mathbf{x} = [x_1, x_2]^T$, $\mathbf{A} = [a_1 \;\; a_{12}/2; \;\; a_{12}/2 \;\; a_2]$ and $\mathbf{b} = [b_1 \;\; b_2]^T$ where $\mathbf{A}$ is symmetric and positive definite. The following expression gives the transformation of the equation in the new co-ordinate system:

$$\mathbf{x} = \mathbf{Q}\bar{\mathbf{x}} + \mathbf{t}, \tag{15.13}$$

where $\mathbf{Q}$ is the matrix for rotational transformation and $\mathbf{t}$ is the vector in the original co-ordinates for a linear shift of the conic. Using Eq. (15.13), the equation of the conic in the transformed co-ordinates can be written as:

$$\bar{\mathbf{x}}^T \mathbf{Q}^T \mathbf{A} \mathbf{Q} \bar{\mathbf{x}} + (2\mathbf{t}^T \mathbf{A} + \mathbf{b}^T)\mathbf{Q}\bar{\mathbf{x}} + \mathbf{t}^T \mathbf{A} \mathbf{t} + \mathbf{b}^T \mathbf{t} + c = 0 \tag{15.14}$$

This can be rewritten in the following simplified form

$$\bar{\mathbf{x}}^T \bar{\mathbf{A}} \bar{\mathbf{x}} + \bar{\mathbf{b}}^T \bar{\mathbf{x}} + \bar{c} = 0, \tag{15.15}$$

where

$$\bar{\mathbf{A}} = \mathbf{Q}^T \mathbf{A} \mathbf{Q} \tag{15.16}$$
$$\bar{\mathbf{b}}^T = (2\mathbf{t}^T \mathbf{A} + \mathbf{b}^T)\mathbf{Q} \tag{15.17}$$
$$\bar{c} = \mathbf{t}^T \mathbf{A} \mathbf{t} + \mathbf{b}^T \mathbf{t} + c. \tag{15.18}$$

Now, $\mathbf{Q}$ can be chosen in a way so that $\bar{\mathbf{A}} = diag(\lambda_1, \lambda_2)$. One approach is to choose $\mathbf{Q}$ as the eigenvector matrix obtained from the eigenvalue decomposition of the matrix $\mathbf{A}$. If the conic is an ellipse with its centre at the origin of the new co-ordinates, then in Eq. (15.15),

$$\bar{\mathbf{b}} = \mathbf{0} \tag{15.19}$$

Therefore, Eq. (15.15) can be simplified as

$$\lambda_1 \bar{x_1}^2 + \lambda_2 \bar{x_2}^2 + \bar{c} = 0 \tag{15.20}$$

or,

$$\frac{\bar{x_1}^2}{\sqrt{\frac{-\bar{c}}{\lambda_1}}} + \frac{\bar{x_2}^2}{\sqrt{\frac{-\bar{c}}{\lambda_2}}} = 1 \tag{15.21}$$

or,

$$\frac{\bar{x_1}^2}{m^2} + \frac{\bar{x_2}^2}{n^2} = 1 \tag{15.22}$$

where,

$$m = \sqrt{\frac{-\bar{c}}{\lambda_1}}, \qquad n = \sqrt{\frac{-\bar{c}}{\lambda_2}} \tag{15.23}$$

The lengths of the axes of the ellipse will be invariant to the transformation. Therefore, in the original co-ordinates the lengths of the axes of the ellipse are $2m$ and $2n$, respectively. The centre of the ellipse is at $\mathbf{t}$, which can be calculated from $\mathbf{t} = -0.5\mathbf{A}^{-1}\mathbf{b}$ (obtained using Eq. (15.19)). The angle of rotation of the ellipse (θ, measured anti-clockwise from the positive horizontal axis) can be calculated using any of the eigenvectors. Since the eigenvectors are of unit length, Eq. (15.13) can be written as

$$x_1 = \bar{x_1}\ cos\theta - \bar{x_2}\ sin\theta + t_1 \tag{15.24}$$

$$x_2 = \bar{x_1}\ sin\theta + \bar{x_2}\ cos\theta + t_2. \tag{15.25}$$

Because apparent stiction is defined as the maximum width of the ellipse in the *op* direction, the distance between two points lying at the intersections of the ellipse and a line parallel to the *op* axis and passing through the centre of the ellipse will be the amount of stiction present in the loop. For any point $P(x,t_2)$ (see Fig. 15.2(f)), Eq. (15.22) can be solved using Eq. (15.24) and (15.25). This gives the (x coordinate) of points A and P in Fig. 15.2(f):

$$x = t_1 \pm \frac{mn}{\sqrt{(m^2\ sin^2\theta + n^2\ cos^2\theta)}}, \tag{15.26}$$

where (t_1,t_2) is the centre of the fitted ellipse, m and n are the length of the major and minor axes of the fitted ellipse, respectively, and θ is the angle of rotation of the ellipse. Therefore, the amount of stiction (length of AP in Fig. 15.2(f)) can be obtained using the following expression

$$Apparent\ \ stiction\ = \ AP\ = \ \triangle x = \frac{2mn}{\sqrt{(m^2\ sin^2\theta + n^2\ cos^2\theta)}} \tag{15.27}$$

Remark. The information obtained by fitting an ellipse in the pv_f vs. op_f plot can be used to quantify hysteresis. Since the hysteresis is measured in the output axis or y-axis, the following equation can be used to quantify apparent hysteresis:

$$Apparent\ \ hysteresis\ = \ \ \triangle y = \frac{2mn}{\sqrt{(m^2\ cos^2\theta + n^2\ sin^2\theta)}} \tag{15.28}$$

15.5 An Illustrative Example

This example explains in detail all steps of the proposed method by applying it to an industrial data set. The data set represents a level control loop in a power plant, which controls the level in a condenser located at the outlet of a turbine by manipulating the flow rate of the liquid condensate from the condenser. In total, 8,640 samples for each tag were collected at a sampling interval of 5 s. Figure 15.2(a) shows the time trends for level (*pv*), set-point (*sp*) and the controller output (*op*). The loop shows oscillatory behaviour. For the bicoherence calculation, 4,096 data points were used. Figure 15.2(b) shows the squared bicoherence plot corresponding to the controller error signal (*sp*–*pv*). The values of NGI and NLI were found to be 0.46 and 0.85, respectively, indicating the presence of significant loop nonlinearity. From the bicoherence plot Fig. 15.2(b), it can be seen that frequencies in the range 0.001–0.1 are the most significant frequencies of the signal responsible for nonlinear interactions. Therefore, the *pv* and *op* signals were filtered using a Wiener filter with frequency boundaries at 0.001 and 0.1. Using the method of Thornhill et al. (2003) described earlier, it was found that the controller output signal was showing regular oscillations with an average period of 19.78 sampling intervals and the maximum r-value of 10.5 for a segment length of 200 data points. The maximum r-value was found for samples 2,801–3,000.

The pv_f–op_f plot is found to be useful for diagnosing the type of nonlinearity. Thus, the filtered pv_f and op_f corresponding to this segment is plotted in Fig. 15.2(c), which shows elliptical patterns, indicating valve stiction. Figure 15.2(d) demonstrates the C-means clustering technique used in the quantification of stiction. The points denoted by empty and filled diamonds are the initial and final centres of the clusters, respectively. This method quantifies the apparent stiction in this loop as 11.3%. Figure 15.2(e) illustrates the use of fuzzy C-means clustering in the quantification of stiction. The trajectories followed by the centres of the clusters during the iteration stages are shown by lines with diamonds directed with arrows. The final centres are again in solid diamonds. The apparent stiction estimated by this method is 11.25%. Figure 15.2(f) shows the algebraic ellipse-fitting technique, and the apparent stiction estimated using this method is 11.40%. All three methods produced approximately the same estimates for the apparent stiction.

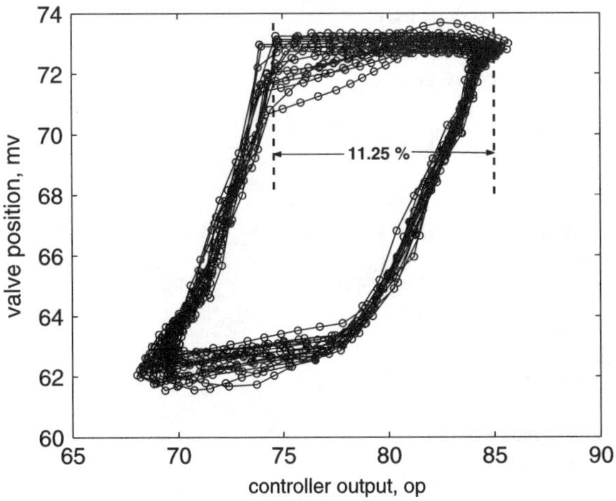

Fig. 15.3 Valve position (*mv*) versus controller output (*op*) plot. This plot confirms that the actual amount of stiction was correctly estimated by the apparent stiction-estimation methods depicted in Fig. 15.2

15.5.1 Validation of the Results

After the results of the analysis were sent to plant engineers, they confirmed that this loop was suffering from stiction. For this loop, the valve positioner data were made available. Figure 15.3 shows the actual valve position (*mv*) vs. controller output (*op*) plot. This plot clearly shows that the valve was sticking during the change of its direction. From this plot, the amount of stiction can be estimated as 11.25%, which is in agreement with the results obtained from the proposed methods.

15.6 Automation of the Method

In order to apply the proposed method to a large number of industrial control loops, it must be automated. The task list of steps required in automating this procedure is described below:

1. Calculate NGI and NLI for the control error signal (*sp–pv*). If both indices are greater than 0, then go to step 2. Otherwise, STOP. Nonlinearity is not a problem. The poor performance may be caused by tight tuning, detuned controller or external oscillatory disturbances (refer to Fig. 15.1).
2. If nonlinearity is detected, obtain frequency coordinates (f_1, f_2) corresponding to the maximum bicoherence peak in step 1. All frequencies are normalized such that the sampling frequency is 1. Let $f_1 = \min(f_1, f_2)$ and $f_2 = \max(f_1, f_2)$.
3. The boundaries of a Wiener filter can be obtained from $[f_L = \max(0.001, f_1 - 0.05), f_H = \min(0.5, f_2 + 0.05)]$. The small increment 0.05 is subtracted

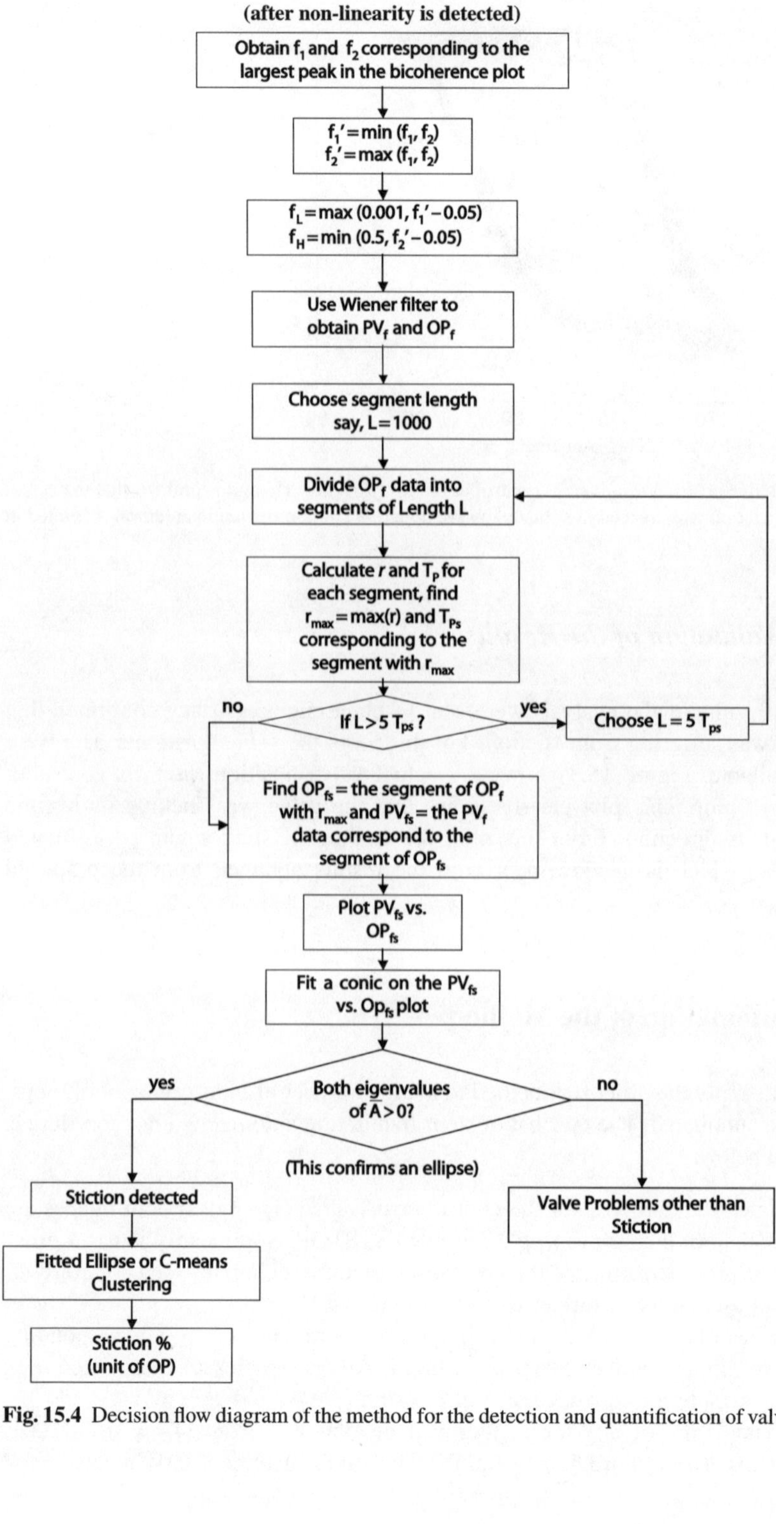

Fig. 15.4 Decision flow diagram of the method for the detection and quantification of valve stiction

or added from the frequencies obtained from the bicoherence plot in order to ensure that the exact location of the significant bicoherence peak does not fall on the filter boundaries.

4. Filter *pv* and *op* data with the Wiener filter designed in step 3 to obtain pv_f and op_f.
5. Obtaining the segment of the data with regular oscillations
   a) Choose a segment length *L*, say $L = 1{,}000$.
   b) Divide the op_f data into segments of length L. Here op_f is chosen instead of pv_f because often the *op* signal is less noisy than the *pv* signal because of the filtering takes place inside the controller.
   c) Calculate *r* and T_p for each segment of op_f data.
   d) Obtain $r_{max} = max\ (r)$.
   e) Obtain T_{ps}, which is equal to the T_p of the segment of *op* with r_{max}.
   f) If $L > 5\ T_{ps}$, then choose $L = 5\ T_{ps}$ and go to step (b).
   g) Now, op_{fs} is the segment of the op_f data that corresponds to the r_{max} and pv_{fs} is the portion of the pv_f data that corresponds to op_{fs}.
6. Use the pv_{fs} and op_{fs} data to get pv_{fs} vs. op_{fs} plot.
7. Fit a conic to the selected pv_{fs} and op_{fs} data. If both eigenvalues of the $\bar{A}$ (see Eq. (15.15)) matrix are greater than zero, then the pv_{fs} vs. op_{fs} plot is an ellipse. Otherwise, output the message, 'Not an ellipse – Other valve problems, not stiction'.
8. Quantify stiction using Formula 15.28 or 15.7, depending on the method chosen for stiction quantification.

The above-mentioned automatic detection and quantification of stiction algorithm has been summarized in Fig. 15.1 and 15.4. Figure 15.1 shows the automatic detection of loop nonlinearity. If nonlinearity is detected, Fig. 15.4 can be used to quantify stiction.

15.7 Simulation Results

15.7.1 A Worked Example

This section demonstrates the applicability of the proposed method for the detection and quantification of valve stiction through a known simulated case of stiction. A simple single-input, single-output (SISO) system in a feedback control configuration (Fig. 15.5) was used for generating simulated data. The first-order process with time delay is given by the following transfer function:

$$G(z^{-1}) = \frac{z^{-3}(1.45 - z^{-1})}{1 - 0.8z^{-1}}. \qquad (15.29)$$

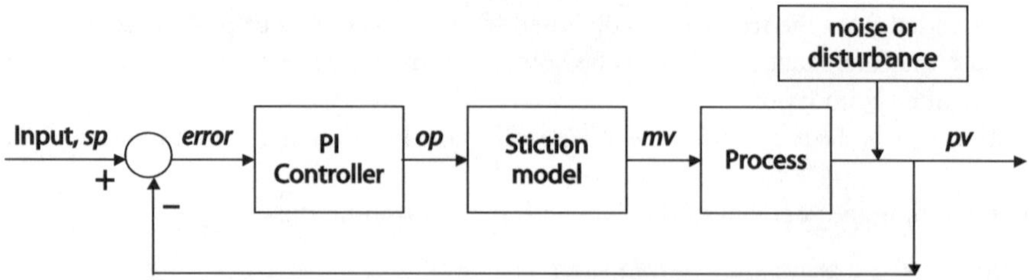

Fig. 15.5 Block diagram of a simple SISO process with stiction nonlinearity in the valve

The process is operating under regulatory control with a PI controller. A random walk disturbance generated by integrating random noise was added to the process output used for feedback. The signal-to-noise ratio defined by the ratio of the variance of the controlled output (*pv*) to the variance of the random noise was 6. The simulation was performed for 6,000 sampling intervals. To remove the effect of transients, the first several hundred data points were discarded and the last 4,096 points of the error signal to the controller (*sp*–*pv*) were analysed to detect the nonlinearity present in the system.

The data-driven stiction model of Chap. 13 was used to introduce the stiction behaviour of the control valve in the closed-loop process. A 3% stiction (*S*) with 3% slip-jump (*J*) were used in simulation. Figure 15.6(a) shows the time trends of *pv*, *op* and *sp*. The presence of stiction produces oscillations in the process. The values of NGI and NLI are 0.28 and 0.59, detecting the presence of nonlinearity in the process signal. The bicoherence plot (Fig. 15.6(b)) shows that the frequency range of interest from a nonlinearity point of view is [0.001 0.28] (using the steps of automation section). After performing Wiener filtering, the segments of the pv_f and op_f data corresponding to the segment of op_f that has the highest oscillation index were chosen to obtain the pv_f–op_f plot. C-means clustering, fuzzy C-means clustering and the fitted ellipse technique were used to quantify stiction. The apparent stiction obtained from these techniques were estimated to be 2.98%, 2.98% and 3.2%, respectively, which is in agreement with the actual stiction used in simulation. Thus, all these methods are capable of correctly quantifying the amount of stiction present in a control loop.

15.7.2 Distinguishing Limit Cycles Caused by Stiction and Those Caused by a Sinusoidal Disturbance

Often, an unmeasured oscillatory disturbance (for example, a sine wave) can initiate cycles in the controlled and manipulated variables. This example illustrates the effect of a sinusoidal disturbance with amplitude 2 and frequency 0.01 Hz on the stiction quantification algorithm. Some measurement noise was also introduced, as can be seen in Fig. 15.5. The stiction model was absent in this simulation study. Thus,

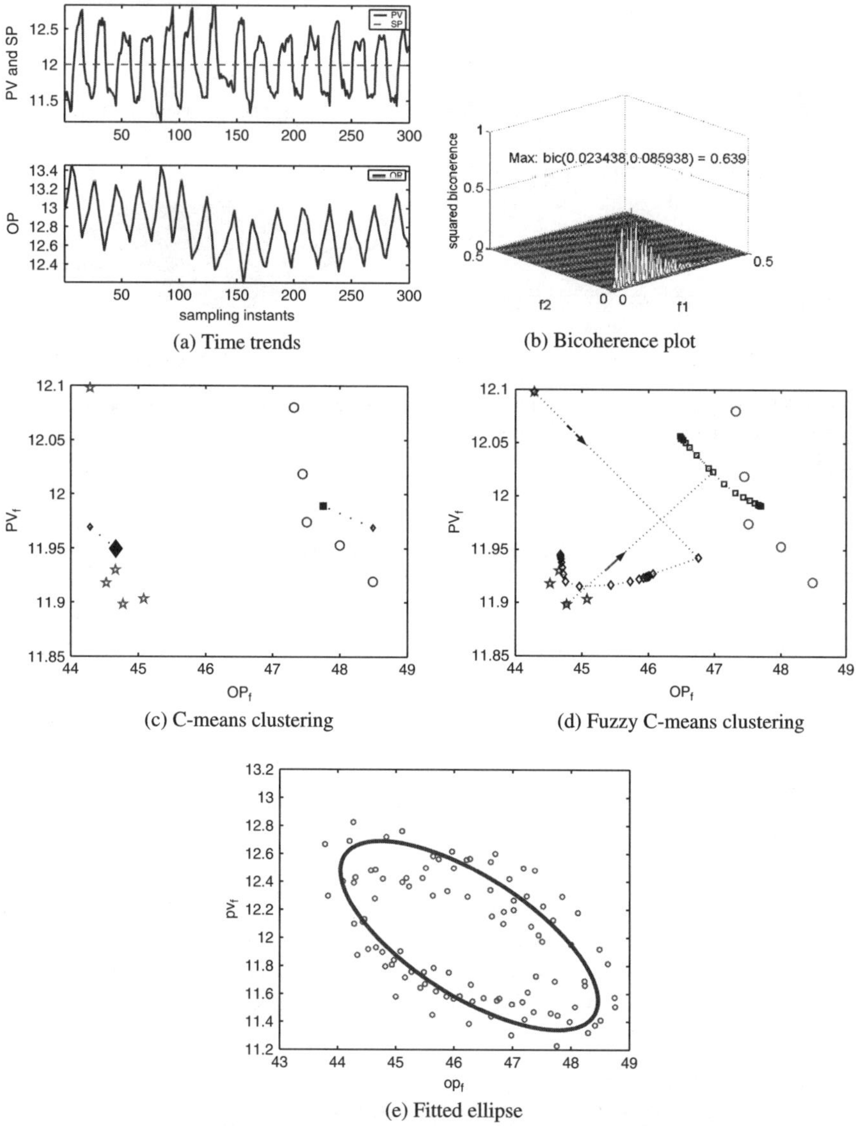

Fig. 15.6 Results of detection and quantification of stiction in a simulated example

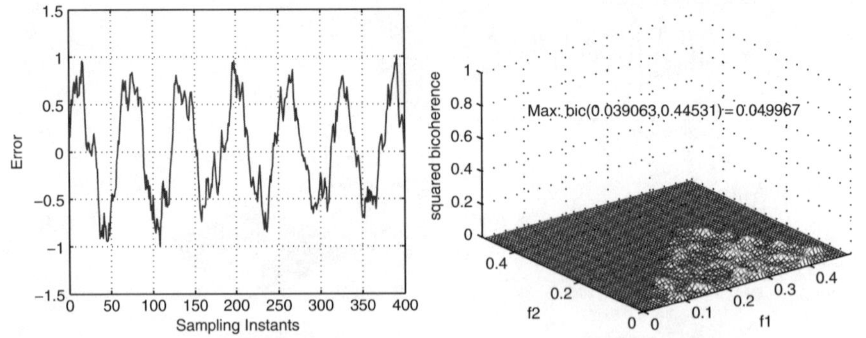

(a) Time trends (b) Bicoherence plot

Fig. 15.7 Stiction detection results for a simulated data set when an external sinusoidal oscillatory disturbance plus noise are entering the control loop

the diagnosis results should not detect stiction or any other nonlinearity. The time trend of the controlled variable (*pv*) in Fig. 15.7(a) shows the oscillatory behaviour of the process output. The NGI value was approximately zero, indicating a linear loop. The bicoherence plot for the error signal to the controller is shown in Fig. 15.7(b). The flatness of the bicoherence plot confirms the linearity of the loop.

15.7.3 Detecting Stiction When Its Impact Propagates as Disturbance

This case demonstrates the root-cause diagnosis of propagation of oscillation(s) from one loop to another. This simulation case has been formulated by feeding the output of a concentration control loop as a disturbance to a level control loop. The transfer function and controllers for both loops are given below.

For the concentration loop:

$$G(s) = \frac{3e^{-10s}}{10s+1} \qquad C(s) = 0.2(1+\frac{1}{10s}) \tag{15.30}$$

For the level loop:

$$G(s) = \frac{e^{-2s}}{4s} \qquad C(s) = 0.1(1+\frac{1}{10s}) \tag{15.31}$$

The simulation block diagram for this study is shown in Fig. 15.8. The data-driven stiction model was used in the composition loop with $S=3$ and $J=2$, and the valve in the level control loop was free from stiction. In reality, it may describe a scenario where the outlet of a mixing chamber in a composition loop is used to feed

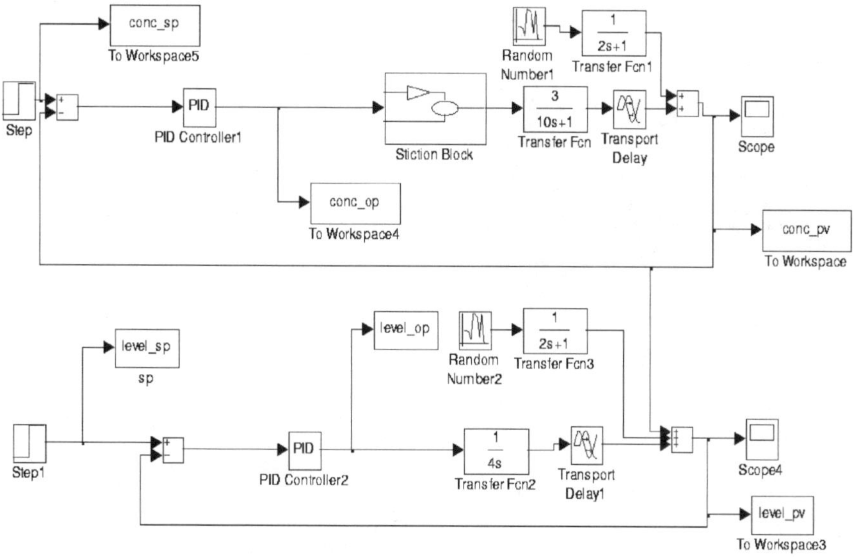

Fig. 15.8 Simulation block diagram implemented in Simulink for the stiction induced oscillation propagation study

another processing unit (e.g. a stock tank), and the level of the unit is controlled by a PI controller. Composition might affect level if the content of the tank is frothy, for example.

Results of the analysis of these two loops are shown in Fig. 15.9. As shown in Fig. 15.9 both loops are oscillating. The NGI and NLI values for the composition loop are 0.004 and 0.57, respectively, and for the level loop, they are NGI = 0.002 and NLI=0.49. NGI values of 0.004 and 0.002 are significant for the given data set, as shown in Table 15.1. The apparent stiction detected in the composition loop is 0.70% whereas in the level loop it is 1%. The method falsely detects stiction in the level loop. The method described in this chapter assumes that the disturbance(s) entering the loop is linear. For the level loop, this assumption has been violated.

For the root-cause diagnosis of plant-wide or unit-wide oscillation(s), the methods described in Chaps. 18 and 19 should be used. For application in industry, it is recommended that the stiction diagnosis method should be used in conjunction with information from the process flow sheet. For example, for this case, these two loops are connected and stiction is detected in both loops. Since there is no recycle, it is most likely that the source of nonlinearity due to stiction is located in the upstream loop, i.e. in the composition loop. The NGI and NLI indices are also higher for the composition loop, confirming this as the root cause of the nonlinearity. Other stiction-detection algorithms (Horch, 1999; Rengaswamy et al., 2001; Stenman et al., 2003) would also produce false-positive results for this kind of propagated disturbance(s) because the other methods also do not consider the interaction among the loops.

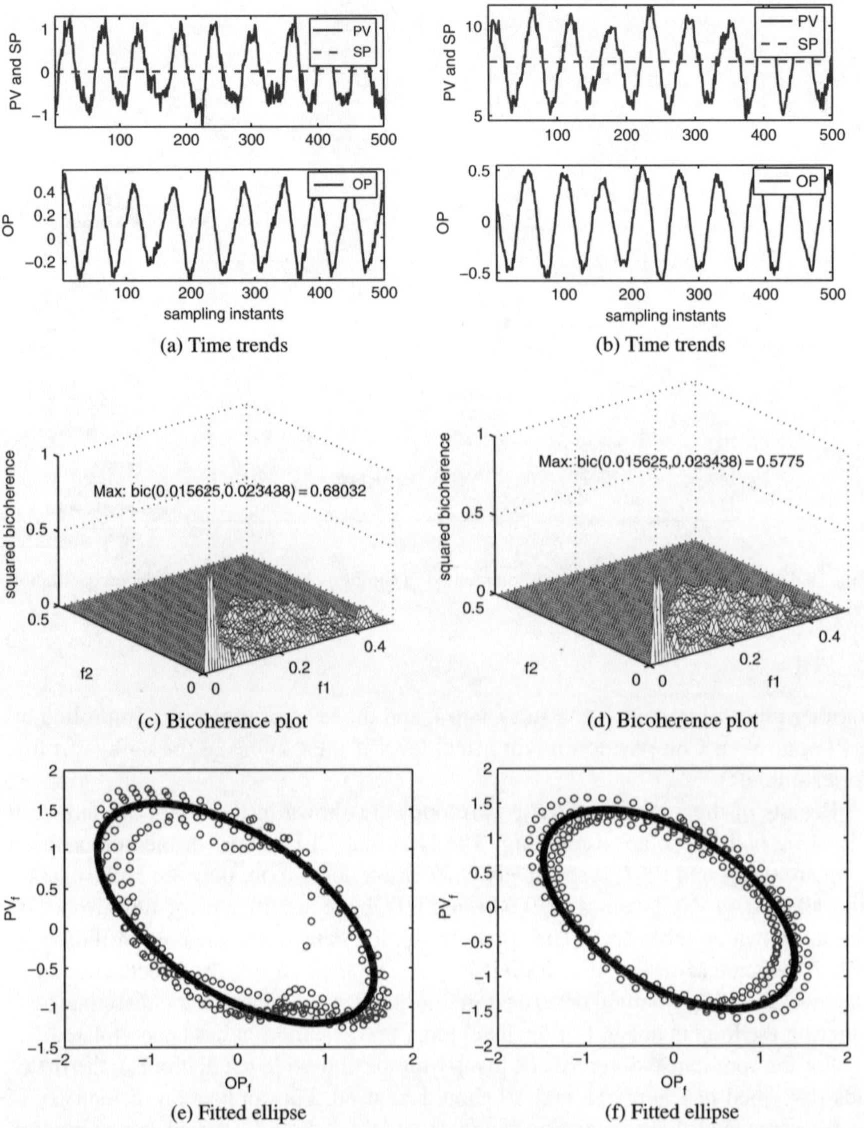

Fig. 15.9 Analysis results for the root-cause diagnosis of a propagated disturbance. Result for the composition loop with a sticky valve (*left*). Results for the level loop affected by the disturbance from the composition loop (*right*)

Table 15.1 Threshold values for NGI and NLI

Data length	NGI_{crit}	NLI_{crit}
4096	0.001	0.01
2048	0.002	0.02
1024	0.004	0.04

15.8 Practical Implementation Issues

For any practical data analysis, a considerable amount of time is spent on data-preprocessing to make the data suitable for analysis. The following sections describe some useful information when analysing data using the method described here.

15.8.1 Bicoherence Estimation

The bispectrum can be normalized in various ways to obtain bicoherence. Some normalization may not deliver bicoherence magnitudes bounded between 0 and 1. For example, the bicoherence calculated using the freely available higher-order statistical analysis (HOSA) toolbox in MATLAB® does not provide bounded values, rather it provides unbounded values of bicoherence. Therefore, the users are recommended to use the normalization provided in Eq. (6.2). More details on bicoherence estimation can be found at Nikias and Petropulu (1993).

15.8.2 Nonstationarity of the Data

Most of the statistical analyses including bicoherence estimation have the assumptions of *stationarity* of the signal. For a slowly drifting or varying signal, the MATLAB® *detrend* command can be used to make it more stationary. If it does not help, the signal can be filtered by a high pass filter. For example, if one likes to remove the signal components at frequencies below 0.001, the signal can be filtered with a Wiener filter with boundaries [0.001 0.5]. For a random walk-type signal, the first differencing of the signal may help.

15.8.3 Problems of Outliers and Abrupt Changes

Bicoherence estimation is very susceptible to outliers or abrupt changes in the signal (Fackrell, 1996). Outliers should be removed and replaced by estimates using a suitable statistical method. Also, care should be taken so that the segment of the signal used for bicoherence calculation does not have step changes or other abrupt feautres.

15.8.4 Dealing with Short Length Data

As discussed in Chap.3, the reliability of the bicoherence calculation improves for long data sets and ideally the data ensemble should contain 4,000 samples or more. Sometimes, however, a practical data set falls short of this ideal, for instance, there may only be short episodes of stationary behaviour that are suitable for the analysis. Practical methods for handling shorter data sets include overlapping the data segments during the calculation of bicoherence using a direct method similar to the Welch periodogram method (Choudhury, 2004). The number of data segments used in the bicoherence estimation should be close to 30 (see Sect. 3.2.2 in Chap. 3). Also, the threshold values used for NGI and NLI should also be changed for obtaining robust decisions with a minimum number of false-positive results. From the experience of the authors, suitable threshold values for NGI and NLI for various data lengths are provided in Table 15.1.

15.8.5 Dealing with Longer Oscillations

Sometimes the oscillation period of the signal is very large, e.g. more than 256 samples. In those cases, the data can be downsampled in such a way that the size of the data ensemble being used to generate the fast Fourier transform for the estimation of bicoherence captures three or four cycles of oscillation. Downsampling of a low-frequency oscillation could lead to aliasing of high- frequency oscillations present in the same time trend, which might become undersampled. If this is likely to be an issue, an antialiasing filter with a cut-off frequency of one half of the new sampling frequency would be applied before downsampling.

15.8.6 Valve Nonlinearity

The valve itself may have a nonlinear characteristic, e.g. a square-root or equal percentage characteristic, which is definitely not a fault. It is therefore necessary to distinguish between the wanted nonlinear characteristic and faults due to nonlinear stiction. The simulation study performed in Chap. 8 showed that that if the movement of the valve stem or the change in input signal to valve is within 30% of the full span (0–100%) of the valve travel, a control loop exhibits linear behaviour under steady-state regulatory control in the presence of a square-root or equal percentage valve (Choudhury, 2004). Therefore, in the beginning of the stiction-detection algorithm, a check on the range of *op* signal can be performed, and if the range of *op* is larger than 30%, a warning can be issued.

15.8.7 Filtering of the Data

In the bicoherence calculations, the main purpose of filtering is to remove the slowly varying trend and high-frequency noise from the data. If the data does not have either of these two, the filtering effect may not be visible in such cases. An example of such data is described in Fig. 15.2, where the data looks very clean and there is no slow drifting of the mean of the data. But this is not the case for the data set described in Fig. 15.10. The description of the data set is provided in Sect. 16.1.3. The plots in the first row of the Fig. 15.10 show that without filtering, the elliptical patterns in the *pv–op* plot are neither clean nor distinct. On the other hand, the lower right plot shows that the elliptical patterns in the *pv–op* mapping of the filtered data are clear and distinct. Here, the filter boundary was $[0.004 \quad 0.1]$. The filtered time trends in the lower left plot show that the filtering has removed the slowly varying mean-shift and high-frequency noise from the *pv* and *op* signals.

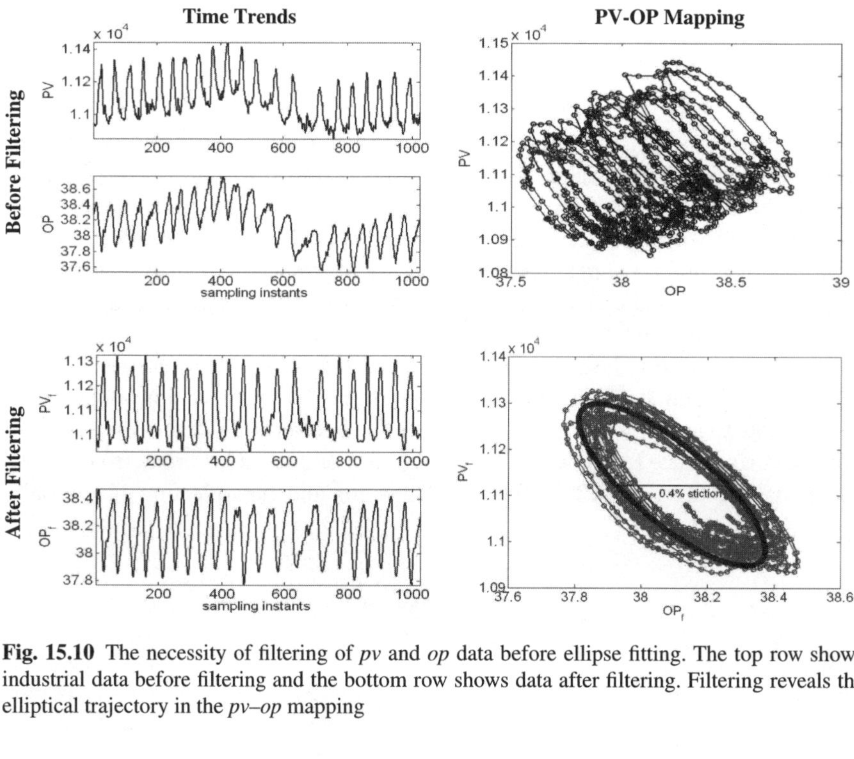

Fig. 15.10 The necessity of filtering of *pv* and *op* data before ellipse fitting. The top row shows industrial data before filtering and the bottom row shows data after filtering. Filtering reveals the elliptical trajectory in the *pv–op* mapping

15.8.8 Segmenting Data for pv–op Plot

If the whole data set has well-defined limit cycles, one can omit this step and plot any part of the *pv* and *op* data. But in reality, there are data sets where the valve may suffer from stiction sometimes, but may not exhibit stiction at other times. Since stiction generally causes limit cycling in closed loop, it is logical to obtain that portion of the data that has a regular oscillation or oscillations. It should be remembered that the estimated stiction using the *pv–op* mapping should be treated as *apparent stiction*, which can be obtained from the available *pv–op* data. As mentioned earlier, the actual amount of stiction should be obtained from the mapping of the valve positioner data (*mv*) and the controller output data (*op*). However, the apparent stiction will provide an indication of the severity of the stiction in the existing operating conditions of the loop.

Further research is required for exact quantification of stiction from routine operating data. Some research results have appeared in Jain et al. (2006), Choudhury et al. (2006b, 2008) on the quantification of actual *S* and *J* from real-life sticky-valve data. This method requires simultaneous identification of the stiction parameters and the closed-loop model of the plant. Also, this method requires extensive computational time and resources. A similar method using a one-parameter stiction model has appeared in Srinivasan et al. (2005b).

15.9 Summary

A non-invasive method for detecting and quantifying stiction in control valves has been presented in this chapter. The method first detects nonlinearity in a control loop by the use of the normalized bispectrum or bicoherence to find the nonlinear interactions that may be present in the control error signal. If nonlinearity is detected, then *pv* and *op* signals are filtered using frequency domain Wiener filter to obtain filtered pv_f and op_f signals. If an ellipse can be fitted satisfactorily onto the pv_f–op_f plot, this can be taken as a signature of valve stiction. Then C-means clustering, fuzzy C-means clustering or fitted ellipse techniques can be used to quantify the amount of stiction. The method has been extensively evaluated on simulated as well as industrial data sets. The algorithm has been commercialized in the Matrikon Control Performance Monitor product from Matrikon Inc. and is protected by an international patent PCT/CA2005/001031 "Detection and quantification of stiction".

Chapter 16
Industrial Applications of the Stiction Quantification Algorithm

This chapter evaluates the stiction detection and quantification algorithm described in Chap. 15 on a variety of control loops obtained from different types of process industries. This chapter shows that the algorithm can detect and quantify stiction blindly without knowing the type of the control loop and type of the control valve. The algorithm has been rigorously and successfully evaluated by applications in more than 20 industrial case studies. In all cases, the algorithm could successfully detect control valves that were suffering from stiction and also quantify the amount of apparent stiction present.

16.1 Industrial Case Studies

Control loops where stiction was detected as the cause of poor performance by the stiction quantification algorithm are described below. For each loop, set point (*sp*), controlled output (*pv*) and controller output (*op*) data were available. Unless otherwise stated, a data ensemble of 4,096 samples was used for the squared bicoherence calculation for each case. For each loop, the time trends of the variables with the corresponding squared bicoherence plot, the C-means clustering plot and the fitted ellipse plot are presented. The numerical results for all loops are provided in Table 16.1. These data were analysed without prior knowledge of the control valve problems, and the results of the analysis were confirmed later by the plant personnel.

16.1.1 Loop 1: A Level Loop

Loop 1 is a level control loop in the same power plant described in the illustrative example in Chap. 15. This particular level loop also controls the level of condenser located at the outlet of a turbine by manipulating the flow rate of the liquid condensate. Figure 16.1a shows the time trends of the *sp*, *pv* and *op* data. Figure 16.1b shows the squared bicoherence plot. The values of NGI and NLI for this loop were found

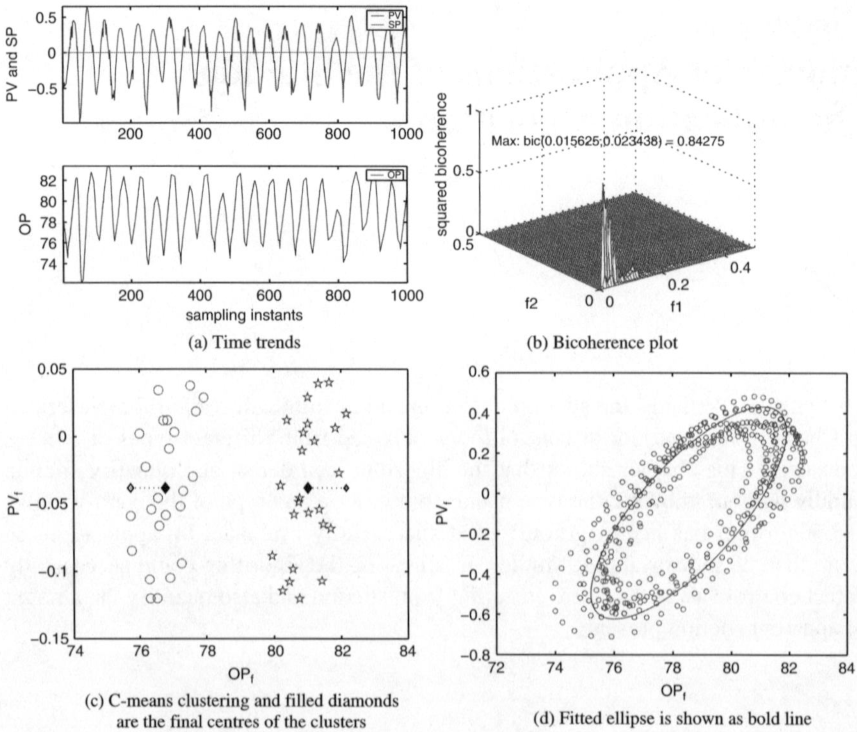

(a) Time trends

(b) Bicoherence plot

(c) C-means clustering and filled diamonds are the final centres of the clusters

(d) Fitted ellipse is shown as bold line

Fig. 16.1 Results for an industrial-level control loop

to be 0.32 and 0.83 and are tabulated in Table 16.1. These indices indicate that nonlinearity was present in this loop. From the position of the maximum peak at the bicoherence plot, the frequency range for the Wiener filter was obtained following the steps described in the automation section (Sect. 15.6). The frequency band for the

Table 16.1 Numerical results for the analysis of selected Industrial loops analyses

Loop no.	Loop type	NGI	NLI	TNLI	f_L	f_H	T_p	r	App. stiction % C-means	Ellipse
1	Level	0.32	0.83	4.45	0.001	0.08	50	22	4.2	4.3
2	Level	−0.02	–	–	–	–	95	3.5	–	–
3	Flow	0.32	0.80	3.5	0.001	0.08	45	8.4	0.35	0.33
4	Flow	0.23	0.32	0.93	0.09	0.20	7	2	0.42	0.42
5	Pres.	0.38	0.36	0.77	0.01	0.25	12.2	12.2	11	11
6	Comp.	0.58	0.78	6.35	0.01	0.15	28.3	11.6	1	1
7	Flow	45	0.44	0.45	0.004	0.14	59	4.6	–	–
8	Temp	0.24	0.38	2.76	0.004	0.28	125	6.5	1	1.14
9	Level	0.24	0.66	1.88	0.001	0.08	61	14.4	9	9.40
10	Flow	0.14	0.16	0.73	0.001	0.49	25	7.3	1.78	1.72

filter is [0.001 0.08] (1,000–12.5 samples/cycle). The filter will therefore remove low frequencies that cause non-stationary behaviour and high-frequency noise. The average period of oscillation was 50 samples for the controller output signal. The segment of the data corresponding to the maximum oscillation index (the magnitude of r was 22 for this case) was selected to quantify stiction. Both C-means clustering and the fitted ellipse technique estimate the amount of apparent stiction as 4%, which is significant and causes poor control performance. The C-means-clustering plot is shown in Fig. 16.1c, while Fig. 16.1d shows the fitted ellipse for quantifying stiction.

16.1.2 Loop 2: A Linear-Level Control Loop

Loop 2 is another level control loop in the same power plant described in the previous section. It also controls the level of a condenser located at the outlet of a different turbine by manipulating the flow rate of the liquid condensate. Figure 16.2a shows

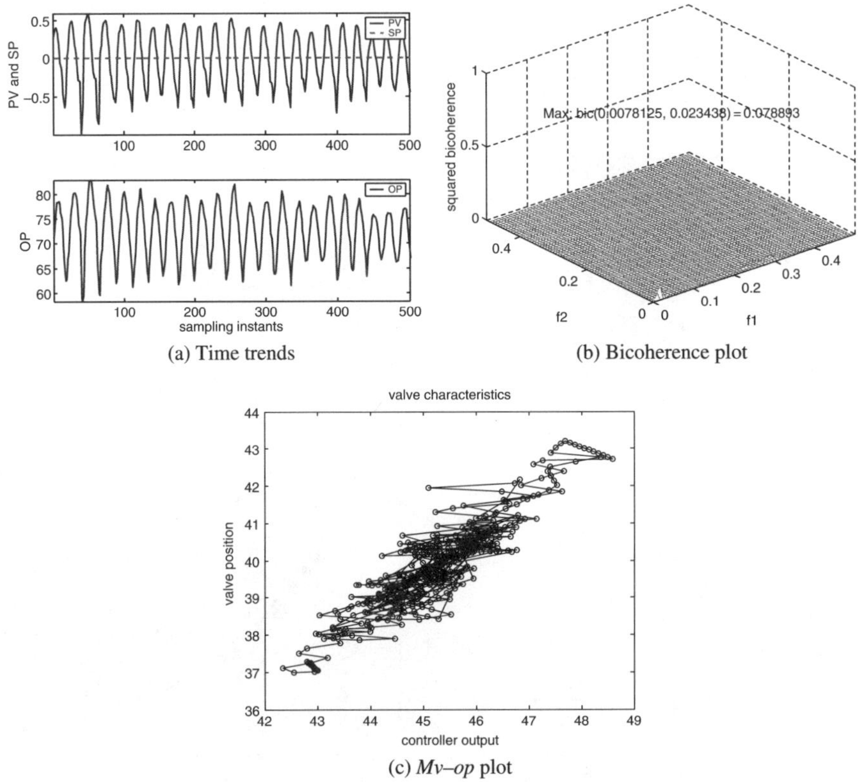

(a) Time trends

(b) Bicoherence plot

(c) *Mv–op* plot

Fig. 16.2 Results for a second industrial level control loop data. The flatness of the bicoherence plot confirms the linearity of this level control valve, which was confirmed in the valve position versus valve input plot (subplot (c)) the time trends of the *sp*, *pv* and *op* data. Figure 16.2b shows the squared bicoherence plot for the control error signal. The magnitude of NGI was −0.02 (which is smaller than zero), indicating that nonlinearity is not present in this loop. Figure 16.2c shows the valve positioner (*mv*) vs. controller output (*op*) plot. This figure shows that the valve has a linear response because the data, though noisy, lie mainly on a straight line.

16.1.3 Loop 3: A Flow Control Loop

Loop 3 is a flow control loop at a refinery. The results of the analysis of this loop are shown in Fig. 16.3 and also in the third row of Table 16.1. The presence of a small amount of apparent stiction (0.35% for loop 3) was causing a large amplitude oscillation (see the magnitude of *pv* axis in Fig. 16.3a) in this loop. The large

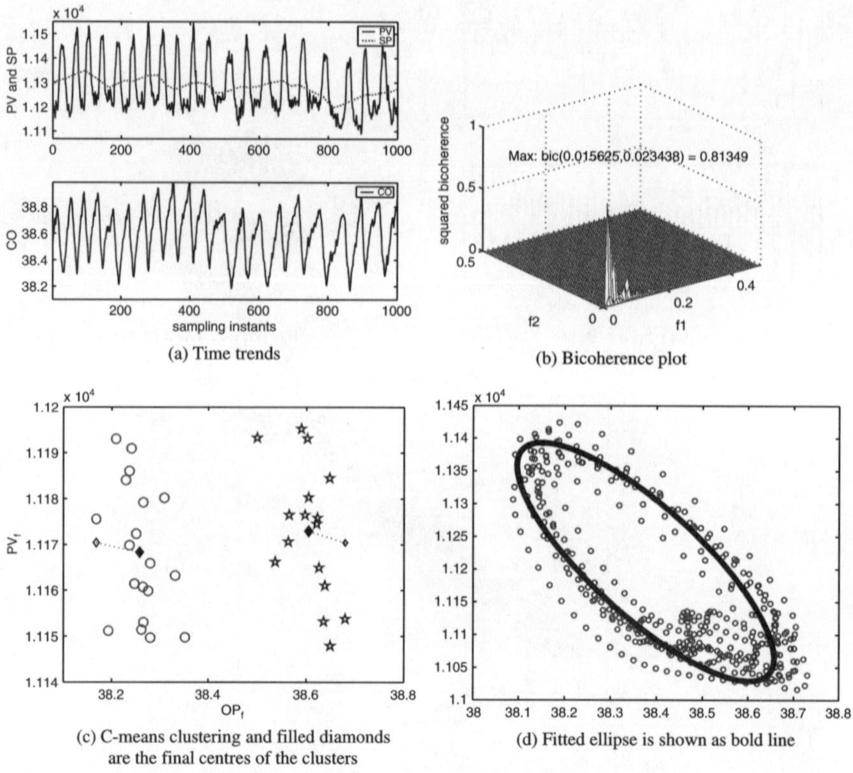

(a) Time trends

(b) Bicoherence plot

(c) C-means clustering and filled diamonds are the final centres of the clusters

(d) Fitted ellipse is shown as bold line

Fig. 16.3 Results of the analysis of a refinery flow control loop. Significant peaks in the bicoherence plot show the nonlinearities in the loop. The presence of an ellipse in the *pv-op* plot confirms stiction amplitude oscillation for this small amount of stiction could be due to a large gain in this process. This data set was also used to illustrate the necessity of using a filter in Sect. 15.8.7.

16.1.4 Loop 4: Flow Control Loop Cascaded with Level Control

Loop 4 is a flow control loop at Celanese Canada Ltd., a chemical complex located in the east of Edmonton, AB, Canada. This flow control loop is cascaded within a level control loop. The stiction quantification algorithm detects stiction in this loop and quantifies it to be approximately 8%. The finding of an apparent stiction of 8% illustrates the benefit of the valve stiction quantification algorithm. Without the algorithm, the valve fault may have gone undetected because the time trends in Fig. 16.4 suggest that the *pv* is generally tracking *sp*, although some deviations can be seen when the *sp* changes the direction. Figure 16.5 shows an *sp*–*pv* plot for the flow loop that highlights the deviations. The flat horizontal segments of the *sp*–*pv*

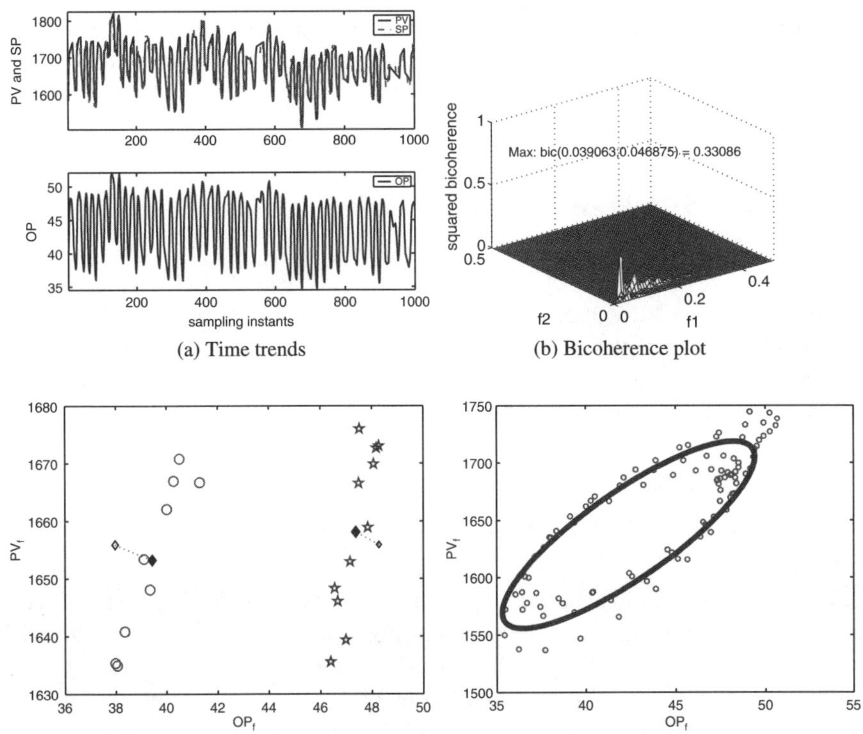

(a) Time trends

(b) Bicoherence plot

(c) C-means clustering and filled diamonds are the final centres of the clusters

(d) Fitted ellipse is shown as bold line

Fig. 16.4 Results for a flow control loop in a chemical plant

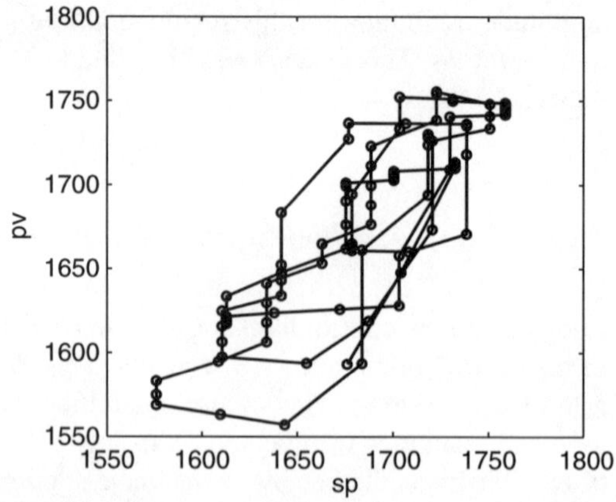

Fig. 16.5 The *pv* vs. *sp* plot for the flow control loop of the cascaded loop in a chemical plant

characteristic indicate that the *pv* sticks whenever the *sp* changes the direction and confirms that stiction is indeed present. It is possible to construct an *sp–pv* plot for this loop because the set point changes in response to the master level controller. Moreover, the flow which is the *pv* of the slave loop is the *mv* for the master level loop, while the *op* of the level controller is the *sp* for the slave flow loop. Therefore the *sp–pv* plot for the flow loop is the *op–mv* plot for the level control loop, which is the preferred mapping for visualization of actuator nonlinearity as discussed in Sect. 11.3 in Chap. 11.

16.1.5 Loop 5: A Pressure Control Loop

Loop 5 is a pressure control loop in a refinery. This data set had only 1,500 data points collected at 20 s sampling intervals. The time trends in Fig. 16.6 show oscillations with 12.2 samples in both *pv* and *op* variables. The detailed results of the analysis are presented in Fig. 16.6 and in the fifth row of Table 16.1. The apparent stiction present in the valve was approximately 11%.

16.1.6 Loop 6: A Composition Control Loop

Loop 6 is a concentration control loop. The data set contains 1,100 points collected at 1 s sampling interval. The time trends in Fig. 16.7a show oscillations with 28.3 samples in both *pv* and *op* variables. The detailed results of the analysis are presented in Fig. 16.7 and also in the sixth row of Table 16.1. The apparent stiction present in the valve of this concentration control loop was approximately 1%.

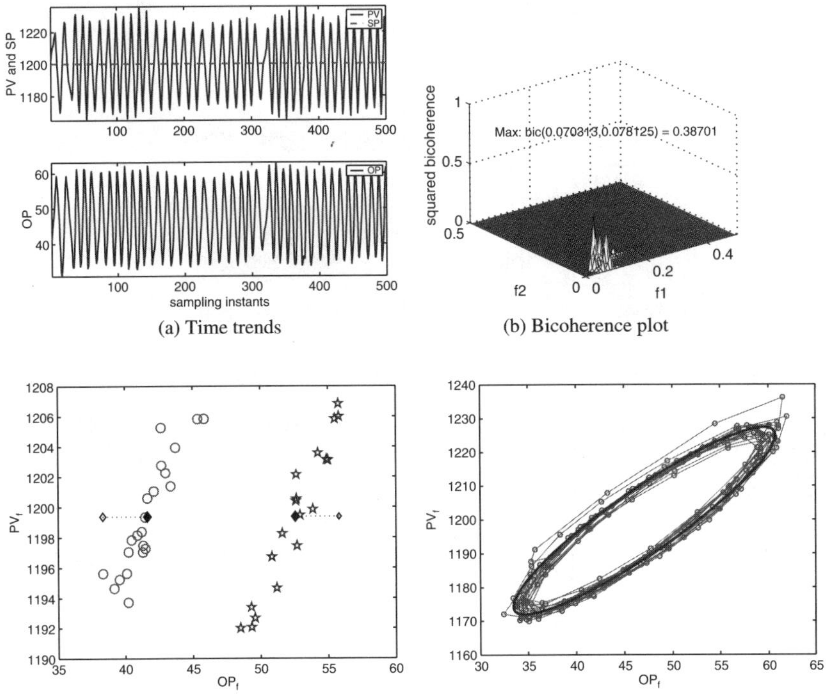

(a) Time trends

(b) Bicoherence plot

(c) C-means clustering and filled diamonds are the final centres of the clusters

(d) Fitted ellipse is shown as bold line

Fig. 16.6 Results of the analysis of a refinery pressure control loop. Approximately 11% stiction was present in the control valve

16.1.7 Loop 7: A Cascaded Flow Control Loop

Loop 7 is a flow control loop cascaded within a level control loop of a drum in an ethylene plant from Mitsubishi Chemical Corporation in Japan. At a sampling rate of 1 sample/min, 7,200 data points were collected. Time trends for the *pv* and *op* variables for the flow control loop are shown in Fig. 16.8a. The detailed results of the analysis are presented in Fig. 16.8 and numerical results are provided in the seventh row of Table 16.1. The bicoherence plot shows the presence of a nonlinearity in this loop. The absence of an elliptical pattern in the *pv–op* plot (Fig. 16.8c) indicates that this nonlinearity is not due to valve stiction. The probable source of this nonlinearity is likely to be a nonlinear disturbance entering this loop or valve problems other than stiction, or nonlinearities in the process itself.

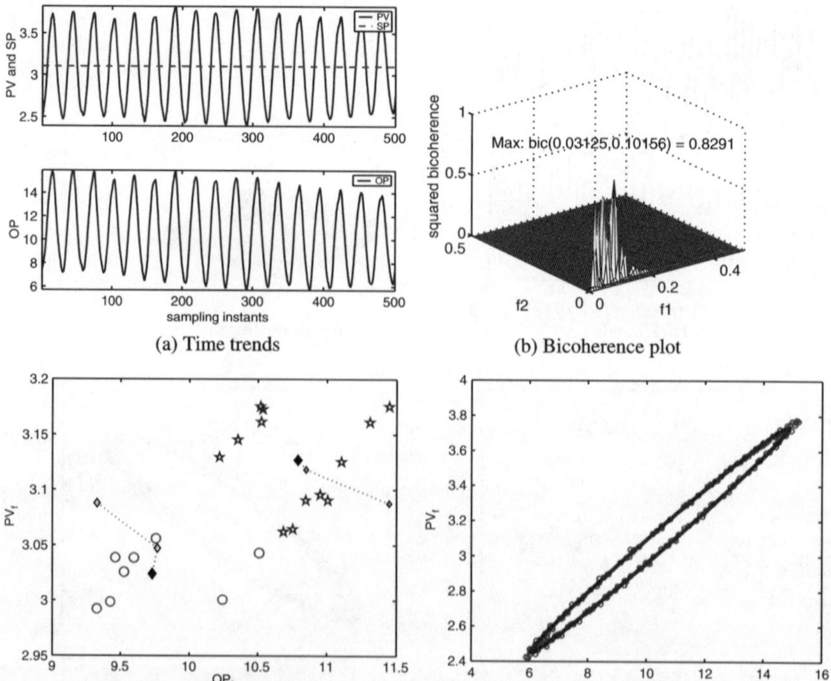

(a) Time trends

(b) Bicoherence plot

(c) C-means clustering and filled diamonds are the final centres of the clusters

(d) Fitted ellipse is shown as bold line

Fig. 16.7 Results of the analysis of an industrial composition control loop. Approximately 1% stiction was present in the control valve

16.1.8 Loop 8: A Temperature Control Loop

Loop 8 is the same temperature loop described in Sect. 9.5.1 of Chap. 9. The temperature of the dryer combustion chamber is controlled by manipulating the flow rate of natural gas to the combustion chamber. The results of the analysis are presented in the eighth row of the Table 16.1. The amount of stiction found in this loop was approximately 1%.

16.1.9 Loops 9 and 10

Both these loops are from a pulp and paper plant. Loop 9 is a level control loop controlling the level of a dilution tank. The detailed diagnosis results are shown in Table 16.1 and Fig. 16.9. This control valve was suffering severely from stiction (approximately 9% stiction), and its poor performance was aggravated by the presence of excessive integral action in the controller. Loop 10 is a flow control loop control-

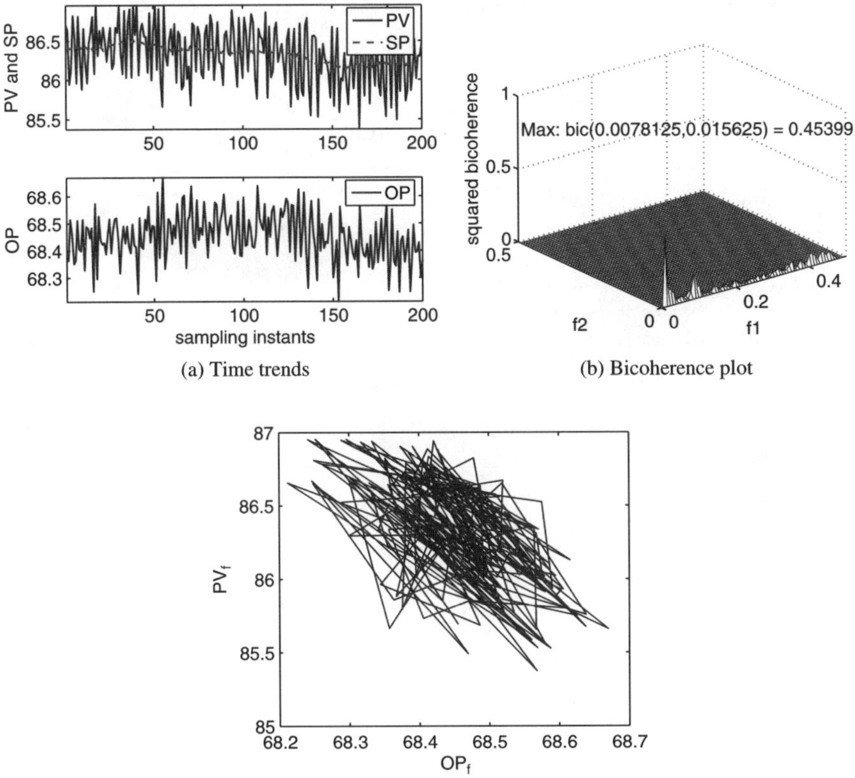

(a) Time trends (b) Bicoherence plot

(c) *pv–op* plot

Fig. 16.8 Results of the analysis of an industrial-cascaded flow loop

ling the flow rate of the water used for washing pulp in a rotary vacuum washer. The tenth row of Table 16.1 and Fig. 16.10 show the results of the analysis. The apparent stiction in this valve was approximately quantified to be 2%.

16.2 Online Compensation for Stiction

After a sticky valve is identified, the best solution is to repair it, which may require taking the valve out of service. In most cases, it is not possible to take the sticky valve out of service for its immediate maintenance because of the absence of a bypass line. A method that can help continuing operation of the plant until the next planned shutdown with minimum loop performance degradation due to stiction can be useful for the process industry. Several online methods to deal with a sticky valve have been suggested (Bergstrom and Dumont, 2003; Armstrong-Hélouvry et al., 1994; Hatipoglu and Ozguner, 1998; Kayihan and Doyle III, 2000; Hägglund,

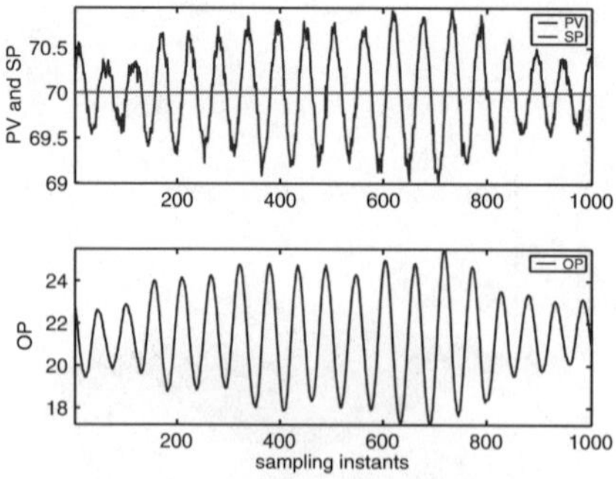

(a) Time trends

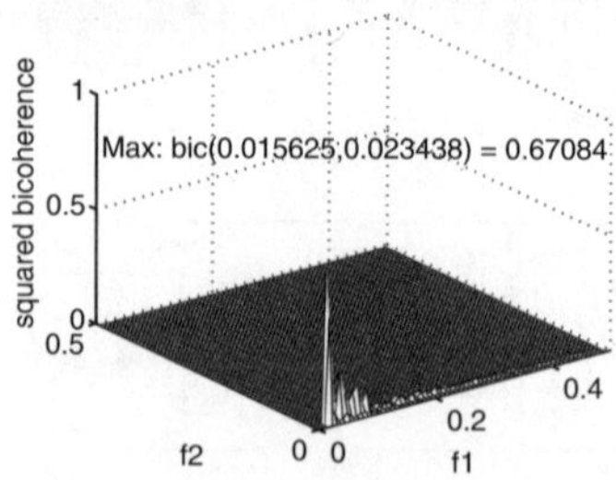

(b) Bicoherence plot

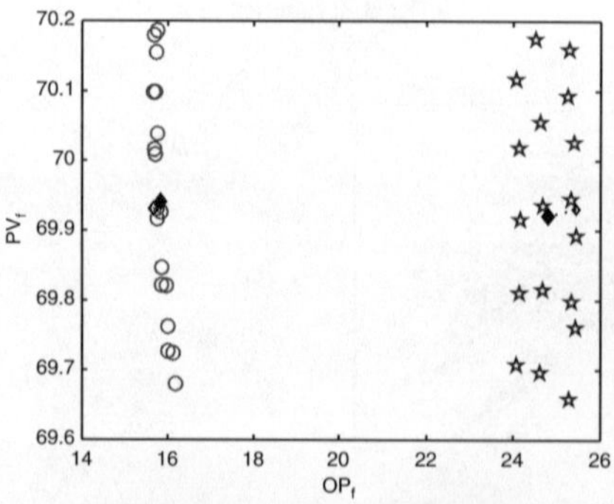

(c) C-means clustering and filled diamonds are the final centres of the clusters

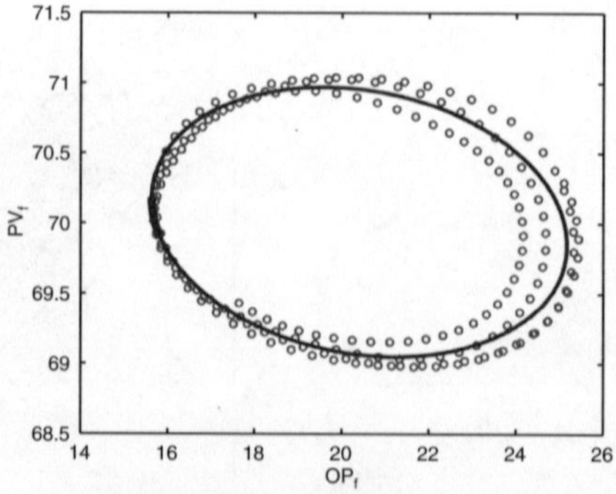

(d) Fitted ellipse is shown as bold line

Fig. 16.9 Results of analysis of a level control loop in a pulp and paper mill

2002; Tao et al., 2002). For online compensation of stiction, Gerry and Ruel (2001) suggested the use of a PI controller where the integral action has variable strength. Hägglund (2002) suggested a method of adding short pulses to the control signal to compensate stiction in pneumatic valves. As claimed by Hägglund (2002), the *knocker* friction compensation method has been patented and implemented in industrial controllers. In (Bergstrom and Dumont, 2003), a special adaptive controller in combination with a PID controller has been suggested for online stiction compensation. It requires a special function for switching between the adaptive controller and the PID controller. Some other methods to compensate stiction in control valves can be found in Armstrong-Hélouvry et al. (1994), Hatipoglu and Ozguner (1998), Kayihan and Doyle III (2000), Tao et al. (2002), Amin et al. (1997).

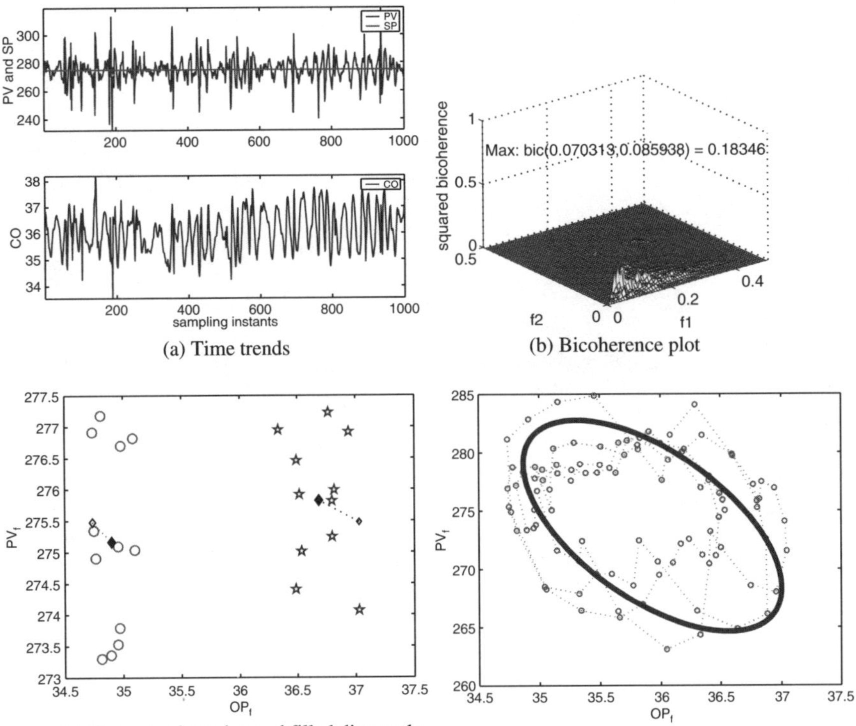

(a) Time trends

(b) Bicoherence plot

(c) C-means clustering and filled diamonds are the final centres of the clusters

(d) Fitted ellipse is shown as bold line

Fig. 16.10 Results of analysis of a flow control loop in a pulp and paper mill

16.3 Summary

This chapter presented the industrial application of the automatic method for detecting and quantifying control valve stiction discussed in Chap. 15. It has been shown that the stiction-detection algorithm can detect and quantify stiction blindly without knowing the type of control valves or type of control loops. The algorithm only requires the set point (*sp*), process variable (*pv*) and controller output (*op*). The method has already been applied to many industrial sites online through the Matrikon Control Performance Monitor product from Matrikon Inc.

Chapter 17
Confirming Valve Stiction

Stiction detection and quantification methods that are based on the assumption of SISO configuration do not take into account the multivariate or MIMO interactions of the control loops, which causes false positive results when applied to an entire plant site. Due to these limitations, usually an invasive valve test is required to confirm if the suspected valves are indeed suffering from stiction. This chapter describes a method for confirming valve stiction based on changing the controller gain under closed-loop control configuration. The method is based on the observation that varying controller gain changes the frequency of the oscillations caused by valve stiction.

17.1 Methods to Confirm Valve Stiction

The automated stiction detection and quantification methods discussed in Chap. 15 can highlight problematic behaviour in a control valve. However, before a maintenance request is issued, it is usual to conduct a valve test to confirm the diagnosis. The various methods for confirming valve stiction are as follows:

1. **Valve travel or bump test:** Stiction in control valves is usually confirmed by putting the valve in manual and increasing the control signal in small increments until there is an observable change in the process output. This test is known as the valve travel or bump test (Taha et al., 1996; Wallén, 1997; Gerry and Ruel, 2001). This method of confirming stiction by putting the loop in manual is not convenient and cost-effective due to the risk of plant upset and production of off-specification products.
2. **Use of valve positioner data:** For smart valves, usually the valve positioner (actual stem position of the valve) data are available. If the plot of valve positioner (*mv*) data against the valve input (*op*) data shows a pattern similar to that described in Fig. 11.2, it can be concluded that the valve has stiction. Unfortunately, most industrial valves (more than 95% cases) are not smart valves. The lack of availability of the valve positioner data restricts the application of this simple test to only a few cases.
3. **Changes in controller gain:** Presence of stiction in a control valve produces limit-cycle oscillations in the process output (*pv*) and the controller output (*op*).

Changes in controller gain cause changes in the amplitude and frequency of these limit-cycle oscillations. Once stiction is detected in a loop, changes in oscillation frequency due to variation in controller gain can help confirm the presence of stiction in the loop.

17.2 Gain Change Method for Confirming Valve Stiction

It appears that the simple test employing changes in controller gain is well known to engineers in the process industries, but is relatively less well known in academia. This simple test was first reported in Thornhill et al. (2003b) where an online closed-loop plant test was conducted to confirm the presence of stiction. The influence of gain change on the amplitude and frequency of oscillations due to stiction has also been described briefly through a simulation study in Choudhury et al. (2005a, 2005b) and Rossi and Scali (2005). Here, a simulation study of this phenomenon has been carried out to distinguish stiction-induced oscillation from an external oscillatory disturbance. A theoretical justification through describing function analysis is also presented.

17.2.1 Distinguishing Stiction from External Oscillatory Disturbance

The two-parameter stiction model described in Chap. 13 has been used to introduce the stiction behaviour of the control valve in a closed-loop process. The two parameters are deadband plus stickband (*S*) and slip–jump (*J*), which are taken as 3% and 1%, respectively, for all simulations. The PI controllers used in all simulations were implemented in the following form:

$$C(s) = K_c\left(1+\frac{1}{\tau_i\, s}\right). \tag{17.1}$$

The controller gains and reset times are reported in the right column of each figure provided for each of the following cases.

17.2.1.1 A Concentration Control Loop with a Sticky Valve

The transfer function model for this loop is similar to the one reported in Horch and Isaksson (1998) except that the time delay has been reduced to 5 for ensuring better control of the process:

$$G(s) = \frac{3e^{-5s}}{10s+1}. \tag{17.2}$$

The model in Eq. (17.2) describes the dynamics of mixing pulp with water from the inlet of a water flow control valve to the outlet of a mixing chamber (Horch and

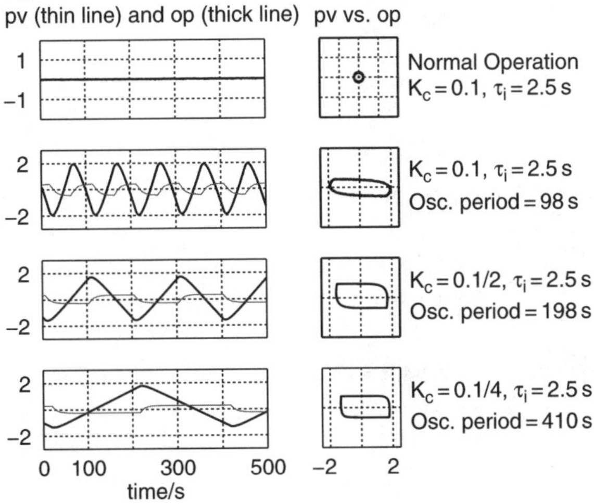

Fig. 17.1 Effect of changing controller gain on the limit cycles generated by a sticky valve in a concentration loop. A decrease in controller gain causes a decrease in the frequency and amplitude of the limit cycles in *pv*

Isaksson, 1998). The top row in Fig. 17.1 shows that during normal operation (no stiction), there is no limit cycle with the existing controller settings reported in the right column of the figure. In the presence of stiction, a limit-cycle oscillation with a period of 98 s is generated in the loop (the plot in the second row of Fig. 17.1). In order to confirm that this oscillation is generated within this loop, the controller gain was reduced by 50%, which in turn decreases the frequency of oscillation. The third row in the figure shows that the oscillation period is now approximately 198 s and the amplitude of the oscillation in *pv* is also decreased. Reducing the gain further decreases the frequency of oscillation. This test confirms that the oscillation originates within the control loop.

17.2.1.2 A Level Control Loop with a Sticky Valve

The transfer function for the level control loop in this example was an integrator of the form of $\frac{1}{5s}$. Figure 17.2 shows the result of changing controller gain, K_c. The plot in the first row of the figure shows that during normal operation, there is no oscillation. The plots from the second to the fourth row in Fig. 17.2 show that, in the presence of stiction, a decrease in controller gain for the level loop decreases the frequency of oscillation but increases the amplitude of the oscillation in *pv*.

17.2.1.3 Connected Concentration and Level Loop

The purpose of this case is to present the technique of distinguishing an oscillation that originates within the control loop from an external oscillatory disturbance. This

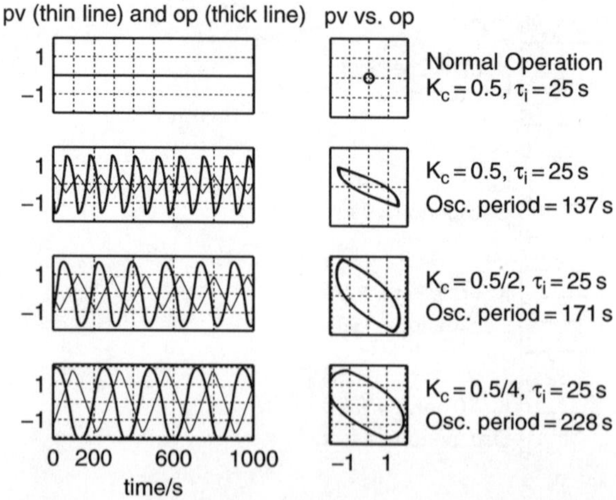

Fig. 17.2 Effect of changing controller gain on the limit cycles generated by a sticky valve in a level loop. A decrease in controller gain causes a decrease in the frequency, but an increase in the amplitude of the limit cycles in *pv*

simulation case has been formulated by feeding the output of the concentration loop as a disturbance to the level control loop. Stiction is now only present in the concentration loop valve, and there is no stiction in the level control valve. It is the same simulation as described in Sect. 15.7.3. Stiction detection algorithms designed for SISO control loops tend to detect stiction in both loops because they do not consider interactions between loops. The method of controller gain change is applied to differentiate between the loops suffering from stiction and the loops oscillating due to propagated oscillatory disturbances.

The results are shown in Fig. 17.3. In the case of normal operation, there were oscillations neither in the concentration loop nor in the level loop (plots in the top row). Then, due to the introduction of valve stiction in the concentration loop control valve, an oscillation with a period of 46 s is generated in the concentration loop and it propagates to the level loop. In order to confirm the presence of stiction in these loops, the proposed gain change method can be applied. If the gain of the concentration loop controller is decreased by 50%, the frequency of the oscillation in the concentration loop decreases to 94 s/cycle, which in turn decreases the frequency of oscillation in the level control loop too (the third row in Fig. 17.3). But when the gain of the level controller is decreased by 50%, the frequency of oscillation in the level loop does not decrease further, rather it remains constant. This confirms that the oscillation in the level control loop does not originate within the loop, but enters the loop as an external disturbance. Therefore, the level control valve is not suffering from stiction.

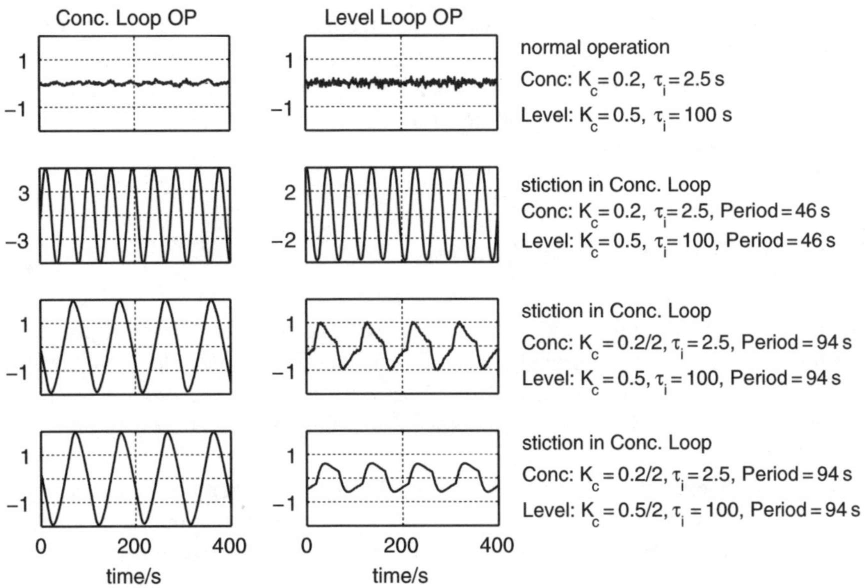

Fig. 17.3 Stiction causes oscillation in concentration control loop and this oscillation propagates to the level control loop. A change in controller gain of the concentration loop causes a change in the frequency of oscillation but a change in controller gain of the level loop does not cause any change in the frequency of oscillation

17.2.1.4 Conclusions from the Simulation Study

For all cases, the limit cycles were present even though the set point to the loop was zero. That is, they originated in and were sustained by the control loop in the absence of any external set-point excitation. Based on this simulation study, the following conclusions can be drawn:

- If a limit cycle originates within a control loop due to valve stiction, a change in controller gain will change the frequency of oscillation.
- If a limit cycle oscillation enters the control loop as an external disturbance, a change in controller gain will not change the frequency of oscillation.

Remark 1. If the oscillations in a loop start due to a tightly tuned controller, a decrease in controller gain will often decrease the frequency of oscillation. This will only prove that this oscillation originated within the loop but would not be able to detect whether it is due to stiction or due to a tightly tuned controller. Therefore, in order to distinguish between these two cases, one must use a stiction detection method. Since the presence of a tightly tuned controller does not introduce any nonlinearity in a control loop, the higher-order statistical method described in Chap. 9 can be used to eliminate this loop as a linear loop in the first diagnostic step. After detection of stiction, the gain test method will help to isolate the valve(s) suffering from stiction.

Remark 2. In practice, it is not necessary to change the controller gains by factors as large as two or four. A change in controller gain by any significant amount should do the job. The central idea is to observe whether a change in controller gain causes a change in the frequency of oscillation. If it does, then it can be concluded that the oscillation originates within the loop.

17.3 Describing Function Analysis

In this section, describing function analysis is used to provide a theoretical justification for the observed closed-loop behavior due to changes in controller gain. For the sake of completeness, a summary of the describing function analysis for the two-parameter stiction model is provided below. For details, refer to Chap. 14 and also Choudhury et al. (2005a).

In the describing function analysis approach, the nonlinearity is modelled by a nonlinear gain N. The inherent assumptions of this approximation are that there are periodic signals present in the system and that the controlled system acts as a low-pass filter and responds principally to the fundamental Fourier component. The condition for oscillations in a negative feedback loop is that the loop gain be -1 or:

$$G_o(i\omega) = -\frac{1}{N(X_m)}, \tag{17.3}$$

where $G_o\,(i\omega)$ is the open-loop frequency response that includes the controlled system and the controller, and $N(X_m)$ is the describing function that depends on the magnitude of the controller output, X_m. When condition Eq. (17.3) is met, the system will spontaneously oscillate with a limit cycle. The variation of the quantity $-1/N(X_m)$ with signal amplitude implies that signals initially present in the loop, e.g. noise, can grow until they are big enough to satisfy the equality and hence provide a self-induced oscillation. The describing function for the two-parameter stiction model as provided in Chap. 14 and also in Choudhury et al. (2005a):

$$N = -\frac{1}{\pi X_m}(A - iB), \tag{17.4}$$

where

$$A = \frac{X_m}{2}\sin 2\phi - 2X_m\cos\phi - X_m\left(\frac{\pi}{2}+\phi\right) + 2(S-J)\cos\phi, \tag{17.5}$$

$$B = -3\frac{X_m}{2} + \frac{X_m}{2}\cos 2\phi + 2X_m\sin\phi - 2(S-J)\sin\phi \tag{17.6}$$

$$\phi = \sin^{-1}\left(\frac{X_m - S}{X_m}\right), \tag{17.7}$$

where X_m is the input to the control valve or the controller output (op), S is the deadband plus stickband and J is the slip-jump.

Figure 17.4 shows the graphical solutions of Eq. (17.3) for the concentration control and level loops. The systems are closed-loop stable and thus both the curves, $G_o(i\omega)$ and $-1/N(X_m)$, intersect the negative real axis between 0 and -1 for both cases. It is clear from Fig. 17.4 that there will be limit cycles for both loops because of the intersection of the $-1/N(X_m)$ curve and $G_o(i\omega)$ curve. With the increase of

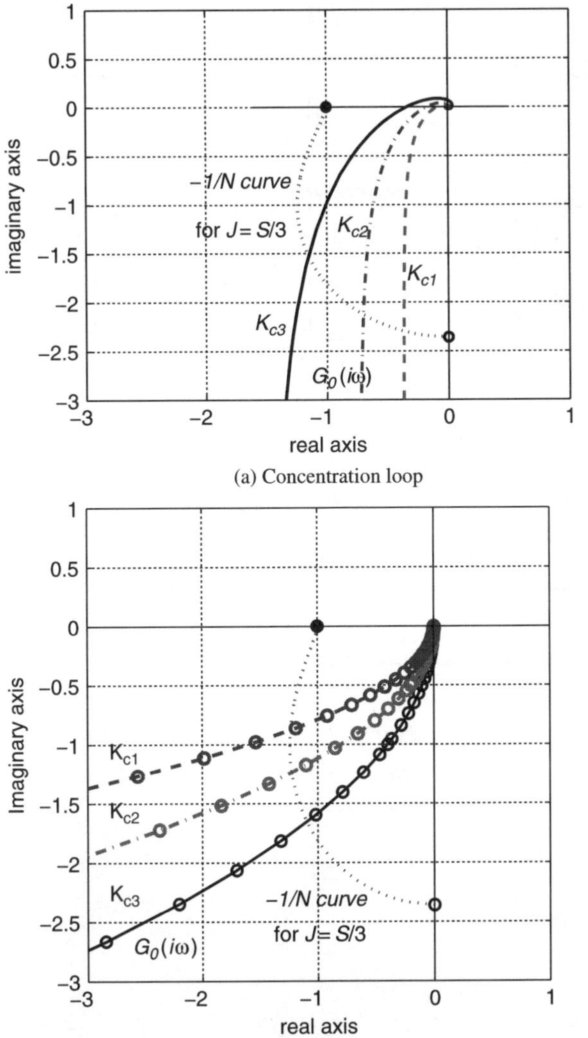

(a) Concentration loop

(b) Level loop

Fig. 17.4 Graphical solutions for limit-cycle oscillations. The *dotted line* is the $-1/N$ curve. The *solid line*, the *dash-dotted line* and the *dashed line* are for the frequency response function of the overall process, $G_o(i\omega)$. Concentration loop: Gains are $K_{c1} = 0.025 < K_{c2} = 0.05 < K_{c3} = 0.1$ and reset time is $\tau_i = 2.5$ s. Level loop: Gains are $K_{c1} = 0.125 < K_{c2} = 0.25 < K_{c3} = 0.5$ and reset time is $\tau_i = 25$ s controller gain, K_c, the intersection point moves to a higher frequency. Thus, the describing function analysis predicts that an increase in controller gain will increase the frequency of limit-cycle oscillation and conversely a reduction in the gain will reduce the frequency of limit-cycle oscillation.

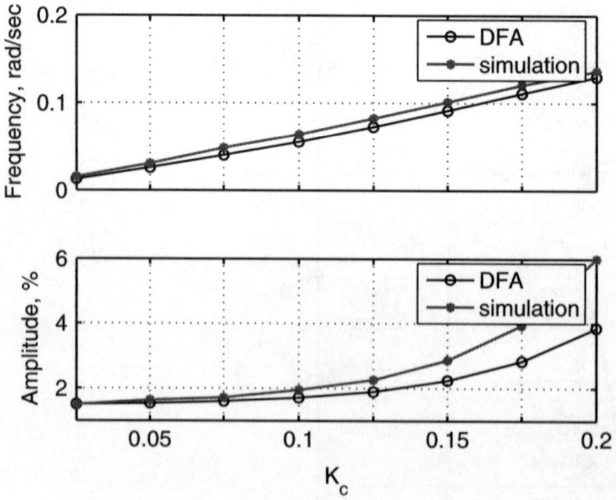

(a) Concentration loop

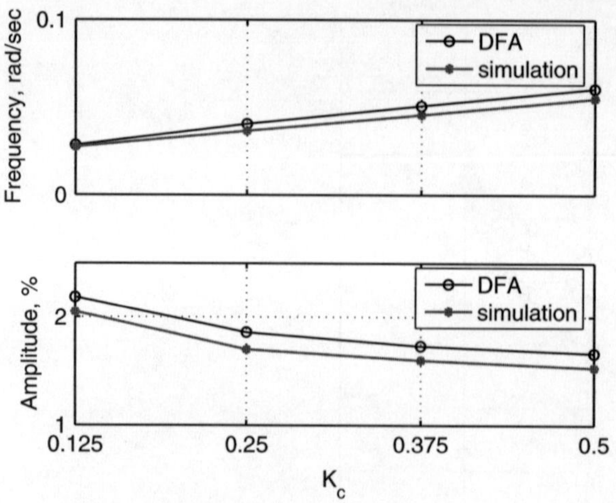

(b) Level loop

Fig. 17.5 Comparison of frequencies and amplitudes of limit-cycle oscillation as predicted by describing function analysis with those obtained from simulation

17.3.1 Comparison of Describing Function Analysis (DFA) Results with Simulation Results

Figure 17.5 compares the frequencies and amplitudes of limit cycles predicted by DFA with those obtained from simulation. Frequencies of limit cycles predicted by DFA are in good agreement with those obtained from simulation for both loops. Amplitudes predicted by DFA for level loop are also very close to those obtained from simulation. However, for the concentration loop, the amplitudes predicted by DFA differs from that obtained from simulation for the cases of higher controller gains. The differences in frequency and amplitudes are probably due to the approximate nature of DFA analysis. The DFA approximates the limit cycles as a pure sinusoid with one fundamental frequency and neglects all other higher harmonics. But in reality, the limit cycles caused by a sticky valve is not a pure sinusoid and its shape changes with the change of controller gains.

17.4 Industrial Example

A case study at Eastman Chemical Company identified a suspected deadband in a level control valve (Thornhill et al., 2003b). The presence of a stiction was confirmed in closed loop by doubling the controller gain following the reasoning given in the previous paragraph. Figure 17.6 shows the plant test data courtesy of Eastman Chemical Company and a simulation of doubling of the gain using the physics-based valve stiction model from Chap. 12. The simulation showed similar waveforms to the industrial data and also a decrease in period and amplitude when the controller gain was doubled.

A feature of interest in the test and simulation of the level control valve is that the shapes of the curves are different from those of Fig. 12.3 for the FOPTD process where *op* was a triangular waveform. *op* is controller output or valve demand, *pv* is the controlled variable or process output, and *mv* is the valve position or valve output. In Fig. 17.6, in contrast, the *pv* is a triangular waveform and *op* has a curved shape. The explanation is that the process is an integrator. In both cases, the *mv* (the flow through the sticking valve) has a square-shaped waveform because of its tendency to stick. In Fig. 12.3 for the FOPTD system, the *pv* (not shown) also has a square-shaped waveform similar to that of the *mv*. The square-shaped *pv* was integrated by the PI controller to become a ramp-shaped *op*.

In Fig. 17.6, by contrast, the controlled system was an integrator and the controller also had an integrator. Therefore the *pv* is ramp-like because it is an integrated square waveform. The *op* is a series of parabolic segments since the result of integration of a ramp is a parabola.

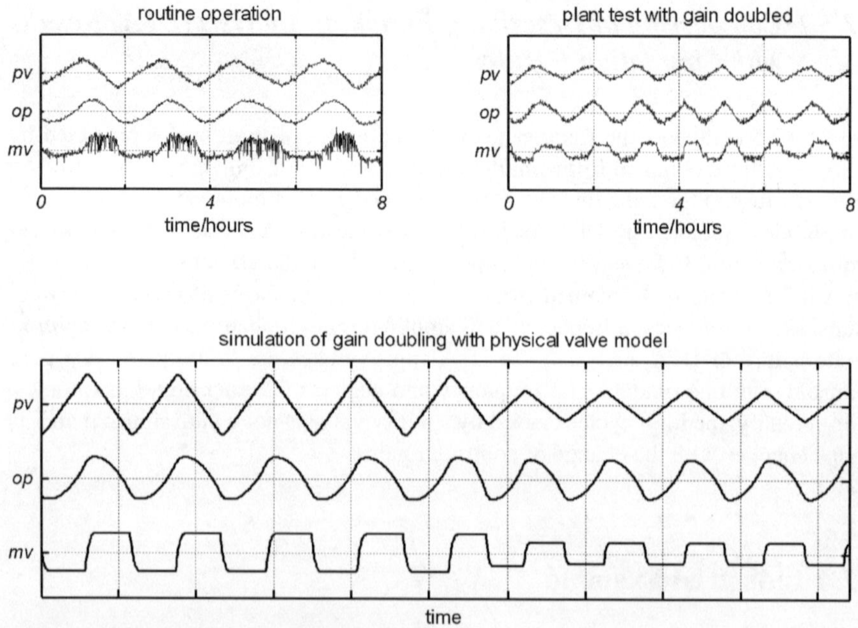

Fig. 17.6 Response to a doubling of controller gain of a level control loop with a sticking valve. *Upper panels*: Plant tests at Eastman Chemical Company, *Lower panel*: Results from simulation

17.5 Summary

A simple closed-loop test for confirming the presence of stiction in control valves has been presented in this chapter. The test is based on the observation that changing controller gain also changes the frequency of the oscillations induced in the control loop due to valve stiction. The method has been evaluated using simulation example and validated by describing function analysis. An industrial case study has been described to illustrate the method. Another industrial example is discussed in Chap. 19.

Part V
Plant-wide Oscillations – Detection and Diagnosis

Chapter 18
Detection of Plantwide Oscillations

Oscillations are a common form of plant-wide disturbances. The presence of oscillations in a control loop increases the variability of the process variables, thus causing inferior quality products, larger rejection rates, increased energy consumption and reduced average throughput. The high level of energy integration in most modern plants is responsible for the propagation of oscillations from one process unit to another. It is important to detect and diagnose the causes of oscillations in a chemical process because a plant running close to product quality limits or operating constraints is more profitable than a plant that has to back away due to oscillations (Martin et al., 1991; Shunta, 1995). This chapter focusses on the detection of oscillations in a single time series as well as in multiple time series.

18.1 Introduction

The well established single-input-single-output (SISO) approach for control-loop performance assessment (CLPA) and benchmarking (Harris, 1989; Huang and Shah, 1999; Desborough and Miller, 2002; Jelali, 2006) has the shortcoming of not considering the connectivity of the control loops. In reality, control loops are not isolated from one another. Specifically, the reason for poor performance in one control loop might be that it is being upset by a disturbance originating elsewhere.

The basic objective in process control is to divert process variability away from key process variables to variables that can accommodate the variability such as buffer tanks and plant utilities (Luyben, et al., 1999). Unfortunately, process variability is often *not* easy to divert. The reason for this is that modern industrial processes have reduced inventory and make use of recycle streams and heat integration. The interactions are strong in such processes because the amount of buffer capacity is small and the opportunities to exchange heat energy with plant utilities are restricted.

A plant-wide approach means that the distribution or the reach of a disturbance throughout the plant has to be explored, and the location and nature of the cause of the disturbance are determined with a high probability of being right the first time. The alternative is a time-consuming procedure of testing each control loop in turn until the root cause is found. Some key requirements for the detection and isolation of plant-wide oscillations are (Qin, 1998; Paulonis and Cox, 2003):

- Detection of the presence of one or more periodic oscillations;
- Detection of non-periodic disturbances and plant upsets;
- Determination of the locations of the various oscillations/disturbances in the plant and their most likely root causes.

A wish-list from Desborough and Miller (2002) included:

- Automated, non-invasive stick-slip detection in control valves;
- Facility-wide approaches including behaviour clustering;
- Automated model-free causal analysis;
- Incorporation of process knowledge such as the role of each controller.

18.2 What is an Oscillation?

If a signal is periodic with well-defined amplitude and frequency, it is called an oscillatory signal. A sinusoidal signal is an example of a simple oscillatory signal. For example, $y(t) = A\sin(\omega t)$ is an oscillatory signal with amplitude A and angular frequency ω.

18.2.1 Units of Frequency

For a continuous time signal, the usual unit of frequency is Hz, which is equivalent to cycles per second or the inverse of the period of the signal expressed in seconds. To show a signal in terms of its angular movement, it may be expressed in rad/s. To obtain the angular frequency in units of rad/s from a given frequency in Hz, the given quantity should be simply multiplied by 2π.

For a discrete time signal, the frequency is often expressed as cycles/sample. To express this in cycles/s, it has to be divided by the sampling intervals or sampling time. It is also convenient to normalize the frequency of a signal by dividing it by the sampling frequency because, in such a case, one can simply obtain the period of the signal from the inverse of the normalized frequency. For example, if the frequency of a signal, f, is 0.01 cycles/s and the sampling frequency, f_s, is 1 sample/s, then the normalized frequency is 0.01. The period of the signal will be 10 samples/cycle ($1/0.01 = 10$).

18.2.2 Examples of Oscillatory Signals

Probably the simplest example of an oscillatory signal is a sine wave. Other examples include saw-tooth signals, rectangular signals and a sum of sines. Figure 18.1

shows four different kinds of oscillatory signals. Oscillations are clearly visible in these signals because they are clean and not corrupted by noise. In reality, oscillations may not be clearly visible from the time domain data. The power spectrum, a frequency domain tool, often proves to be useful in detecting and visualizing oscillation(s) in signals.

18.3 Detection of Oscillation(s) in a Single Time Series

18.3.1 The Power Spectrum

The power spectrum is an useful tool to visualize the energy or strength of various frequency components present in a signal. Figure 18.2 shows the power spectrum of the signals shown in Fig. 18.1. Various frequency components of the signals appear as peaks in the power spectrum.

18.3.2 Hägglund's IAE Method

The classic SISO real-time oscillation-detection method of Hägglund (1995) calculates the integrated absolute deviation (IAE) between successive zero crossings of the controller error signal. Its motivation is that when the controller error is oscillatory rather than random, such deviations are large and the intervals between them are long. An oscillatory signal therefore has larger IAE values than a random one. The IAE is defined by the following expression:

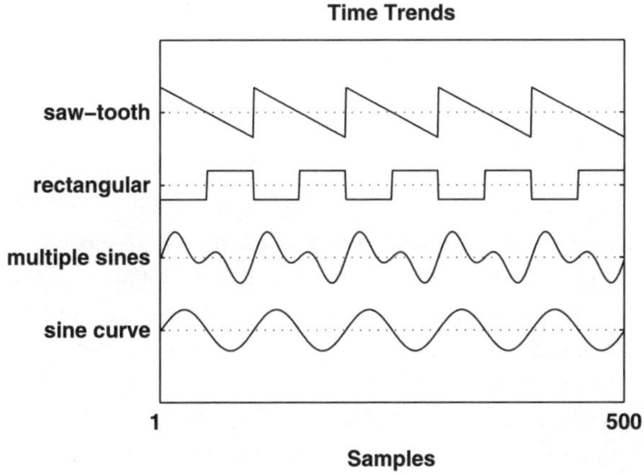

Fig. 18.1 Various oscillatory signals

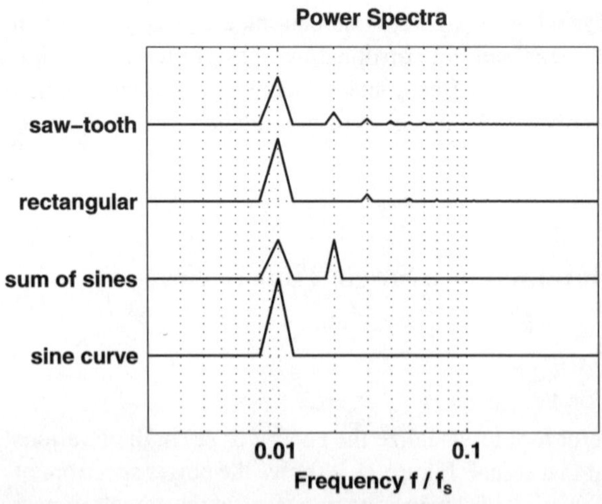

Fig. 18.2 Power spectra of various oscillatory signals

$$IAE_i = \int_{t_i}^{t_{i+1}} |Y(t)|\,dt,$$

where $Y(t)$ is the controller error signal and t_i and t_{i+1} are times of successive zero crossings of $Y(t)$. These integrated deviations are compared to a threshold value, which is the *IAE* value of a sinusoidal oscillation having an amplitude a and with a presumed oscillation having angular frequency equal to $2\pi/\tau_i$, where τ_i is the controller integration time. An oscillation is detected when the integrated deviations persistently exceed the threshold, with a set to 1% of the controller range.

18.3.3 Autocovariance (ACF) Based Method

18.3.3.1 Period of Oscillation or Frequency?

In the detection of plantwide oscillations, one is often interested in the period of oscillation rather than its frequency for the pragmatic reason that it is easier to visualize and talk about a 30 min oscillation than an oscillation of 0.00056 Hz (for example). It is therefore desirable to estimate the period of oscillation accurately.

Zero crossings allow a direct estimation of the oscillation period T_p, which is more accurate than an estimate determined from taking the reciprocal of the frequency of the main peak in the power spectrum. The heart of the matter is that Δf errors in determining the exact position of a spectral peak convert to ΔT_p errors in the estimate of the period of oscillation, which is large at low frequencies. *Low frequency* in this context broadly means oscillations characterized by 10 samples/

cycle or more. The fundamental reasoning behind this statement can be seen from the propagation of errors from Δf to ΔT_p:

$$T_p = \frac{1}{f} \Rightarrow \Delta T_p = -\frac{1}{f^2}\Delta f.$$

For low-frequency oscillations, the ΔT_p uncertainty from inversion of the estimated spectral peak frequency is much larger than the uncertainty arising when the period is determined directly from zero crossings.[1]

18.3.3.2 The Autocovariance Function (ACF)

It is recommended to use the zero crossings of the autocovariance function (ACF) for determination of the period of oscillation. The ACF of an oscillating signal is itself oscillatory with the same period as the oscillation in the time trend. Oscillations can therefore be detected if the zero crossings of the autocovariance function are regular. The benefit of using the ACF is that it is less noisy than the time trend. For instance, in the case of white noise all the noise appears in the zero lag channel. Therefore noise does not disrupt the zero crossing of the autocovariance function in the same way as it disrupts zero crossings in the time domain, and the pattern of zero crossings reveals the presence of an oscillation more clearly than the zero crossings of the time trend. The phenomenon is illustrated in Fig. 18.3.

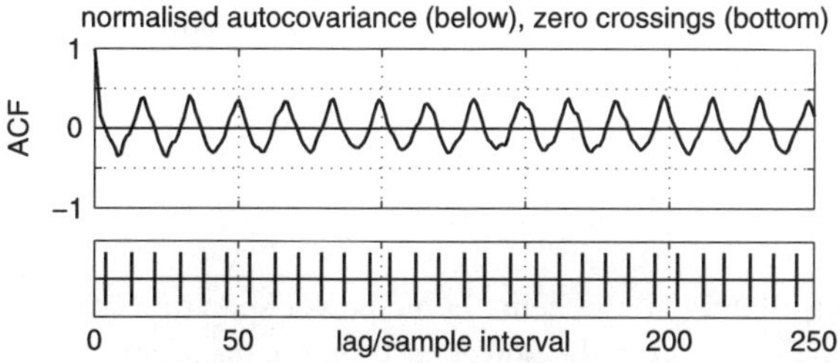

Fig. 18.3 Oscillation detection in the presence of noise. *Upper plot*: portion of time trend. *Middle*: autocovariance function. *Bottom*: regularly spaced zero crossings

[1] The converse is also true. It is generally more accurate to estimate the frequency from the position of the spectral peak than to calculate the reciprocal of the oscillation period, especially for high-frequency oscillations when the period of oscillation is small.

The horizontal axis of the autocovariance functions is a linear axis indicating lag, and an estimate of the autocovariance at lag ℓ may be determined directly from the data as

$$ACF(\ell) = \frac{1}{N-\ell-1}\sum_{i=\ell+1}^{N} \tilde{x}(i)\tilde{x}(i-\ell),$$

where $\tilde{x}$ is the mean-centred data scaled to unit standard deviation. The Wiener–Khinchin theorem shows that the ACF is the inverse Fourier transform of the two-sided power spectrum (see, e.g. Proakis and Manolakis, 1996), so the ACF may also be computed via the power spectrum.

18.3.3.3 Assessment of Period and Regularity of Oscillation

Each oscillation has two zero crossings and hence the average interval between zero crossings is:

$$interval = \frac{1}{2}\left(\overline{T_p} \pm \Delta T_p\right)$$

where $\overline{T_p}$ is the mean period and ΔT_p a random variation in the period. Thus $\overline{T_p}$ is twice the mean value of the intervals, and the standard deviation of the period is $\sigma_{T_p} = 2 \times \sigma_{intervals}$, where $\sigma_{intervals}$ is the standard deviation of the intervals.

The interval from lag zero up to the first zero crossing is excluded from the calculation because it corresponds to only one half of a completed deviation. An oscillation is considered to be regular if the standard deviation of the period is less than one third of the mean value. A regularity statistic was discussed and defined in Thornhill et al., (2003a) such that values of $r > 1$ indicate a regular oscillation with a well-defined period:

$$r = \frac{1}{3} \times \frac{\overline{T_p}}{\sigma_{T_p}}.$$

The statistic r has an interpretation as a three-sigma test of the null hypothesis that the intervals between zero crossings are random with an equal probability of arrival in each sampling interval.

18.3.3.4 Multiple Oscillations

It is easy to detect multiple oscillatory components in the power spectrum because they appear as multiple peaks. For instance, the power spectrum in the second panel of Fig. 18.4 shows that the time trend in the top panel consists of two superimposed oscillations. The two oscillations can be seen in the autocovariance function, but their superposition destroys the regularity of the zero crossings, as shown in the bottom panel of Fig. 18.4.

The ACF may be used for estimating the periods of multiple oscillations if the interfering spectral peaks are removed from the power spectrum before calculation

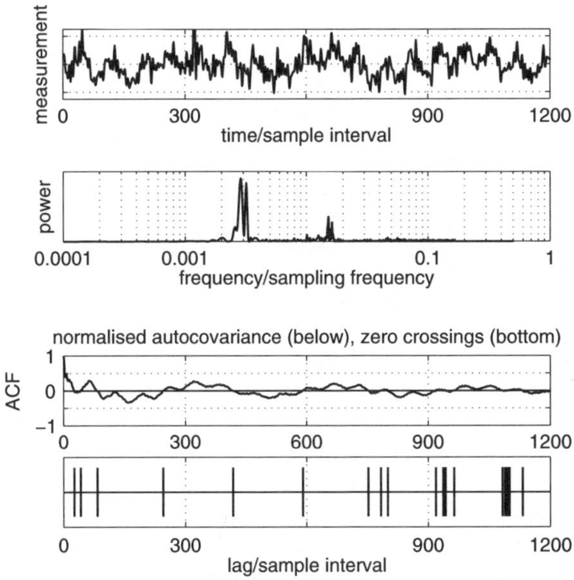

Fig. 18.4 Multiple oscillations. *Upper plots*: time trend and power spectrum. *Lower plots*: autocovariance function and zero crossings

of the ACF. Figure 18.5 shows the effect of the removal of the spectral content above 0.01 on the normalized frequency axis. Frequency content above 0.01 has been removed leaving only the left-hand peak. The period of oscillation can be determined from zero crossings of the ACF as 344 ± 3.6 samples/cycle. In Fig. 18.6, the right-hand spectral peak has been isolated. The oscillation has a period of 64.3 ± 3.1 samples/cycle.

18.3.3.5 Implementation of the Frequency Domain Filter

The filter that removes unwanted spectral content operates by re-setting spectral power in all the unwanted channels to zero. The ACF is then calculated from the filtered power spectrum.

Care is needed in coding of the filter because the ACF calculation requires the two-sided power spectrum to use frequency channels from 0 to $N-1$. The two-sided spectrum includes the aliased part of the spectrum at frequencies above the Nyquist frequency, which resides between 0.5 and 1 on the normalized frequency axis. The filter must remove the unwanted peaks and *their aliases*. Further details and a worked example can be found in Thornhill et al., (2003a).

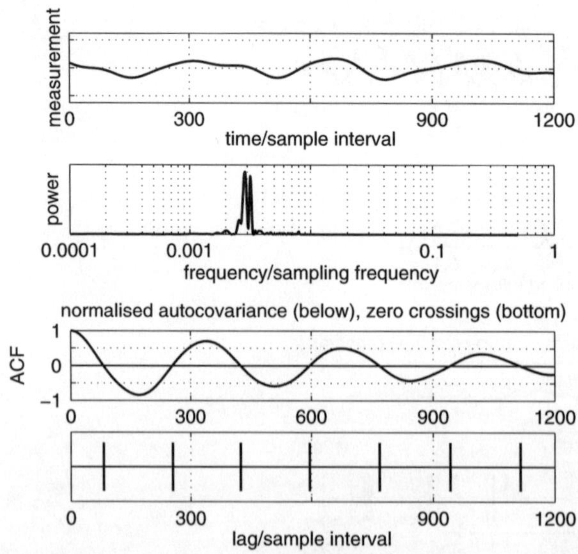

Fig. 18.5 Isolation via signal filtering and ACF analysis of the low-frequency oscillation in Fig. 18.4

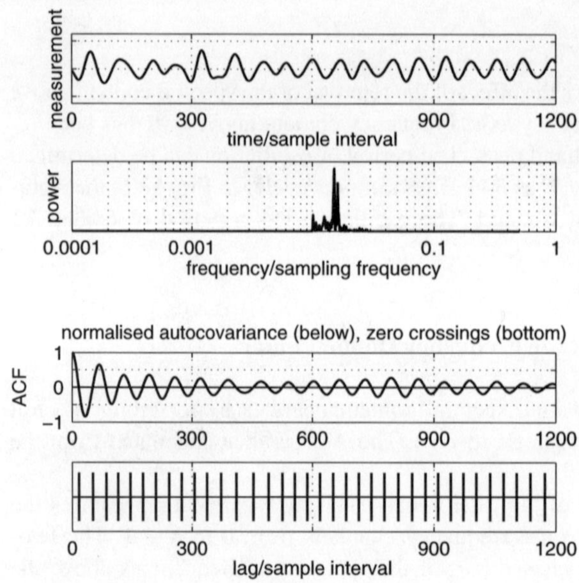

Fig. 18.6 Isolation via signal filtering and ACF analysis of the high-frequency oscillation in Fig. 18.4

18.3.4 Other Methods

Other methods for oscillation detection (e.g. Forsman and Stattin, 1999; Miao and Seborg, 1999; Thornhill and Hägglund, 1997) are off-line auditing methods and exploit the advantages of working with historical data using the time domain, auto-covariance function (ACF) or spectral peak detection. They achieve the detection of an oscillation in one measurement at a time; however, more than that is needed for the detection of common oscillation(s) in multiple time series for the detection of plant-wide oscillations.

18.4 What are Plant-wide Oscillations?

When an oscillation is generated somewhere in the plant and propagates throughout the whole plant or to some units of the plant, such oscillations are termed plant- or unit-wide oscillations. An oscillation may propagate through many units because of the tight heat and mass integration in the plant as well as the presence of recycle streams. Figure 18.9 shows an example of a plant-wide oscillation problem. The upper-left panel of Fig. 18.9 shows the time trends of 37 variables representing a plant-wide oscillation problem in a SE Asian refinery. The upper-right panel of Fig. 18.9 shows the power spectra of these variables, indicating the presence of an oscillation with a frequency of 0.06 or approximately 17 samples/cycle. The presence of such plant-wide oscillations causes significant performance degradation in the overall plant economy. Therefore, they must be detected, isolated and eliminated to ensure a good overall profitability of the plant.

18.5 Classification of Plant-wide Oscillations or Disturbances

18.5.1 Time scales

The first distinction in a classification of plant-wide disturbances concerns the time scale, which may be (a) slowly developing, e.g. catalyst degradation or fouling of a heat exchanger, (b) persistent and dynamic and (c) abrupt, e.g. a compressor trip. The focus in this book is on (b), dynamic disturbances that persist over a time horizon of hours to days. The approach is typically one of process auditing in which a historical data set is analysed off-line. The off-line approach gives opportunities for advanced signal analysis methods such as integral transforms and non-causal filtering.

18.5.2 Oscillating and Non-oscillating Disturbances

Figure 18.7 shows a family tree of methods for the detection of plant-wide disturbances that was first presented by Thornhill and Horch (2007). The main subdivision is between oscillating and non-oscillating behaviours. An oscillation significant enough to cause a process disturbance can be seen both in the time domain and as a peak in the frequency domain, suggesting that either might be exploited for detection. The time trends at measurement points affected by a non-oscillating disturbance, by contrast, often look somehow similar but in a way that is hard to characterize because of the multiple frequencies that are present. The frequency domain readily reveals the similarities in the spectral content, however, and therefore spectra are useful for the detection of non-oscillating disturbances. Some dynamic disturbances are not stationary. For instance, an oscillation may come and go or may change in magnitude. This localization in time suggests that wavelet methods would be best for such cases.

Plant-wide oscillation detection requires the recognition that an oscillation in one measurement is the same as the oscillation in another measurement, even though the shape of the waveform may differ and when interferences such as other oscillations are present. A characterization and clustering step is needed in addition to oscillation detection. Over the last few years, many studies have been carried out to detect plant-wide oscillations (Thornhill et al., 2003a; Tangirala et al., 2005; Jiang et al., 2006) and to group similar oscillations together. The following is a brief description of some of these methods.

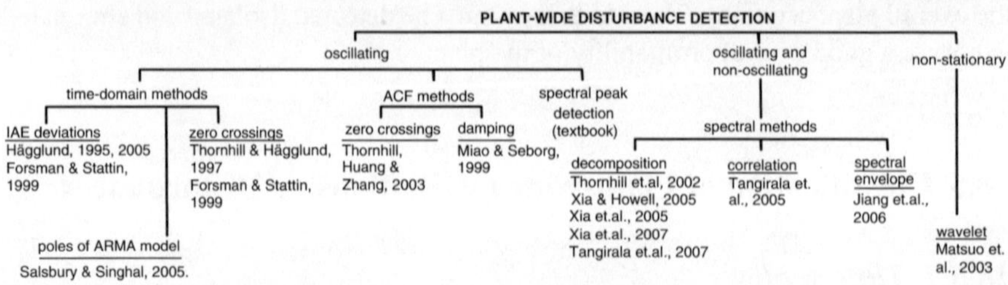

Fig. 18.7 Family tree of methods for data-driven plant-wide disturbance detection

18.6 Detection of Plant-wide Oscillations

18.6.1 High-Density Plots

High-density plots provide an off-line visualization tool. This plot presents time series data of all the concerned tags or variables and their spectra in a compact form. From this plot, one may easily visualize the nature of the data and common oscillation(s) present in the data. However, this method has not been automated, and therefore it does not provide automatically a list of the variables that oscillate together. Figure 18.9(a) is an example of a high-density plot.

18.6.2 ACF-Based Method

Thornhill et al., (2003a) automated the detection of clusters of similar oscillations using the ACF. The mean value of the intervals between zero crossings of the ACF gave the period of oscillation, while the regularity of the oscillation was also assessed using the standard deviation of the intervals. A fully automated agglomerative classification algorithm detects clusters of measurements oscillating with the same frequency and issues a report. All comparisons are referenced to the period and standard deviation of the most regular oscillation in the group, and a measurement belongs to a group if its period is no more different from that of the most regular oscillation in the group by less than one standard deviation.

18.6.3 Power Spectral Correlation Map (PSCMAP)

The power spectral correlation index (PSCI) is defined as the correlation between the power spectra of two different measurements (Tangirala et al. (2005)). It is a measure of the similarity of spectral shapes, i.e. a measure of the common frequencies of oscillations in all variables. The procedure to calculate the correlation is illustrated in the block diagram shown in Fig. 18.8.

The DFTs used to calculate the spectrum are calculated after removal of means from the time series data. However, the correlation used in the calculation of PSCI is calculated without the removal of mean of the spectra. The PSCI for any two spectra $|X_i(\omega)|^2$ and $|X_j(\omega)|^2$ is calculated as

$$correlation(|X_i(\omega)|^2, |X_j(\omega)|^2) = \frac{\sum_{\omega_k} |X_i(\omega_k)|^2 |X_j(\omega_k)|^2}{\sqrt{|X_i(\omega_k)|^4 |X_j(\omega_k)|^4}} \tag{18.1}$$

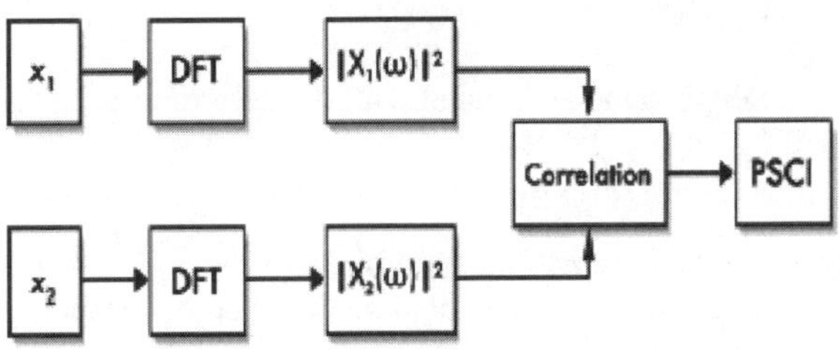

Fig. 18.8 The procedure for calculation of PSCI using discrete Fourier transforms or DFTs

As a result, PSCI always lies between 0 and 1. As shown in the schematic, the phase information is excluded because only the magnitude of the DFT of the measurements are considered. In the detection of plant-wide oscillations, the objective is to collect variables with similar oscillatory behaviour.

For multivariate processes, the PSCI is a matrix of size $m \times m$, where m is the number of measured variables. In order to provide an effective interpretation of the PSCI, the matrix is plotted as a colour map, which is termed as the power spectral correlation map (PSCMAP). The intensity as well as the colour in the map is assigned in proportion to the value of the correlation index. An important aspect of this colour map is its ability to automatically re-arrange and group variables together with similar shapes, i.e. variables, which oscillate at a common frequency and have therefore similar values of PSCI. Tangirala et al. (2005) have provided a detailed description of the method.

18.6.4 Spectral Envelope Method

The spectral envelope method is a frequency domain technique that was first introduced by Stoffer et al. (1993) to explore the periodic nature of categorical time series. In 1997, McDougall et al. (1997) extended the concept of spectral envelope to real-valued series. The key idea is to select optimal transformations of a real-valued series that emphasize any periodic nature in the frequency domain.

Let $\mathbf{X}$ be a data matrix of dimension $n \times m$, where n is the number of samples and m the number of variables. If the covariance matrix of $\mathbf{X}$ is $\mathbf{V_X}$ and the power spectral density (PSD) matrix of $\mathbf{X}$ is $\mathbf{P_X}(\omega)$, then the spectral envelope of $\mathbf{X}$ is defined as:

$$\lambda(\omega) \triangleq \sup_{\beta \neq 0}\{\frac{\beta^* \mathbf{P_X}(\omega)\beta}{\beta^* \mathbf{V_X}\beta}\}, \tag{18.2}$$

where ω represents angular frequency $\lambda(\omega)$ is the spectral envelope at angular frequency ω, $\beta(\omega)$ is the optimal scaling vector that maximizes the power (or variance) at the frequency ω and the '*' represents conjugate transpose. The optimal scaling vector $\beta(\omega)$ is different for each frequency, ω.

With the optimal scaling vector $\beta(\omega)$, Eq. (18.2) can be rewritten as:

$$\lambda(\omega)\mathbf{V_X}\beta(\omega) = \mathbf{P_X}(\omega)\beta(\omega) \tag{18.3}$$

It follows that $\lambda(\omega)$ is the largest eigenvalue associated with the determinant equation:

$$|\mathbf{P_X}(\omega) - \lambda(\omega)\mathbf{V_X}| = 0 \tag{18.4}$$

$\beta(\omega)$ is the corresponding eigenvector satisfying Eq. (18.3).

A simplified version of the spectral envelope can also be obtained by replacing $\mathbf{V_X}$ with $\mathbf{V}$ in the previous equations, where $\mathbf{V} = \mathbf{diag}\{\mathbf{V_X}\}$.

It is common to limit β to the constraint $\beta^* \mathbf{V_X} \beta = 1$ or $\beta^* \mathbf{V} \beta = 1$, which makes the calculated spectral envelope more interpretable and explicable. For such cases, the quantity $\lambda(\omega)$ represents the largest power (variance) that can be obtained at the frequency ω for any scaled series with unit variance.

18.6.4.1 Using the Spectral Envelope to Detect Plant-wide Oscillation

Jiang et al. (2006) described the viability of the spectral envelope for detecting common oscillation(s). If the univariate time series $x_i(t)$, $1 \leq i \leq m$, $-\infty < t < \infty$, is in the form of a common signal plus independent white noise, say $x_i(t) = s(t) + \varepsilon_i(t)$, then in terms of power spectra, $P_{x_i}(\omega) = P_s(\omega) + \sigma_\varepsilon{}^2$, where $var(\varepsilon_i) = \sigma_\varepsilon{}^2$. A simple linear combination of $x_i(t)$, say $\bar{x}(t) = m^{-1} \sum_{i=1}^{m} x_i(t)$, will have as its power spectrum, $P_{\bar{x}}(\omega) = P_s(\omega) + m^{-1} \sigma^2{}_\varepsilon$. The signal-to-noise ratio of $\bar{x}(t)$ has increased by a factor of m over the individual $x_i(t)$. Therefore, this indicates that the right linear combination of the original time series will enhance the signal and attenuate the noise. The spectral envelope method selects the optimal linear combination that can enhance the signal spectra and dampen the noise spectra at each frequency ω. This feature makes the spectral envelope particularly suitable for analysing noise-corrupted data.

18.6.5 Spectral Decomposition Methods

18.6.5.1 Introduction

Spectral decomposition methods can also be used to detect and classify spectral features in multivariate data sets. The basic decomposition method is spectral principal component analysis (spectral PCA). The m rows of the data matrix $\mathbf{X}$ are the power spectra $P(f)$ of the signals from m measurement and a spectral decomposition reconstructs the $\mathbf{X}$ matrix as a sum over p basis functions $\mathbf{w}'_1$ to $\mathbf{w}'_p$, which are spectrum-like functions each having N frequency channels arranged as a row vector (Thornhill et al., 2002).

$$\mathbf{X} = \begin{pmatrix} t_{1,1} \\ \dots \\ t_{m,1} \end{pmatrix} \mathbf{w}'_1 + \begin{pmatrix} t_{1,2} \\ \dots \\ t_{m,2} \end{pmatrix} \mathbf{w}'_2 + \dots + \begin{pmatrix} t_{1,p} \\ \dots \\ t_{m,p} \end{pmatrix} \mathbf{w}'_p + \mathbf{E}$$

The error matrix $\mathbf{E}$ contains the residual parts of the spectra that are not captured by the p basis functions, while a suitable value for p is assessed by a standard method such as the average eigenvalue criterion or other methods reviewed by Valle et al. (1999). Figure 18.9 shows the time trends and spectra of the error signals (*sp–pv*) in the SE Asian refinery data set that was introduced in Sect. 18.4 and a set of seven basis functions from which the spectra in the rows of $\mathbf{X}$ can be approximately reconstructed. In spectral PCA, the basis functions are orthogonal with peaks and/or

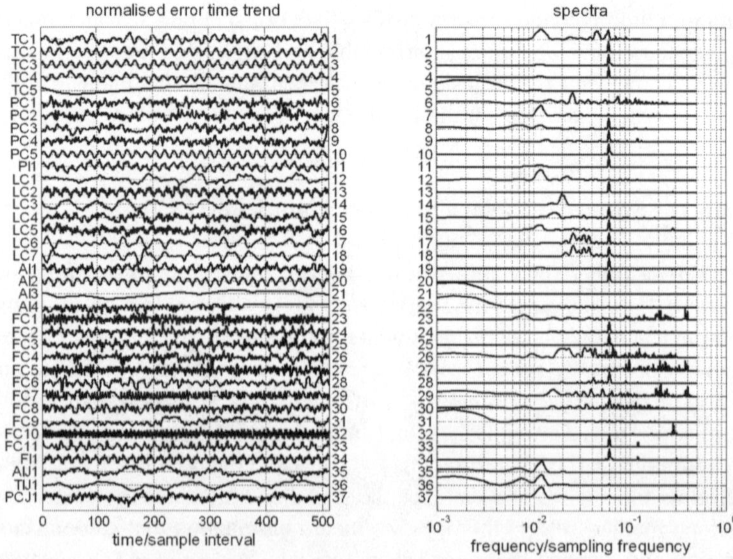

(a)

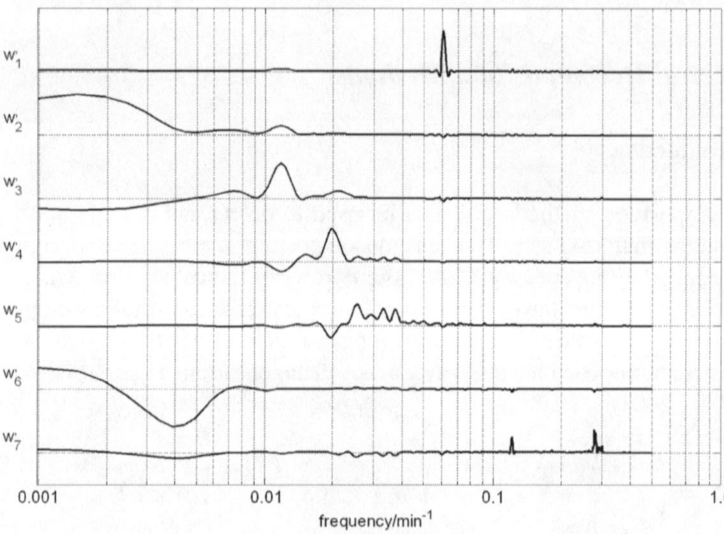

(b)

Fig. 18.9 *(a)* time trends and spectra of the error signals (*sp–pv*) in the SE Asian refinery data set. *(b)* A set of seven spectral PCA basis functions from which the spectra can be approximately reconstructed valleys at one or more frequencies. Some of the basis functions have negative spectral values. It is for this reason that several researchers have imposed the condition of non-negativity in SPCA analysis (Tangirala et al., 2007; Xia et al., 2007; Lee and Seung, 1999). This aspect is discussed in more detail in Sect. 18.6.5.5.

The spectrum in the ith row of $\mathbf{X}$ is characterized by its coordinates $\mathbf{t'}_i = (t_{i,1}, t_{i,2}, \ldots t_{i,p})$ in the spectral PCA score space. Figure 18.10 shows the score space for a three-PC decomposition of the spectra in the SE Asia data set. The scores indicate the weighting for each basis function in the reconstruction of the spectra and can be used for clustering because similar spectra have similar t-coordinates. The plant-wide oscillation detection problem requires (i) a suitable distance measure by which to detect similarity, and (ii) determination and visualization of clusters of measurements with similar spectra in a multidimensional space.

Section 18.6.5.2 discusses a suitable measure while Sect. 18.6.5.3 shows how a hierarchical tree can be used to visualize the clusters in a high-dimensional space (i.e. when $p > 3$). A generalization of power spectral correlation (Sect. 18.6.5.4) gives an alternative means of visualization of the spectral clusters within a data set.

Spectral independent component analysis and spectral non-negative matrix factorization are post-processing methods that give the basis functions more insightful properties than those of spectral PCA for the purposes of diagnosis. They aim to find linear combinations such that each basis function represents just one spectral peak. These methods are described in Sect. 18.6.5.5.

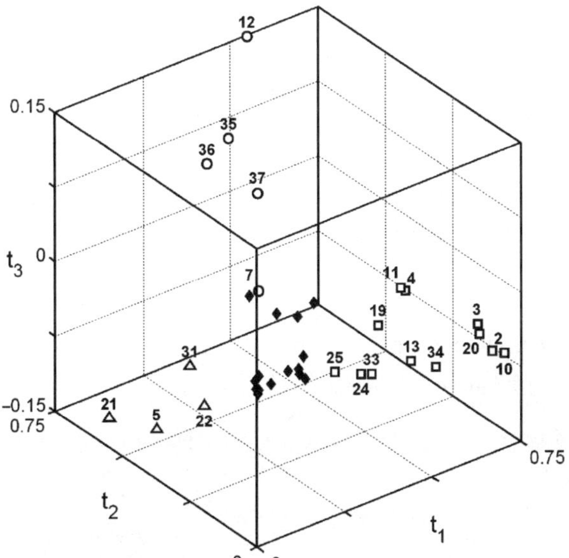

Fig. 18.10 Results from spectral PCA represented as a three-PC score plot. Major clusters (*circles, triangles* and *squares*) form plumes radiating from the origin. Many tags remain undifferentiated near the origin when using only three PCs (*small black diamonds*)

18.6.5.2 A Distance Measure for Clustering

The vector $\mathbf{t}'_i = (t_{i,1},\ t_{i,2},\ \dots\ t_{i,p})$ for the ith spectrum represents a vector joining the origin to a position or spot in a p-dimensional spectral PCA score space. In spectral PCA, clusters frequently take the form of plumes radiating from the origin, as can be seen in Fig. 18.10 where the clusters represented by triangles, squares and circles each have the shape of an elongated plume. The physical interpretation of a plume can be understood with reference to the white squares in Fig. 18.10, where the plume extends along the $\mathbf{t}_1$-axis of the score plot and the $\mathbf{t}_2$ and $\mathbf{t}_3$ values are close to zero. It means that the spectra in this cluster closely resemble the $\mathbf{w}'_1$ basis function, which has a single spectral peak corresponding to an oscillation with a period of 16.7 min. Their position along the $\mathbf{t}_1$-axis indicates their strength. Tags 2, 3, 10 and 20 with strong, well-defined oscillations are far out towards the end of the plume while tags 24 and 25 in which the oscillation is much more noisy are closer to the origin.

Figure 18.10 is a three-PC representation of the score space, but these three basis vectors may not be enough to accurately reconstruct the spectra in the rows of $\mathbf{X}$. What is needed is a means of visualization of the structure of the score space of arbitrary dimension. An angular measure (Duda et al., 2000) is well suited for finding plume-shaped clusters in any dimensions. It uses $\theta_{i,j}$, the angle between $\mathbf{t}'_i$ and $\mathbf{t}'_j$ determined through calculation of the scalar product:

$$\cos(\theta_{i,j}) = \frac{\mathbf{t}'_i\,(\mathbf{t}'_j)^{\mathbf{T}}}{|\mathbf{t}'_i|\,|\mathbf{t}'_j|},$$

where

$$\mathbf{t}'_i\,(\mathbf{t}'_j)^{\mathbf{T}} = \sum_{k=1}^{p} t_{i,k}t_{j,k} \text{ and } |\mathbf{t}'_i| = \sqrt{\sum_{k=1}^{p} t_{i,k}^2}.$$

Two $\mathbf{t}'$ vectors joining the origin to spots in the score plot have a small angular separation when they are roughly parallel. An algorithm for the determination of a matrix of $\theta_{i,j}$ values is given below:

Algorithm: Calculation of the angle measures (from Thornhill et. al., 2006)

Step 1: Create a normalized matrix $\tilde{\mathbf{T}}$ from $\mathbf{T}_p$ whose row vectors are of unit length. Each element in the ith row of $\mathbf{T}_p$ is divided by $\sqrt{\sum_{k=1}^{p} t_{i,k}^2}$, where p is the number of principal components in use.
Step 2: Determine the matrix $C = \tilde{\mathbf{T}}\tilde{\mathbf{T}}^{\mathrm{T}}$. The elements of C are $\cos(\theta_{i,j})$.
Step 3: Create the matrix of angles from $\mathbf{A} = \arccos(C)$, which returns results in the range 0–180°.

18.6.5.3 Finding and Visualizing the Clusters – a Hierarchical Tree

The matrix **A**, whose elements are $\theta_{i,j}$, has to be analysed to find plume-shaped clusters in a PCA score plot. The fully automated agglomerative hierarchical-clustering algorithm is given below. At the core is step 3, which has the following possible outcomes: (i) two items are combined to start a new cluster, (ii) another item is added to a cluster already identified or (iii) two clusters are combined to form a larger cluster.

Algorithm: Agglomerative classification (from Thornhill et al., 2006; based on Chatfield and Collins, 1980)

Step 1: The starting point is the matrix **A** of angular distances with elements $\theta_{i,j}$. A text vector of row and column headings is also defined, which initially is $(1\ 2\ 3\ 4\ 5\ \ldots.)$ to keep track of the items in the data set. For a spectral PCA application the items are the spectra in the data set.
Step 2: At the kth iteration, the smallest non-zero value $\theta_{i,j}$ in the matrix $\mathbf{A}_{k-1}$ from the previous iteration is identified. Its row and column indexes i and j indicate the items with the smallest angular separation, which are to be agglomerated to form a cluster.
Step 3: A smaller matrix $\mathbf{A}_k$ is then generated from $\mathbf{A}_{k-1}$. It does not have rows and columns for the two similar items identified at step 2. Instead, it has one row and column that give the distances of all the other items from the cluster created from the agglomerated items. The distances are $\min\left\{\theta_{i,n}, \theta_{j,n}\right\}$, i.e. the angular distance between the nth item and whichever member of the cluster was closer.
Step 4: The row and column headings are redefined. The heading for the new row created at step 3 indicates the items that have been combined.
Step 5: The results of the kth step are written to a report showing the cluster size, the row heading for the cluster formed at iteration k and the two subclusters within it.
Step 6: Steps 2–5 are repeated until all the items have been clustered.

A graphical representation in the form of a hierarchical tree can be extracted from the report generated by the above algorithm. The tree is a dendogram because it represents the sizes of the clusters on the vertical axis. It utilizes an algorithm that starts at the top and systematically searches down the left and then the right branches and sub-branches to parse the structure of the tree. The algorithm is recursive, meaning it calls itself over and over again in a nested way until it reaches a leaf of the tree. The end result is a set of y-coordinates representing the cluster sizes and tracing the path that joins each individual item on the horizontal axis to the master node at the top of the tree, which are then plotted using a staircase plot.

Algorithm: Path Search (from Thornhill and Melbø, 2006)
At the current node,

Step 1: Search left if the next node to the left is not done
find description of the next node to the left
if the next node to the left is a leaf of the tree
set *label* equal to the item number
mark the path to that leaf as *done*
return (back to the next highest level of recursion)
else if the next node to the left is not done yet
call Path Search (recursive call)
build the path by adding the *y*-coordinate of the node to the path
(the path starts empty)
else
mark the left node to the left as done.

Step 2: Search right if the next node to the right is not done:
find the next node to the right
If the next node to the right is a leaf of the tree
set*label* equal to the item number
mark the path to that leaf as *done*
return (back to the next highest level of recursion)
else if the next node to the right is not done yet
call Path Search (recursive call)
build the path by adding the *y*-coordinate of the node to the path
(the path starts empty)
else
mark the node to the right as *done*
mark the current node as *done*.

Step 3: Plot paths for each leaf as a stairs plot to construct the tree from the leaf of the tree

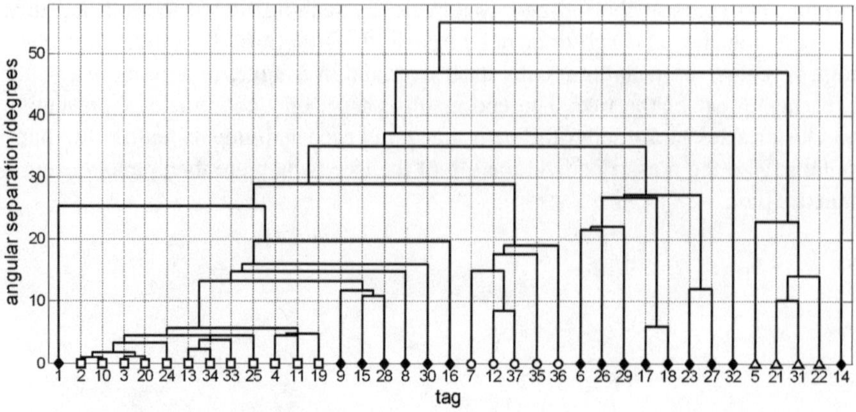

Fig. 18.11 Hierarchical classification tree based on angle measures for a seven-PC model. The same symbols have been used as in the score plot of Fig. 18.10

Figure 18.11 is a hierarchical tree for the SE Asian refinery data set using seven PCs. The vertical axis represents the angular separation of the tags within a cluster. For example, the spectrum of tag 24 joints the subcluster comprising tags 2, 3, 10 and 20 at a node with a y-axis coordinate of about 4°. This means the $\mathbf{t}'$ vectors for the spectra of these five tags all lie within 4° of one another in the 7-dimensional space.

It is easy to recognize clusters visually, for instance, all the tags represented by squares lie together and well separated from the rest of the tree. The overall angular size of the cluster of tags with the square symbols is about 6°, while the next highest node is at 13°. The seven-PC plot shows the relationships among the spectra represented by black diamonds that could not be resolved in the three-PC plot. It is clear that tags 1, 14 and 32 have spectra that are unique in the data set and not similar to any other spectra. It also shows that tags 17 and 18 are closely related.

18.6.5.4 Finding and Visualizing the Clusters – Spectral Correlation Tree

Spectral PCA gives a geometrical insight into the power spectral correlation map introduced in Sect. 18.6.3. The element in the ith row and jth column of the m-by-m power spectral correlation matrix SC is:

$$\mathbf{SC}_{i,j} = \frac{\sum\limits_k X_{ik}X_{kj}}{\sqrt{\sum\limits_k X_{ik}^2 \sum\limits_k X_{kj}^2}}.$$

When all m principal components are retained in the model, the spectral correlation matrix is identical to the matrix $\mathbf{C}$ whose elements are $\cos(\theta_{i,j}) = \frac{\mathbf{t}'_i (\mathbf{t}'_j)^{\mathbf{T}}}{|\mathbf{t}'_i| |\mathbf{t}'_j|}$, as can be demonstrated by writing the spectra in the ith and jth rows of $\mathbf{X}$ in their decomposed form and exploiting the orthonormality of the $\mathbf{w}'$ vectors:

$$\mathbf{SC}_{i,j} = \frac{\sum\limits_{k=1}^{N} X_{ik}X_{kj}}{\sqrt{\sum\limits_{k=1}^{N} X_{ik}^2 \sum\limits_{k=1}^{N} X_{kj}^2}}$$

$$= \frac{(t_{i,1}\mathbf{w}'_1 + t_{i,2}\mathbf{w}'_2 + \ldots t_{i,m}\mathbf{w}'_m)(t_{j,1}\mathbf{w}'_1 + t_{j,2}\mathbf{w}'_2 + \ldots t_{j,m}\mathbf{w}'_m)^{\mathbf{T}}}{\sqrt{(t_{i,1}\mathbf{w}'_1 + \ldots t_{i,m}\mathbf{w}'_m)(t_{i,1}\mathbf{w}'_1 + \ldots t_{i,m}\mathbf{w}'_m)^{\mathbf{T}}}\sqrt{(t_{j,1}\mathbf{w}'_1 + \ldots t_{j,m}\mathbf{w}'_m)(t_{j,1}\mathbf{w}'_1 + \ldots t_{j,m}\mathbf{w}'_m)^{\mathbf{T}}}}$$

$$= \frac{(t_{i,1}\mathbf{w}'_1 + t_{i,2}\mathbf{w}'_2 + \ldots t_{i,m}\mathbf{w}'_m)(t_{j,1}\mathbf{w}'_1 + t_{j,2}\mathbf{w}'_2 + \ldots t_{j,m}\mathbf{w}'_m)^{\mathbf{T}}}{\sqrt{\sum_{\ell=1}^{m} t_{i,\ell}^2}\sqrt{\sum_{\ell=1}^{m} t_{j,\ell}^2}}$$

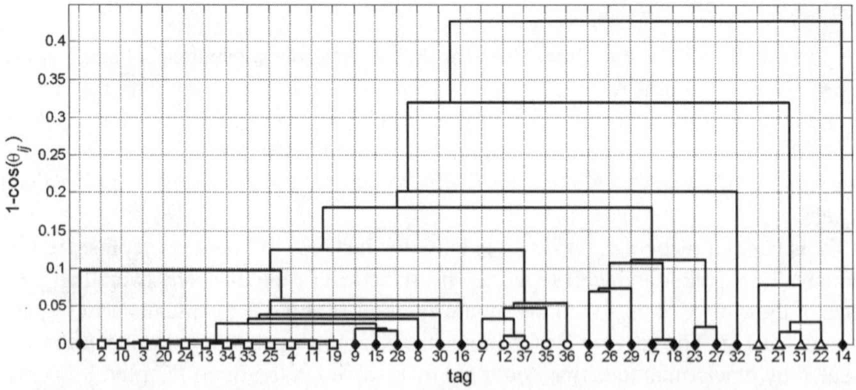

Fig. 18.12 Hierarchical classification tree based on power spectral correlation for a seven-PC model

$$= \frac{\sum_{k=1}^{m} t_{i,k} t_{j,k}}{\sqrt{\sum_{\ell=1}^{m} t_{i,\ell}^2}\sqrt{\sum_{\ell=1}^{m} t_{j,\ell}^2}} = \frac{\mathbf{t'}_i\,(\mathbf{t'}_j)^{\mathbf{T}}}{|\mathbf{t'}_i|\,|\mathbf{t'}_j|}$$

A power spectral correlation can also be defined using p rather than m PCs (where $p < m$) as follows, with $\hat{X}_i = t_{i,1}\mathbf{w'}_1 + t_{i,2}\mathbf{w'}_2 + \cdots t_{i,p}\mathbf{w'}_p$ being the approximation to the ith spectrum reconstructed from p basis vectors. The quantity $\mathbf{SC}_{ij}^{(p)}$ represents the power spectral correlation between the reconstructed vectors without interference from the other spectral content that is discarded in the unused $m - p$ PCs.

$$\mathbf{SC}_{i,j}^{(p)} = \frac{\sum_k \hat{X}_{ik}\hat{X}_{kj}}{\sqrt{\sum_k \hat{X}_{ik}^2 \sum_k \hat{X}_{kj}^2}} = \frac{\sum_{k=1}^{p} t_{i,k} t_{j,k}}{\sqrt{\sum_{\ell=1}^{p} t_{i,\ell}^2}\sqrt{\sum_{\ell=1}^{p} t_{j,\ell}^2}} = \frac{\mathbf{t'}_i\,(\mathbf{t'}_j)^{\mathbf{T}}}{|\mathbf{t'}_i|\,|\mathbf{t'}_j|} = \cos(\theta_{ij})$$

It is convenient in the construction of a hierarchical tree to use $1 - \mathbf{SC}^{(p)}$ to give a distance measure that is zero when two $\mathbf{t'}$ vectors are parallel. The corresponding hierarchical tree is shown in Fig. 18.12. Its advantage is that the nonlinearity in the cos function makes the clusters more sharply defined with longer branches; however the interpretation of the vertical axis scale is perhaps less intuitive.

18.6.5.5 Post-processing of the Basis Vectors – Spectral ICA and NMF

A desirable property of any decomposition of a spectral matrix would be that the various spectral peaks distribute themselves among the basis functions ideally with one peak per basis function. If that were the case, then further analysis would be facilitated including (i) a one-to-one identification of each basis function with a different plant-wide oscillation and (ii) an easy algorithmic analysis of which spectrum contains which spectral peaks, and hence a mapping of the distribution of each plant-wide oscillation.

The basis functions used in spectral PCA to reconstruct the spectra are shown in the lower panel of Fig. 18.9. As observed by Xia (2003), their complicated shapes do not lend themselves to an intuitive interpretation of the character and distribution of plant-wide disturbances. The matrix decompositions proposed to tackle this problem start from the PCA basis functions and apply further transforms to create a better set of basis functions and a new matrix of scores.

An ICA decomposition is $\mathbf{X} = \mathbf{A}\mathbf{S}' + \mathbf{F}$ where the rows of $\mathbf{S}'$ are independent and matrix $\mathbf{F}$ is the error in reconstruction. Independence means that, if s_i is an element of row vector $\mathbf{s}'_i$ and s_j is an element of row vector $\mathbf{s}'_j$, then

$$\Pr(s_i, s_j) = \Pr(s_i)\Pr(s_j),$$

where $\Pr(s_i)$ and $\Pr(s_j)$ are the probability density functions of s_i and s_j and $\Pr(s_i, s_j)$ is the joint probability density function. Hyvarinen and Oja (2000) showed that statistical independence can be approximated when the kurtosis (the fourth moments) of the distributions $\Pr(s_i)$ of the row vectors are maximized and gave an algorithm called FastICA that generates an $\mathbf{S}'$ matrix whose rows have probability density functions with maximized kurtosis. The algorithm requires initialization, which is provided by the $\mathbf{W}'$ matrix from spectral PCA.

Xia (2003) and Xia and Howell (2005) gave the first application of ICA to process spectra and a demonstration of what can be achieved with ICA as a post-processing step following a spectral PCA analysis. ICA produced a useful set of spectral-like basis functions with just one spectral peak in each independent component, and the reasons for this is that kurtosis is maximized when just one spectral peak is present, as proved in Xia et al., (2005).

Figure 18.13 shows the basis functions from spectral ICA. The fact that each ICA basis function tends to have one strong spectral peak aids the characterization of the plant-wide oscillations present in the data set. Moreover, inspection of the rows of the $\mathbf{A}$ matrix indicates at which measurement points the various oscillations can be observed and where they are absent, which aids in the diagnosis of the spread of a disturbance through the plant. A hierarchical tree can be created based on the rows of the $\mathbf{A}$ matrix in the same way as described above for spectral PCA using the $\mathbf{T}$ matrix, but the penalty for cleaning up the basis vectors is that some spectral content which was useful in separating clusters in spectral PCA has been removed and the clusters in the score space tend to be less distinct.

Non-negative matrix factorization (NMF) was introduced in the area of image recognition (Lee and Seung, 1999) where it proved effective in decomposing images into their parts, e.g. noses, eyes and mouths in the case of images of faces. In NMF, every element in every basis function is either positive or zero, making it a good choice for the analysis of power spectra, while the decomposition into parts tends

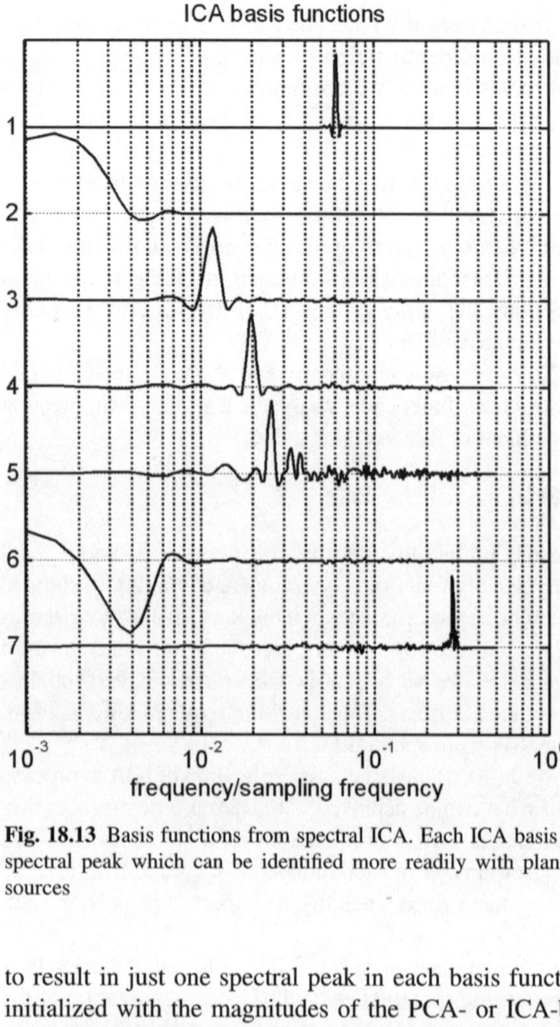

Fig. 18.13 Basis functions from spectral ICA. Each ICA basis function tends to have one strong spectral peak which can be identified more readily with plant-wide oscillations from different sources

to result in just one spectral peak in each basis function. The search algorithm is initialized with the magnitudes of the PCA- or ICA-basis vectors. Studies on the application of NMF to process spectra for plant-wide disturbance analysis have been presented by Tangirala et. al., (2007) and Xia et. al., (2007).

18.7 Summary

This chapter has addressed the topic of oscillations in chemical processes. It described and demonstrated methods for finding oscillations in the time series from a single measurement point, and also for detecting plant-wide oscillations.

Oscillation-detection methods include a zero-crossing technique, which is greatly enhanced by the use of autocovariance functions, as well as spectral methods. The position of a peak in the spectrum indicates the frequency of oscillation, and the spectra are thus useful inputs to algorithms for plant-wide oscillation detection, which compare the spectral peaks present in time series collected from different measurement points around the plant.

The aim of a plant-wide approach is to map out the distribution of an oscillating disturbance. The next chapter will extend the work to the diagnosis of the root causes, where the basic idea is that just one of the oscillating measurements points is the cause of the secondary-propagated oscillations appearing at other measurement points in the plant.

Chapter 19
Diagnosis of Plant-wide Oscillations

This chapter discusses a systematic procedure for the detection and diagnosis of plant-wide oscillations and demonstrates its efficacy through industrial case studies. The oscillation-detection methods described in Chap. 18 have been used for the detection of oscillations in process measurements. Signals having common oscillations are diagnosed for finding and isolating the root cause(s).

19.1 Root Cause Diagnosis of Plant-wide Oscillation

Oscillations are a common form of plant-wide disturbance. The high level of energy integration in most modern plants facilitates the propagation of oscillations from one process unit to another. To operate the plant with a good overall profitability, it is important to detect and diagnose the root cause(s) of such oscillations in a chemical plant. Figure 19.1 is a family tree of methods for the diagnosis of the root cause of a plant-wide disturbance that was first presented by Thornhill and Horch (2007). It focusses on data-driven methods using signal analysis of the measurements from routine operation. The main distinction in the family tree is between *nonlinear* and *linear* root causes.

The diagnosis problem consists of two parts. First the root cause of each plant-wide disturbance should be distinguished from the secondary propagated disturbances, which will be solved without any further work when the root cause is addressed. The second stage is testing of the candidate root cause loop to confirm the diagnosis.

19.1.1 Finding a Nonlinear Root Cause of a Plant-Wide Disturbance

Nonlinear root causes of plant-wide disturbances include:

- Control valves with excessive static friction;
- On–off and split-range control;

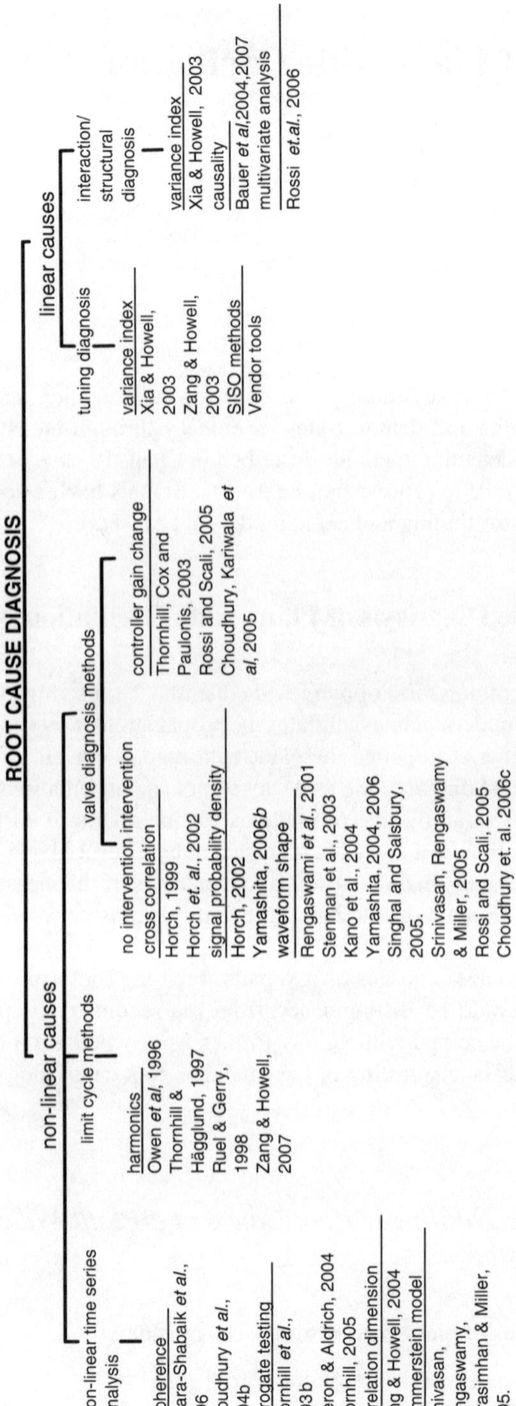

Fig. 19.1 Family tree of methods for data-driven plant-wide root cause diagnosis

- Sensor faults;
- Process nonlinearities;
- Oscillatory disturbances
- Hydrodynamic instabilities such as slugging flows.

Sustained limit cycles are common in control loops having nonlinearities. Examples include the stop–start nature of flow from a funnel feeding molten steel into a rolling mill (Graebe et al., 1995) and variations in the consistency of pulp in a mixing process (Ruel and Gerry, 1998). Thornhill (2005) described a hydrodynamic instability caused by foaming in an absorber column. These examples show that disturbances due to nonlinearity are not just confined to control valve problems.

19.1.1.1 Nonlinear Time Series Analysis

A nonlinear time series means a time series that was generated as the output of a nonlinear system, and whose distinctive characteristic is the presence of phase coupling between different frequency bands. Nonlinear time series analysis uses concepts that are quite different from linear time series methods and are covered in the textbook of Kantz and Schreiber (1997). For example, surrogate data are times series having the same power spectrum as the time series under test but with the phase coupling removed by randomization of the phase. A key property of the test time series is compared with that of its surrogates and nonlinearity is diagnosed if the property is significantly different in the test time series. Another method of nonlinearity detection uses higher- order spectra because these are sensitive to certain types of phase coupling. The bispectrum and the related bicoherence have been used to detect the presence of nonlinearity in process data (Choudhury, 2004; Choudhury et al., 2004b). Root cause diagnosis based on nonlinearity has been reported on the assumption that the measurement with the highest nonlinearity is closest to the root cause (Thornhill et al., 2003b; Thornhill, 2005; Zang and Howell, 2005).

19.1.1.2 Disturbance Propagation

The reason why nonlinearity is strongest in the time trends of measurements nearest to the source of a disturbance is that the plant acts as a mechanical filter. As the limit cycle propagates to other variables such as levels, compositions and temperatures, the waveforms generally become more sinusoidal and more linear because plant dynamics destroys the phase coupling and removes the spectral harmonics that characterize a limit-cycle oscillation. Empirically, nonlinearity measures do very well in the isolation of nonlinear root causes. However, a full theoretical analysis is missing at the present as to why and how the various nonlinearity measures change as a disturbance propagates, and this remains an open research question.

19.1.1.3 Limit Cycles and Harmonics

The waveform in a limit cycle is periodic but non-sinusoidal and therefore has harmonics that can be used to detect nonlinearity.

It is not always true, however, that the time trend with the largest harmonic content is the root cause because the action of a control loop may split the harmonic content of an incoming disturbance between the manipulated variable and the controlled variable. Insight into the distribution of harmonic content is gained from the frequency responses of the control-loop sensitivity and complementary sensitivity functions (Zang and Howell, 2005). Section 5.3.2.3 of this book showed an example from Mitsui Chemical Company (Matsuo et al., 2004). A level control loop had an incoming disturbance from upstream, which comprised a fundamental oscillation of about 46 samples/cycle and a second harmonic with 23–24 samples/ cycle. The controlled variable (level) had a strong second harmonic at 23–24 samples/cycle while the manipulated variable contained only the fundamental oscillation with a period of 46 samples. Harmonic analysis would wrongly suggest the level controller as the root cause because of the strong second harmonic in the controlled variable. Nonlinearity assessment, by contrast, correctly found the time trend of the disturbance to be more nonlinear than those of the manipulated and controlled variables.

19.1.2 Finding a Linear Root Cause of a Plant-wide Disturbance

Common root causes, besides nonlinearity, are poor controller tuning, controller interaction and structural problem involving recycles. The detection of poorly tuned SISO loops is routine using commercial CLPA tools, but the question of whether an oscillation is generated within the control loop or is external has not yet been solved satisfactorily using only signal analysis of routine operating data. Promising approaches to date require some knowledge of the transfer function (Xia and Howell, 2003).

There has been little academic work to address the diagnosis of controller interaction and structural problems using only data from routine process operations. Some progress is being made, however, by the cause and effect analysis of the process signals using a quantity called *transfer entropy*, which is sensitive to directionality to find the origin of a disturbance (Schreiber, 2000; Bauer et al., 2004, 2007; Bauer, 2005). Transfer entropy uses joint probability density functions and is sensitive to time delays, attenuation and the presence of noise and further disturbances that affect the propagating signals. The outcome of the analysis is a qualitative process model showing the causal relationships between variables.

19.1.2.1 An Example

Bauer et al., (2004), used the method of transfer entropy with data from a process with a recycle stream (Fig. 19.2). None of the time trends was nonlinear and the cause and effect analysis implicated the recycle stream because all the variables in the recycle stream were present in the order of flow. Knowing that the problem involves the recycle stream rather than originating with any individual control loop suggested the need for an advanced control solution. The time trends of the data set for this example are presented in Fig. 7.10.

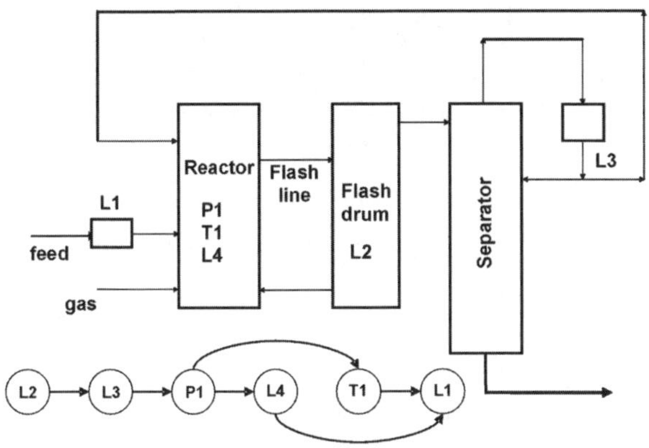

Fig. 19.2 Cause and effect analysis in a process with recycle (courtesy of A. Meaburn and M. Bauer)

19.2 Industrial Case Study 1 – Eastman Chemical Plant

An industrial data set was provided by the Advanced Control Technology group of Eastman Chemical Company. Figure 19.3 shows the process schematic of the plant, which contains three distillation columns, two decanters and several recycle streams. There are 15 control loops and 15 indicators on the schematic. There are eight flow controllers. Six of them are in cascade configuration. The Advanced Control Technology group had identified a need for the diagnosis of a common disturbance with an oscillation period of about 2 h. Thornhill et al., (2003b) completed a plant-wide diagnosis using nonlinearity detection by surrogate analysis and confirmed that the root cause was an excessive deadband in the control valve of LC2. In this section, the procedure described in Chap. 18 and in the beginning of this chapter is applied to this data set to demonstrate its efficacy in the detection and diagnosis of the plant-wide disturbance in this benchmark example.

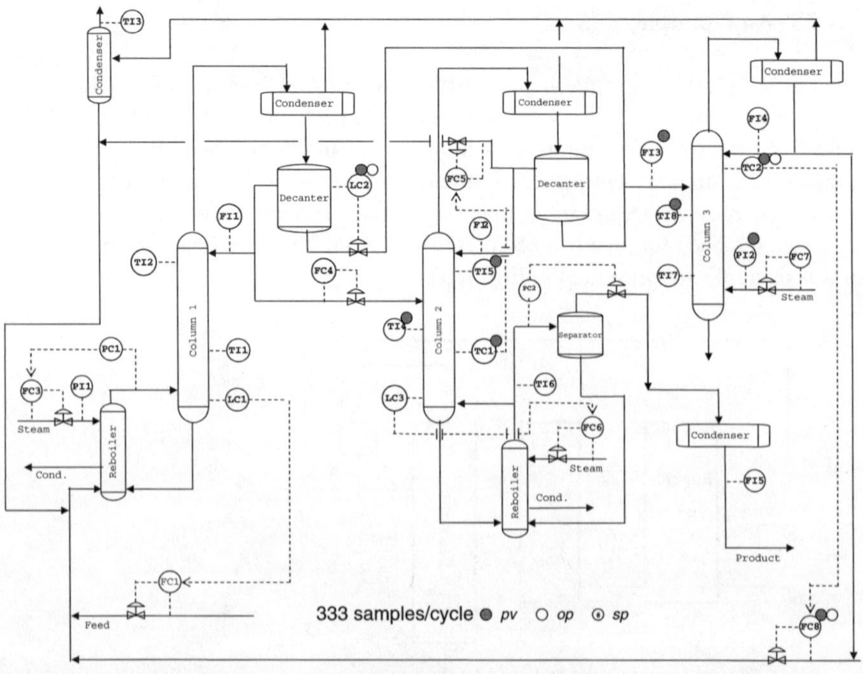

Fig. 19.3 Process schematic (courtesy of Eastman Chemical Company)

19.2.1 Data Description

Uncompressed plant data were collected at a sampling period of 20 s for the period of 2 days. Therefore, there are 8,640 samples for each tag. The data set contains 48 variables: 14 process variables (*pvs*), 14 controller outputs (*ops*), 15 indicator variables and 5 cascade-loop set points (*sps*). The time trends and power spectra are plotted in Figs. 19.4, 19.5 and 19.6.

Figure 19.4 shows the time trends and power spectra of the the 15 variables which are *pvs*. The power spectra have a peak showing the presence of an oscillation at 0.003 on the normalized frequency axis corresponding to oscillations with a period of 333 samples/cycle (i.e. 111 min or nearly 2 h). This oscillation had propagated throughout the plant and affected many variables in the process.

19.2.2 Reduction of the Problem Size

Thornhill et al., (2003b) carried out a plant-wide oscillation detection followed by nonlinearity analysis on the time trents of the measurements from 18 control loops and indicators that were affected by the plant-wide oscillation. Examining the con-

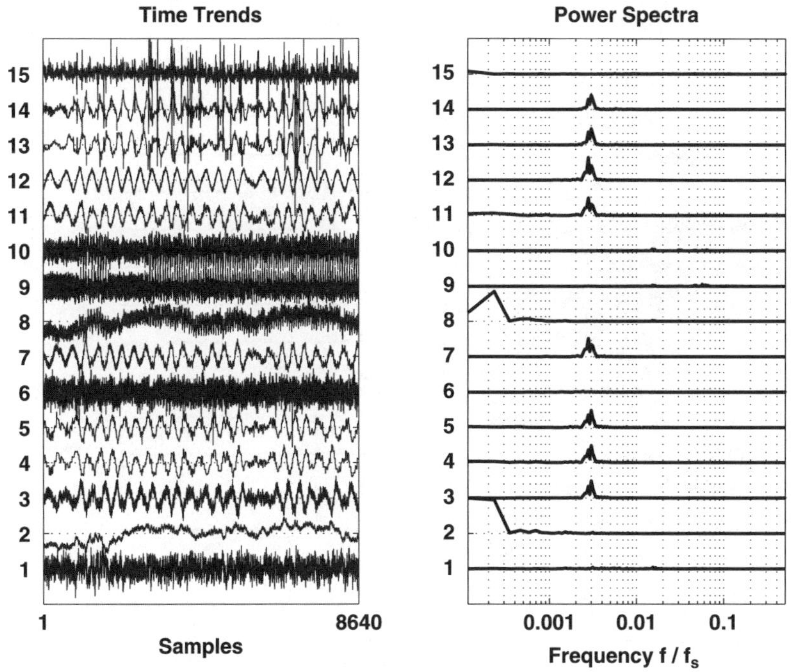

Fig. 19.4 Time trends and power spectra of 15 process variables

trol error signal for each control loop makes the problem easier to analyse because, in this case, oscillation-detection algorithm will not detect oscillations for those loops whose *sp* and *pv* both are oscillating in phase. For example, Fig. 19.4 or 19.5 shows that eight control loops are oscillating at a frequency of 0.003 cycles/sample (or about 333 samples/cycle, nearly a period of 2 h). However, the spectral peaks at 0.003 on the normalized frequency axis for the control error signal in Fig. 19.7 show that only four loops (LC1, TC1, LC2 and TC2) have oscillations at a frequency of 0.003 cycles/sample in the controller error signal. Therefore, as a first attempt to diagnose the root cause of this oscillation, one should examine these four control loops.

19.2.3 Detection of Plant-wide Oscillation by PSCMAP

Figure 19.8 shows the power spectral correlation map (PSCMAP) of the 15 error variables (see Chap. 18 for more details on PSCMAP). The block in the lower-left corner confirms that the spectra of LC1, TC1, LC2 and TC2 (tag numbers 3, 7, 12 and 14) are strongly correlated and have similar oscillations. The spectra of PC2, LC3 and FC5 (tag numbers 9, 10 and 11) are also well correlated with each other

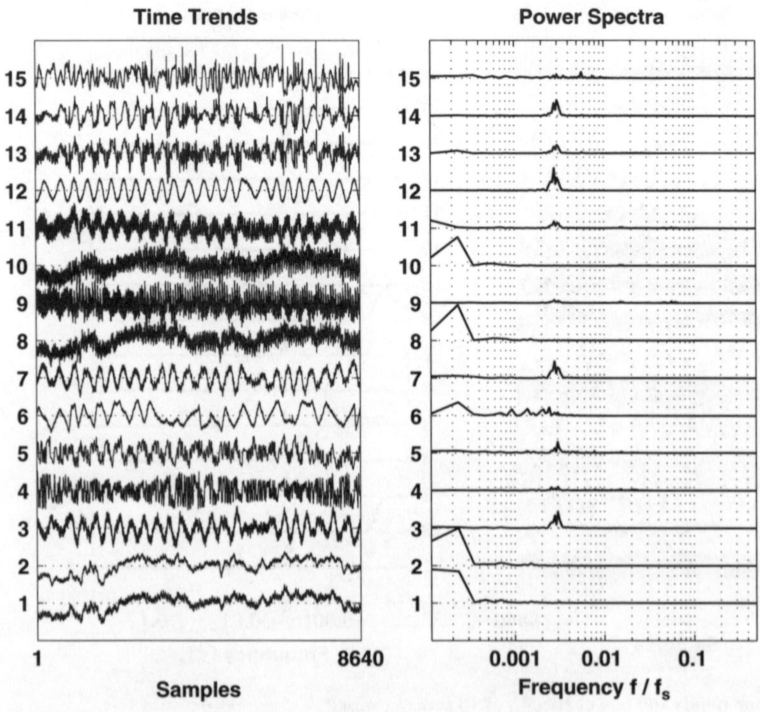

Fig. 19.5 Time trends and power spectra of 15 controller outputs

but are not considered further because these three loops do not have the oscillation frequency for which this troubleshooting exercise is intended, as can be seen from Fig. 19.7.

19.2.4 Nonlinearity Analysis Using Bicoherence-Based Indices

Nonlinearity analysis was performed on the four controller error signals that show the oscillation at 333 samples/cycle. The data were filtered using a band-pass Wiener filter with boundaries (0.001–0.1) cycles/sample in order to exclude any effects from other oscillations. Then, the filtered data were down-sampled by a factor of 10 in order to make them suitable to use with the parameters of the bicoherence calculation algorithm. The direct method of bicoherence estimation using an approach similar to Welch's average periodogram estimation method is used where a 128-point FFT is implemented using 128 samples for each data segment. By down-sampling the data by a factor of 10, the oscillation period reduces to approximately 34 samples/cycle. Therefore, a 128 sample data segment contains enough number of cycles of oscillation to perform higher-order statistical analysis. The results of this analysis

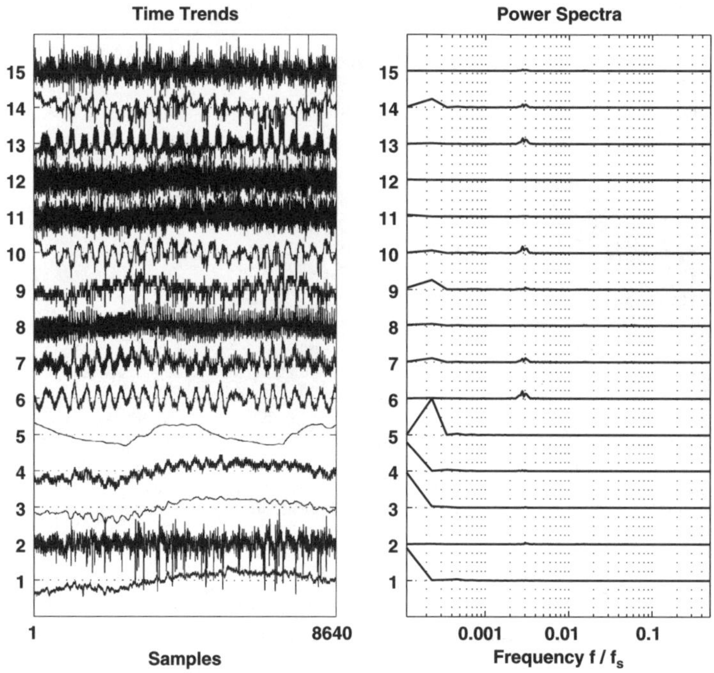

Fig. 19.6 Time trends and power spectra of 15 indicator variables

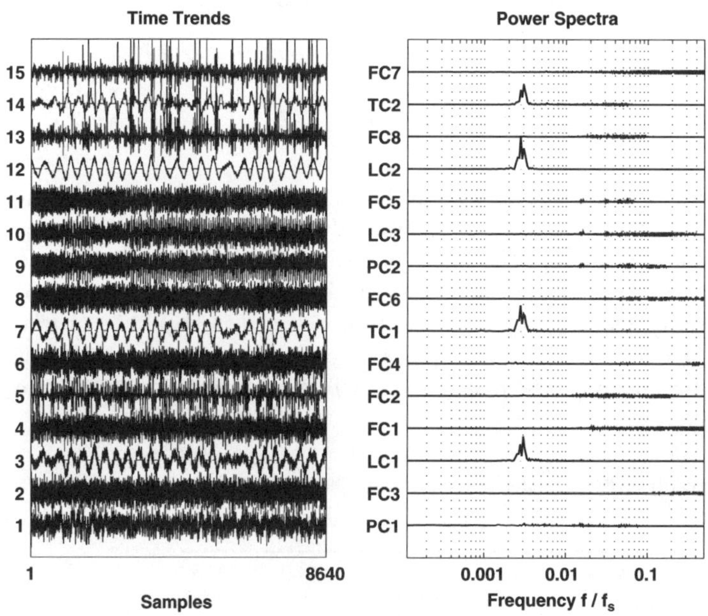

Fig. 19.7 Time trends and power spectra of 15 controller error signals

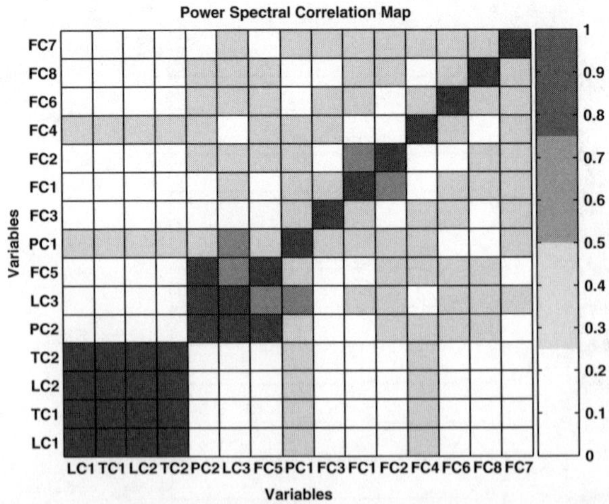

Fig. 19.8 Power Spectral Correlation Map (PSCMAP) for 15 controller error signals

are shown in Table 19.1. Since NGI and NLI values are greater than 0, all four loops have nonlinearities. The total nonlinearity index (TNLI) shows the extent of nonlinearity in each loop. The tag LC2 has the highest nonlinearity. Therefore, with the assumption of having highest nonlinearity near the root cause, it can be concluded that this loop is the root cause for this plant-wide oscillation. This is the same conclusion as was found in the analysis performed in Thornhill et al. (2003b).

19.2.5 Diagnosis of the Problem in Loop LC2

Figure 19.9 shows the diagnostic plots for the LC2 loop that had the highest value of TNLI. Figure 19.9*a* shows that there are significantly large peaks in the bicoherence plot indicating a nonlinear loop. The values of NGI and NLI for this loop reported in Table 19.1 are 0.20 and 0.57, respectively, which clearly indicates that the loop exhibits nonlinearity. Once a loop nonlinearity is detected, it should be checked whether this is due to stiction or other process nonlinearity. Figure 19.9b shows the *pv–op* plot for this loop. The plot shows an elliptical trajectory indicating the

Table 19.1 Nonlinearity analysis results for the Eastman Chemical Plant data

Tag name	NGI	NLI	TNLI
LC1	0.048	0.3	0.297
PC1	0.178	0.43	0.428
LC2	0.202	0.57	0.902
PC2	0.068	0.32	0.636

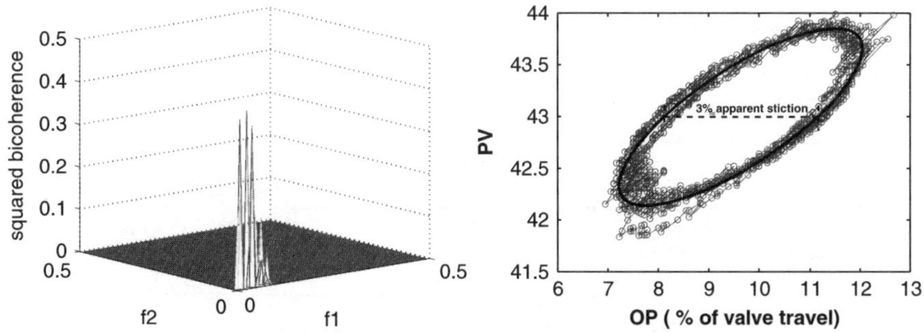

(a) Bicoherence plot (b) LC2.*pv*–LC2.*op* plot

Fig. 19.9 Oscillation diagnosis plots for LC2 loop

presence of stiction in the control valve or a phase lag between *pv* and *op* in a linear loop. The NGI and NLI indices showed that it is not a linear loop. Therefore, the elliptical pattern in the *pv–op* plot is due to valve stiction. The apparent stiction is quantified to be approximately 3% using the method described in Chap. 15. Note that the *pv–op* plot is investigated only after a successful nonlinearity detection in the loop. That is why the *pv–op* plot should not be used alone to detect stiction. This must be used in conjunction with the nonlinearity test.

Similar results of root cause diagnosis have also been discussed in Thornhill et al. (2003b). It was reported that the control valve of loop LC2 suffered from a high static friction or stiction problem (Thornhill et al., 2003b). It has been confirmed that the control valve caused the variable LC2.*pv* to oscillate, and the oscillation passed through the feedback controller and made the controller output LC.*op* also to oscillate. After that, the oscillations propagated to the temperature control loop TC1 in the second distillation column and caused the temperature to oscillate. This is the reason why temperature indicator TI4.*pv* and control variable TC1.*pv* had oscillations too. The paper by Thornhill et al. (2003b) gives more information about the diagnosis of the LC2 valve.

19.3 Industrial Case Study 2 – SE Asia Refinery Data Analysis

This case study describes the method applied to a refinery data set, courtesy of a SE Asian Refinery. A simplified schematic of the refinery process is shown in Fig. 19.10. The data set consist of 512 samples of 37 measurements sampled at 1-min intervals. It comprises measurements of temperature, flow, pressure and level loops along with composition measurements from the gas analysers. The process contains a recycle loop from the PSA unit to the reformer unit. Controller errors $(sp - pv)$ are analysed. The time trends of the controller errors and the corresponding power specta are shown in the upper panel of Fig. 18.9 in Chap. 18. Both figures show that there is very dominant oscillation at 0.06 on the normalized frequency

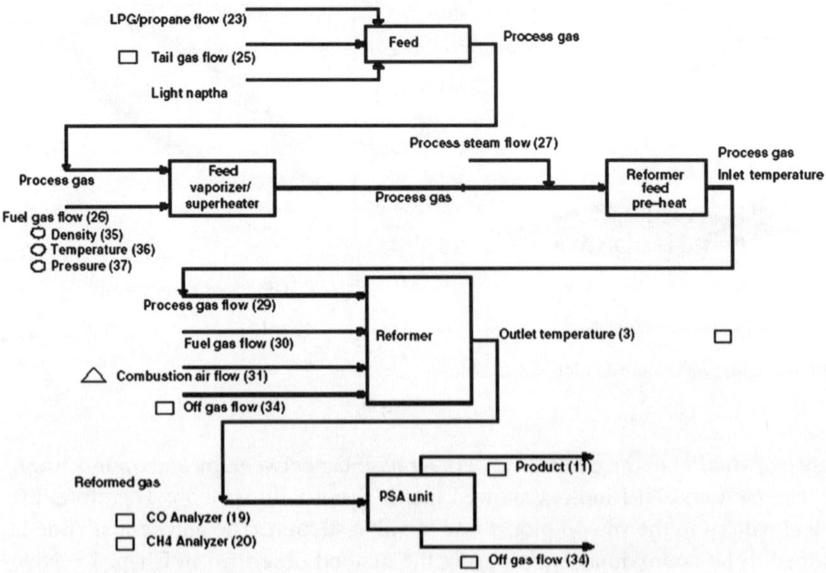

Fig. 19.10 Schematic of the SE Asian refinery process

axis or approximately 17 samples/cycle. This particular oscillation was present all over the plant and affected the plant operation adversely.

19.3.1 Oscillation Detection by PSCMAP

The rearranged spectral correlation colour maps is shown in Fig. 19.11. This clearly identifies that the tags 2, 3, 4, 8, 9, 10, 11, 13, 15, 16, 19, 20, 24, 25, 28, 33 and 34 have similar spectral shape and are affected with the above-mentioned oscillation.

The spectral map shows three main groups of correlated spectra. The largest group is 2, 3, 4, 8, 9, 10, 11, 13, 15, 16, 19, 20, 24, 25, 28, 33 and 34 in the lower-left corner. An examination of the spectra shows that this is the group with the oscillation at 17 samples/cycle. Another two groups are also identified by the PSCMAP. The group comprising tags 7, 12, 35–37 is characterized by a lower-frequency oscillation close to 0.012 on the normalized frequency axis while the group consisting of 6, 21, 22 and 31 has spectral content at low frequencies.

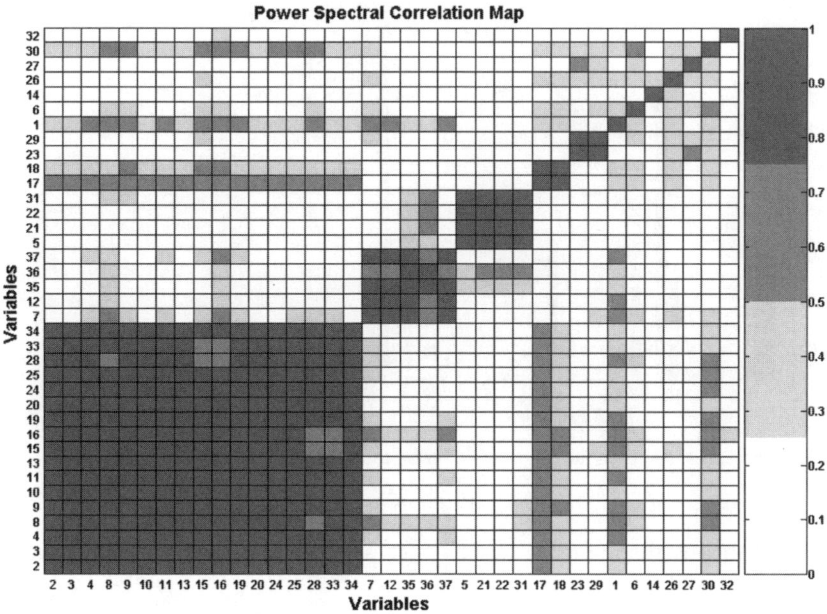

Fig. 19.11 Power Spectral Correlation Color Map (PSCMAP) for the SE Asian refinery

19.3.2 Oscillation Detection by Spectral Envelope

Figure 19.12 shows the spectral envelope for the refinery data set. It also shows that there is a strong oscillation at 0.06 on the normalized frequency axis. Figure 19.13 shows the variables that are oscillating at this frequency. It can also estimate the

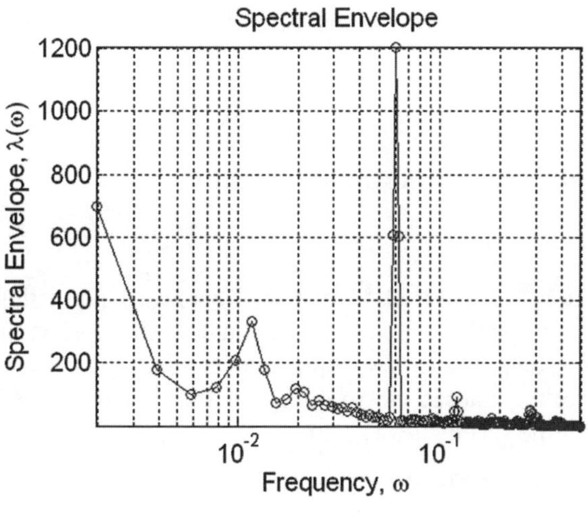

Fig. 19.12 Spectral envelope of the SE Asian refinery process

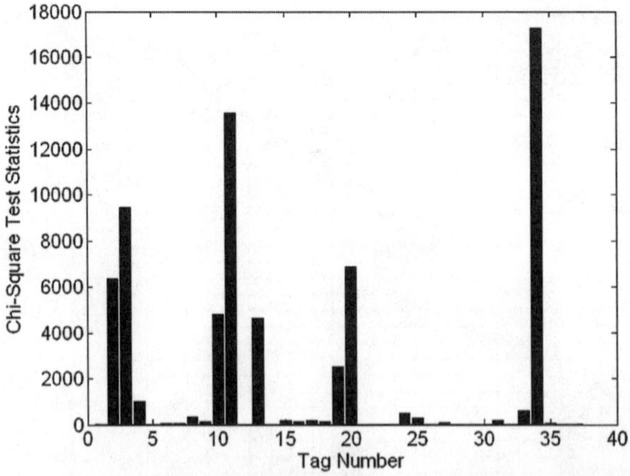

Fig. 19.13 χ^2 Test statistics at the frequency 0.06 for the SE Asian refinery process

strength of this oscillation using a chi-square statistic. Jiang et al. (2006) gave a detailed analysis of the method.

19.3.3 Oscillation Diagnosis

Both the spectral envelope and the power spectral correlation map find that tags 2, 3, 4, 8, 9, 10, 11, 13, 15, 16, 19, 20, 24, 25, 28, 33 and 34 are oscillating with the frequency 0.06. The total nonlinearity test was applied to these tags to troubleshoot this oscillation. Table 19.2 shows the results of this analysis. The table shows that the tag 34 has the highest total nonlinearity index (TNLI). Therefore, tag 34 is identified as the root cause of this plant-wide oscillation problem. This finding is in agreement with other previous studies performed on this data set (Thornhill et al. 2001; Thornhill, 2005).

19.4 Industrial Case Study 3 – Mitshubishi Chemical Corporation

The process under investigation is a plant at Mitsubishi Chemical Corporation, Mizushima, Japan. The plant personnel reported large amplitude oscillations in the condenser level of a distillation column causing suboptimal operation and large economic losses. The schematic of the plant is shown in Fig. 19.14. The condenser is located at the top left corner of the schematic. The aim of the study was to find the root cause of this oscillation.

Table 19.2 Nonlinearity analysis results for SE Asian refinery process

Tag name	NGI	NLI	TNLI
2	0.15	0.99	2.71
3	0.14	0.94	2.65
4	0.06	0.81	0.81
8	0.06	0	0.81
9	0	0.64	0
10	0.04	0.79	0.79
11	0.20	0.96	2.84
13	0.15	0.96	1.80
15	0.05	0.80	0.80
16	0.09	0.49	1.68
19	0.13	0.88	0.88
20	0.13	0.94	1.76
24	0.01	0.76	0.76
25	0	0	0
28	0.01	0.76	0.76
33	0.08	0.87	3.32
34	0.20	0.99	7.63

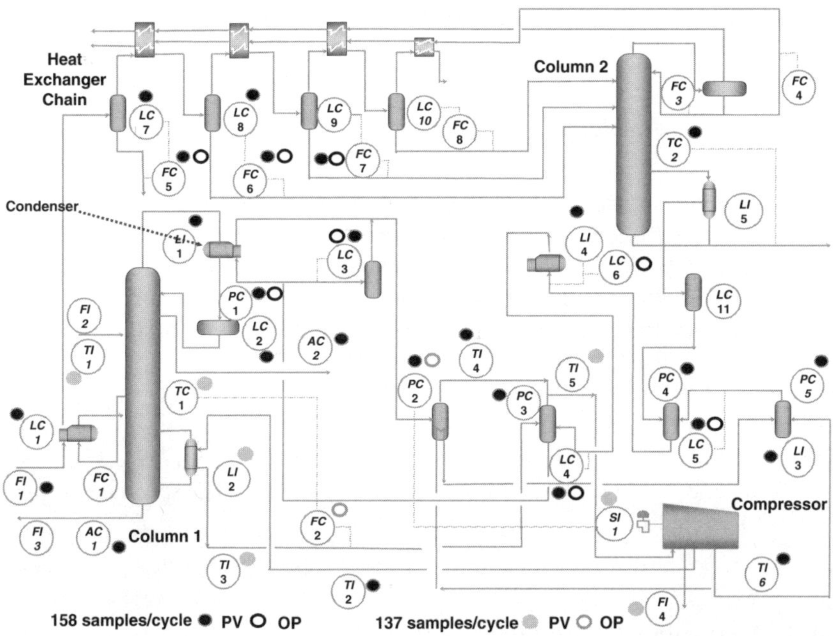

Fig. 19.14 Process flow diagram showing oscillations detected in the low frequency range (88–183 samples/cycle)

19.4.1 Scope of the Analysis and Data Set

There were many variables in the process and a sequential approach was used to define the scope of the analysis. Starting from the condenser level, the scope of the study was expanded based on mass, energy and information connectivity of the process until no more oscillations were detected. For this preliminary analysis, visual inspection of time trends and spectra were found to be sufficient.

Based on the preliminary analysis, the final data set was chosen to consist of 59 tags. These tags were taken from various process units shown in Fig. 19.14. The condenser which is at the centre of the study is located at the top-left of the figure. Fifteen of these variables are controlled using PI(D) controllers, and the controller outputs for these variables are also included in the study. For these variables, a sample data set consisting of 2,880 samples was collected at the rate of 1 sample/min, which corresponds to data for 2 days of operation. Figure 19.15 shows the time trends and power spectra of some of the tags.

In Fig. 19.14 and the ensuing discussion, the notations used for tags are AC, FC, LC, PC and TC to represent controller for composition, flow, level, pressure and temperature tags, respectively. Similarly, FI, LI, SI and TI represent the flow, level, rotor speed and temperature indicators. The set point, process variable and controller output are denoted as sp, pv and op, respectively.

19.4.2 Oscillation-Detection Results

For this industrial case study, the data set is filtered using a band-pass Wiener filter with normalized frequency boundary [0.002 0.5] in order to remove the low-frequency non-stationary trends. Then, PSCMAP was used to detect and group the plant-wide oscillations. Figure 19.16 shows that there are two major oscillation groups in this data set. The condenser level (tag 10) belongs to the first group consisting of the tags 4, 9, 10, 11, 12, 14, 20, 24, 25, 28, 29, 31, 32 and 36. The condenser level (LI1) oscillated with a period of approximately 158 samples/cycle. Thus, the oscillation with a period of 158 samples/cycle was of primary interest of this study. The other oscillation group consists of the tags 5, 7, 13, 17, 23, 27, 37, 39, 48, 49, 50 and 53. This oscillation is at a very low frequency. The spectral envelope method also offered similar results.

19.4.3 Oscillation Diagnosis

The bicoherence-based NGI, NLI and TNLI indices were calculated for all variables. The results for the first oscillation group containing the condenser-level oscillation (tag 10) have been reported in Table 19.3. The highest TNLI index indicates tag 14, which has the maximum nonlinearity. Tag 14 is a composition variable whose data came from laboratory analysis and was only available at every 12 min. A further investigation of the data for this variable showed that in the data historian, each data point for this tag was repeated 12 times until the next lab result became available. Therefore, this tag was eliminated from the possible candidate list of the root cause of this plant-wide oscillation. The next highest TNLI is for tag 12, which corresponds to the control error signal (*sp–pv*) for loop PC1. The controller PC1 was controlling the pressure of the reflux drum by manipulating the coolant flow rate to the condensers. Therefore, this loop was suspected to be the candidate of the root cause for this plant-wide oscillation.

A further analysis of PC1 loop using the techniques described in Chap. 15 shows that the valve that was manipulating the coolant flow rate has approximately 0.5% apparent stiction. The presence of 1.25% stiction was also detected in the valve for TC2 loop. Based on this analysis, we conclude that one or both of the loops PC1 and TC2 is most likely the root cause of the oscillations. The results of the analysis was communicated to the plant personnel. They replied that the pressure loop PC1 is critical for the plant operation and it is not possible to put this loop in manual to perform a valve travel or bump test for confirming stiction. Therefore, the method of changing controller gain described earlier in this book and also in Choudhury et al. (2005b, 2007) was applied to confirm the presence of valve stiction.

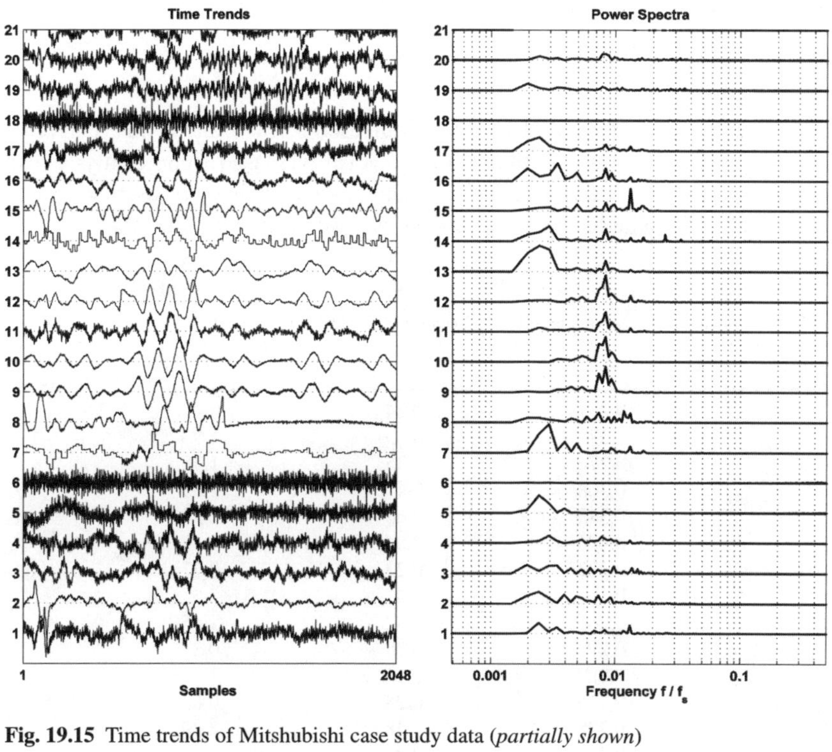

Fig. 19.15 Time trends of Mitshubishi case study data (*partially shown*)

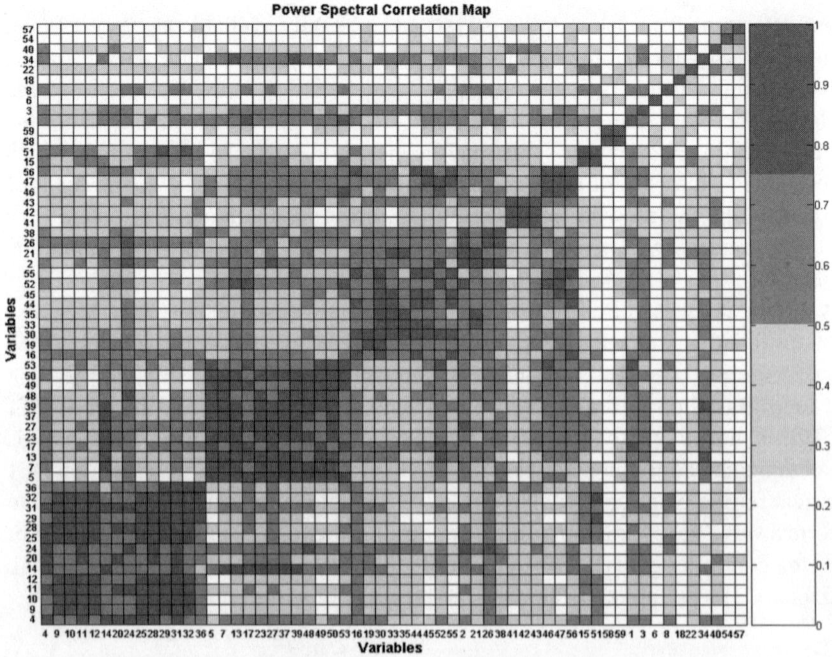

Fig. 19.16 Spectral correlation map for the Mitshubishi case study data

The controller gain of the PC1 loop was increased by 50% and it was found that the frequency of oscillations in the condenser level increased significantly, as can be seen from Fig. 19.17. Before the gain change, the frequency of oscillation was approximately 154 samples/cycle or 0.0068 cycles/sample and after the gain was

Table 19.3 Nonlinearity analysis results for Mitshubishi Refinery data set

Tag name	NGI	NLI	TNLI
4	0.237	0.24	0.47
9	0.278	0.33	1.95
10	0.276	0.35	1.38
11	0.308	0.41	1.85
12	0.345	0.66	10.34
14	0.406	0.87	15.42
20	0.272	0.25	0.27
24	0.3	0.43	3.3
25	0.236	0.32	1.65
28	0.295	0.28	0.3
29	0.227	0.25	1.36
31	0.275	0.28	0.83
32	0.278	0.48	3.89
36	0.217	0.2	0.43

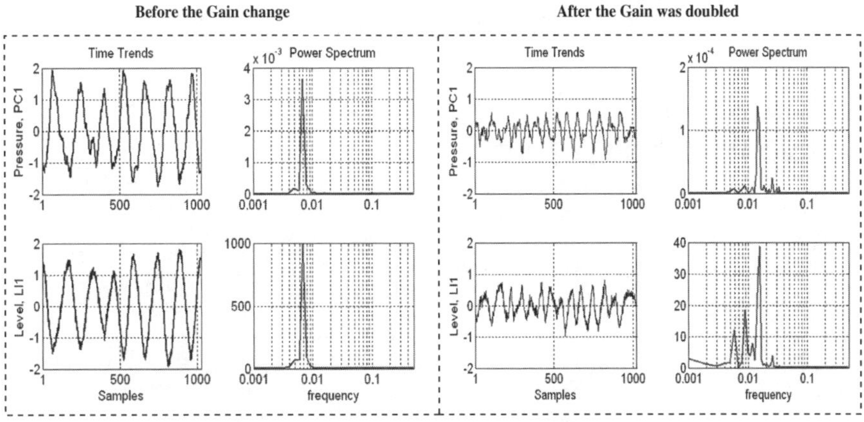

Fig. 19.17 Effect of changes in controller gain on the oscillatory behaviour for the Mitsubishi case study data. The plots in the left column are for before the gain-change case and the plots in the right column correspond to after the gain-change case

doubled the frequency increased to 63 samples/cycle or 0.016 cycles/sample. The power spectra in Fig. 19.17 show the result. The amplitude of the oscillation also decreased with the increase of controller gain, as can be seen from the time trends shown in Fig. 19.17. This phenomenon serves as a confirmation of the presence of valve stiction and is discussed in detail in Chap. 17.

19.4.4 The Results of Maintenance on the PC1 and LI1 Loops

During the shutdown maintenance period of summer 2005, the control valve for the PC1 loop was repaired and cleaned. After the resumption of the plant operation, the MPC controllers were also retuned. Figure 19.18 shows the new time trends and the power spectrum of the data corresponding to PC1 and LI1 loops. The figure shows that the previous oscillation at the frequency 63 samples/cycle or 0.016 cycles/sample is now gone, as can be seen from the power spectra in Fig. 19.17. Both loops are operating satisfactorily. By comparing the power spectra of Figs. 19.17 and 19.18, it is clear that the power of all oscillations has been reduced significantly. The small peak in the power spectrum of LI1 in Fig. 19.18 is due to a low-frequency oscillation (200 samples/cycle) probably caused by interaction from other loops. This peak was also present in Fig. 19.17, but somewhat overshadowed by larger peaks.

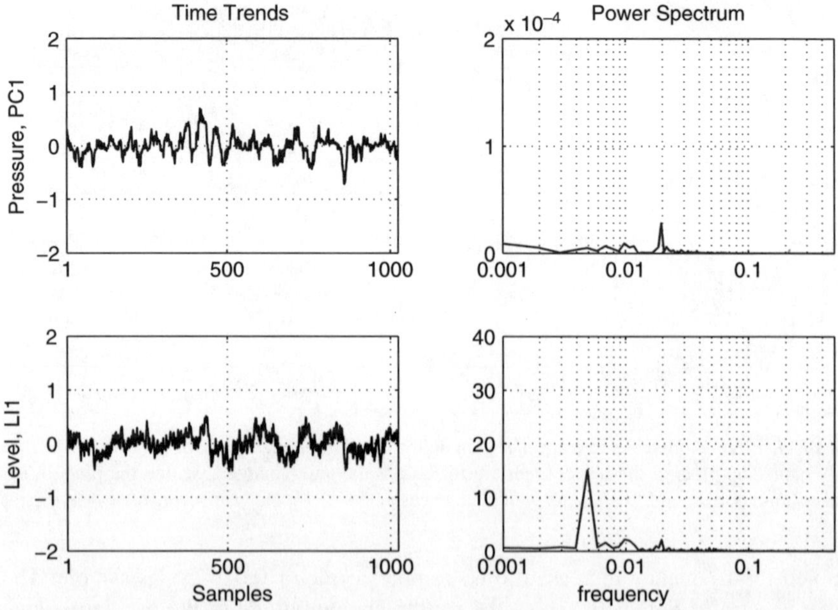

Fig. 19.18 The pressure PC1 and the level LI1 after maintenance of the sticky valve of loop PC1

19.5 Summary

This chapter discussed a systematic procedure to find the root cause of plant-wide oscillations. The variables affected by the frequency of concern should be detected and grouped. If the reason for oscillation is suspected to be nonlinearity, the higher-order statistical-based nonlinearity measures or surrogate data-based nonlinearity measures can be applied to detect the probable candidates of the root cause. Thereafter additional process information combined with further analysis or test should be performed to confirm the root cause.

References

Adeleye, O. (2006). Design and evaluation of optimal PID controllers. Master's thesis. University of Alberta. Canada.

Agrawal, P. and S. Lakshminarayanan (2003). Tuning PID controllers using achievable performance indices. *Industrial and Engineering Chemistry Research* **42**, 5576–5582.

Aldrich, C. and M. Barkhuizen (2003). Process system identification strategies based on the use of singular spectrum analysis. *Minerals Engineering* **16**, 815–826.

Amin, J., B. Friedland and A. Harnoy (1997). Implementation of a friction estimation and compensation technique. *IEEE Control Systems Magazin* **17**(4), 71–76.

Armstrong-Hélouvry, B., P. Dupont and C. C. De Wit (1994). A survey of models, analysis tools and compensation methods for the control of machines with friction. *Automatica* **30**(7), 1083–1138.

Aspentech (2001). Analysis of data storage technologies for the management of real-time process manufacturing data. Accessed Apr 2nd 2003, from http://www.advanced-energy.com/Upload/symphony_wp_infoplus.pdf.

Aubrun, C., M. Robert and T. Cecchin (1995). Fault detection in control loops.*Control Engineering Practice* **3**, 1441–1446.

Barnard, J. P., C. Aldrich and M. Gerber (2001). Identification of dynamic process systems with surrogate data methods. *AIChE Journal* **47**, 2064–2075.

Bauer, M. (2005). Data-Driven Methods for Process Analysis. PhD thesis. University of London.

Bauer, M., J. W. Cox, M. H. Caveness, J. J. Downs and N. F. Thornhill (2007). Finding the direction of disturbance propagation in a chemical process using transfer entropy. *IEEE Transactions on Control Systems Technology* **15**, 12–21.

Bauer, M., N. F. Thornhill and A. Meaburn (2004). Specifying the directionality of fault propagation paths using transfer entropy. In: *DYCOPS 7 Conference*. Boston, USA.

Baumann, H. D. (1994). *Control Valve Primer – A users guide*. Instrument Society of America. USA.

Bergstrom, J. and G. A. Dumont (2003). Adaptive control of valves with friction. *Pulp and Paper Canada* **104**(6), 23–27.

Bezdek, J. C. (1981). *Pattern Recognition with Fuzzy Objective Function Algorithms*. Plenum. New York.

Bialkowski, W. L. (1992). Dreams vs. reality: A view from both sides of the gap. In: *Control Systems*. Whistler, BC, Canada. pp. 283–294.

Birgitta, K. (2003). PID Controllers: Design and Evaluation. Ph.D. thesis. Chalmers University of Technology. Sweden.

Brillinger, D. R. and M. Rosenblatt (1967*a*). A symptotic theory of estimates of *k*th order spectra. *in Spectral Analysis of Time Series*, 153–188. B. Harris, ed.

Brillinger, D. R. and M. Rosenblatt (1967*b*). Computation and interpretation of kth order spectra. *in Spectral Analysis of Time Series*, 189–232. B. Harris, ed.

Bristol, E. H. (1990). Swinging door trending: adaptive trend recording. In: *Proceedings of the ISA National Conf.*, 749–753.

Chandaran, V. and S. L. Elgar (1991). Mean and variance of estimates of the bispectrum of a harmonic random process – an analysis including leakage effects. *IEEE Transactions on Signal Processing* **39**, 2640–2651.

Chatfield, C. and A. J. Collins (1980). *Introduction to Multivariate Analysis*. Chapman and Hall. London, UK.

Choudhury, M. A. A. S. (2004). Detection and Diagnosis of Control Loop Nonlinearities, Valve Stiction and Data Compression. Ph.D. thesis. Universityof Alberta, Canada.

Choudhury, M. A. A. S., M. Jain and S. L. Shah (2006*b*). Quantification of valve stiction. In: *Proceedings of the ACC 2006*. Minneapolis, USA.

Choudhury, M. A. A. S., M. Jain and S. L. Shah (2008). Stiction – definition, modeling, detection and quantification, Journal of Process Control, **18**, 232–243.

Choudhury, M. A. A. S., V. Kariwala, S. L. Shah, H. Doukec, H. Takada and N. F. Thornhill (2005*b*). A simple test to confirm control valve stiction. In: *16th IFAC World Congress*. Praha.

Choudhury, M. A. A. S., V. Kariwala, N. F. Thornhill, H. Doukec, S. L. Shah, H. Takadac and J. F. Forbes (2007). Detection and diagnosis of plant-wide oscillations. *Canadian Journal of Chemical Engineering* **85**, 208–219.

Choudhury, M. A. A. S., S. L. Shah and N. F. Thornhill (2002). Detection and diagnosis of system nonlinearities using higher order statistics. In: *15th IFAC World Congress*. Barcelona, Spain.

Choudhury, M. A. A. S., S. L. Shah and N. F. Thornhill (2004b). Diagnosis of poor control loop performance using higher order statistics. *Automatica* **40**(10), 1719–1728.

Choudhury, M. A. A. S., S. L. Shah, N. F. Thornhill and D. Shook (2006*c*). Automatic detection and quantification of stiction in control valves.*Control Engineering Practice* **14**, 1395–1412.

Choudhury, M. A. A. S., D. Shook and S. L. Shah (2006*a*). Linear or nonlinear? a bicoherence based metric of nonlinearity measure. In: *In the Proceedings of the* 6th *IFAC Symposium of SAFEPROCESS*. Beijing, China.

Choudhury, M. A. A. S., N. F. Thornhill and S. L. Shah (2004*a*). A data-driven model for valve stiction. In: *The Proceedings of ADCHEM 2003, January 11–14, 2004*. Hong Kong.

Choudhury, M. A. A. S., N. F. Thornhill and S. L. Shah (2005*a*). Modelling valve stiction. *Control Engineering Practice* **13**, 641–658.

Collis, W. B., P. R. White and J. K. Hammond (1998). Higher-order spectra: The bispectrum and trispectrum. *Mechanical Systems and Signal Processing* **12**, 375–394.

Cook, P. A. (1986). *Nonlinear Dynamical Systems*. Prentice-Hall International. London.

Desborough, L. and T. Harris (1992). Performance assessment measures for univariate feedforward feedback control. *The Canadian Journal of Chemical Engineering* **70**, 1186–1197.

Desborough, L. and R. Miller (2002). Increasing customer value of industrial control performance monitoring – Honeywell's experience. In: *AIChE Symposium Series*. number 326. 153–186.

Desborough, L., R. Miller and P. Nordh (2000). Regulatory control survey. Honeywell, unpublished manuscript.

Dewit, C. C., H. Olsson, K. J. Åström and P. Lischinsky (1995). A new model for control of systems with friction. *IEEE Transactions on Automatic Control* **40**, 419–425.

Duda, R.O., P.E. Hart and D.G. Stork (2000). *Pattern Classification*. 2ed.. Wiley-Interscience. UK.

Dulyakarn, P. and Y. Rangsanseri (2001). Fuzzy c-means clustering using spatial information with application to remote sensing. In: *Proceedings of the 22nd Asian Conference on Remote Sensing*. Singapore.

Elgar, S. and R. T. Guza (1988). Statistics of bicoherence. *IEEE Transactions on Acoustics, Speech and Signal Processing* **36**, 1667–1668.

Emara-Shabaik, H. E., J. Bomberger and D. E. Seborg (1996). Cumulant/bispectrummodel structure identification applied to a pH neutralization process. In: *IEE Proceedings of UKACC International Conference on Control 96*. Exeter, UK. 1046–1051.

EnTech (1998). *EnTech Control Valve Dynamic Specification (version 3.0)*.

Fackrell, J. W. A. (1996). Bispectral Analysis of Speech Signals. PhD thesis. The University of Edinburgh. UK.

Fackrell, J. W. A., P. R. White, J. K. Hammond, R. J. Pinnington and A. T. Parsons (1995). The interpretation of the bispectra of vibration signals – 1. theory. *Mechanical Systems and Signal Processing* **9**, 257–266.

Fedenczuk, P., P. Fountain and R. Miller (1999). Loop Scout, RPID and Profit Controller team up to produce significant benefits for bp. Accessed Apr 2nd 2003, from http://loopscout.com/loopscout/info/bpamoco.pdf.

Fisher-Rosemount (1999). *Control Valve Handbook.* Fisher Controls International, Inc. Marshalltown, Iowa, USA.

Fitzgerald, B. (1995). *Control Valves for the Chemical Process Industries.* McGraw-Hill, Inc. New York.

Forsman, K., and A. Stattin (1999) A new criterion for detecting oscillations in control loops, *European Control Conference*, Karlsruhe, Germany.

Gander, W., G. H. Golub and R. Strebel (1994). Fitting of circles and ellipses least squares solution. *BIT Numerical Mathematics* **34**, 558578.

Gerry, J. and M. Ruel (2001). How to measure and combat valve stiction online. In: *ISA 2001 conference*. Instrumentation, Systems and Automated Society. Houston, Texas, USA. http://www.expertune.com/articles/isa2001/StictionMR.htm.

Graebe, S. F., G. C. Goodwin and G. Elsley (1995). Control design and implementation in continuous steel casting. *IEEE Control Systems Magazine* **15**, 64–71.

Guay, M. (1996). Measurement Of Nonlinearity in Chemical Process Control. PhD thesis. Queens University at Kingston. ON, Canada.

Hägglund, T. (1995). A control loop performance monitor. *Control Engineering Practice* **3**(11), 1543–1551.

Hägglund, T. (2002). A friction compensator for pneumatic control valves. *Journal of Process Control* **12**(8), 897–904.

Hägglund, T. (2005). Industrial implementation of on-line performance monitoring tools. *Control Engineering Practice*, **13**, 1383–1390.

Hale, J. C. and H. L. Sellars (1981). Historical data recording for process computers. *Chemical Engineering Progress* **77**(11), 38-43.

Harris, T. J. (1989). Assessment of control loop performance. *The Canadian Journal of Chemical Engineering* **67**, 856–861.

Hatipoglu, C. and U. Ozguner (1998). Robust control of systems involving non-smooth nonlinearities using modifed sliding manifolds. In: *Proceedings of American Control Conference*. Philadelphia, PA. 2133–2137.

Hegger, R., H. Kantz and T. Schreiber (1999). Practical implementation of nonlinear time series methods: The TISEAN package. *Chaos* **9**, 413–435.

Hegger, R., H. Kantz and T. Schreiber (2000). *TISEAN 2.1 Surrogates Manual, Periodicity Artefacts*, Accessed May 20th 2008, from http://www.mpipks-dresden.mpg.de/~tisean/TISEAN_2.1/index.html.

Helbig, A., W. Marquardt and F. Allgöwer (2000). Nonlinearity measure: definition, computation and applications. *Journal of Process Control* **10**, 113–123.

Hinich, M. J. (1982). Testing for gaussianity and linearity of a stationary time series. *Journal of Time Series Analysis* **3**, 169–176.

Horch, A. (1999). A simple method for detection of stiction in control valves. *Control Engineering Practice* **7**, 1221–1231.

Horch, A. (2000). Condition Monitoring of Control Loops. PhD thesis. Royal Institute of Technology. Stockholm, Sweden.

Horch, A. (2002). Patents WO0239201 and US2004/0078168.

Horch, A. (2007). Benchmarking control loops with oscillations and stiction. In: *Process Control Performance Assessment; From Theory to Implementation* (A. Ordys, D. Uduehi and M.A. Johnson, Eds.). Chap. 7. Springer-Verlag(London) Ltd.

Horch, A. and A. J. Isaksson (1998). A method for detection of stiction in control valves. In: *Proceedings of the IFAC workshop on On line Fault Detection and Supervision in the Chemical Process Industry*. Session 4B. Lyon, France.

Horch, A., A. J. Isaksson and K. Forsman (2000). Diagnosis and characterization of oscillations in process control loops. In: *Proceedings of the Control Systems 2000*. Victoria, Canada. 161–165.

Horch, A., A .J. Isaksson, and K. Forsman (2002). Diagnosis and characterization of oscillations in process control loops. Pulp & Paper-Canada, **103**(3), 19–23.

Huang, B. and S. Shah (1999). *Performance Assessment of Control Loops: Theory and Applications*. Springer-Verlag. Germany.

Hyvarinen, A. and E. Oja (2000). Independent component analysis: algorithms and applications. *Neural Networks* **13**, 411–430.

Ifeachor, E. C. and B. W. Jervis (1993). *Digital Signal Processing: A Practical Approach*. Addison-Wesley.

Imtiaz, S. A., M. A. A. S. Choudhury and S. L. Shah (2007). Building multivariate models from compressed data. *Industrial Engineering and Chemistry Research* **46**(2), 481–491.

ISA Committee SP51 (1996). Method of evaluating the performance of positioners with analog input signals and pneumatic output. Technical Report ANSI/ISA-75.13-1996. Instrument Society of America.

ISA Committee SP51 (2001). Control valve terminology. Technical Report ANSI/ISA-75.05.01-2000. Instrument Society of America.

ISA Subcommittee SP75.05 (1979). Process instrumentation terminology. Technical Report ANSI/ISA-S51.1-1979. Instrument Society of America.

Jain, M., M. A. A. S. Choudhury and S. L. Shah (2006). Quantification of valve stiction. In: *Proceedings of the IFAC- ADCHEM Symposium*. Gramado, Brazil.

Jamsek, J., A. Stefanovska, P. V. E. McClintock and I. A. Khovanov (2003). Time-phase bispectral analysis. *Physical Review E* **68**, 016201–1–12.

Jelali, M. (2006). An overview of control performance assessment technology and industrial applications. *Control Engineering Practice* **14**, 441–466.

Jiang, H., M. A. A. S. Choudhury and S. L. Shah (2007). Detection and Diagnosis of Plant-wide Oscillations From Industrial Data using the Spectral Envelope Method, *Journal of Process Control*, **17**(2), 143–155

Jiang, H., M. A. A. S. Choudhury, S. L. Shah, J. W. Cox and M. A. Paulonis (2006). Detection and diagnosis of plant-wide oscillations via the spectral envelope method. In: *In the Proceedings of ADCHEM 2006*. Gramado, Brazil 1139–1144.

Johnson, R. A. and D. W. Wichern (1998). *Applied Multivariate Statistical Analysis*. Prentice-Hall. New Jersey.

Kano, M., H. Maruta, H. Kugemoto and K. Shimizu (2004). Practical model and detection algorithm for valve stiction. In: *Proceedings of the Seventh IFAC-DYCOPS Symposium*. Boston, USA.

Kantz, H. and T. Schreiber (1997). *Nonlinear Time Series Analysis*. Cambridge University Press. Cambridge.

Karnopp, D. (1985). Computer simulation of stic-slip friction in mechanical dynamical systems. *Journal of Dynamic Systems, Measurement, and Control* **107**, 420–436.

Kayihan, A. and F. J. Doyle III (2000). Friction compensation for a process control valve. *Control Engineering Practice* **8**, 799–812.

Kim, Y. C. and E. J. Powers (1979). Digital bispectral analysis and its applications to nonlinear wave interactions. *IEEE Transactions on Plasma Science* **PS-7**, 120–131.

Kennedy, J.P., Data treatment and applications – Future of the desktop, *Proceedings of FOCAPO*, CACHE, 1993.

Lai, Y. C. and D. Lerner (1998). Effective scaling regime for computing the correlation dimension from chaotic time series. *Physica D* **115**, 1–18.

Lee, D. D. and H. S. Seung (1999). Learning the parts of objects by non-negative matrix factorization. *Nature* **401**, 788–791.

Luyben, W. and D. C. Hendershot (2004). Dynamic disadvantages of intensification in inherently safer process design. *Industrial and Engineering Chemistry Research* **43**, 384–396.

Mah, R. S. H., A. C. Tamhane, S. H. Tung and A. N. Patel (1995). Process trending with piece wise linear smoothing. *Computers and Chemical Engineering* **19**, 129–137.

Marlin, T. E. (1995). *Process Control: Designing Processes and Control Systems for Dynamic Performance*. McGraw Hill. New York.

Martin, G. D., L. E. Turpin and R. P. Cline (1991). Estimating control function benefits. *Hydrocarbon Processing* **70**, 68–73.

Matsuo, T., H. Sasaoka and Y. Yamashita (2003). Detection and diagnosis of oscillations in process plants. *Lecture Notes in Computer Science* **2773**, 1258–1264.

Matsuo, T., I. Tadakuma and N. F. Thornhill (2004). Diagnosis of a unit-wide disturbance caused by saturation in a manipulated variable. In: *Proceedings of IEEE Advanced Process Control Applications for Industry Workshop 2004*. Vancouver, Canada.

McDougall, A. J., D. S. Stoffer and D. E. Tyler (1997). Optimal transformations and the spectral envelope for real-valued time series. *Journal of Statistical Planning and Inference* **57**, 195–214.

McMillan, G. K. (1995). Improve control valve response. *Chemical Engineering Progress* **91**(6), 77–84.

Mendel, J. M. (1991). Tutorial on higher order statistics (spectra) in signal processing and systems theory: Theoretical results and some applications. *Proceedings of the IEEE* **79**, 278–305.

Miao, T. and D. E. Seborg (1999). Automatic detection of excessively oscillatory feedback control loops, *IEEE Conference on Control Applications*, Hawaii, 359–364.

Nikias, C. L. and J. M. Mendel (1993). Signal processing with higher order spectra. *IEEE Signal Processing Magazine* **10**, 10–37.

Nikias, C. L. and A. P. Petropulu (1993). *Higher-Order Spectra: A Nonlinear Signal Processing Framework*. Prentice Hall. New Jeresy.

Nikias, C. L. and M. R. Raghuveer (1987). Bispectral estimation: A digital signal processing framework. *Proceedings of the IEEE* **75**, 869–891.

Olsson, H. (1996). Control Systems with Friction. PhD thesis. Lund Institute of Technology. Sweden.

OSI Software Inc. (2002). PI data storage component overview. Accessed Apr 2nd 2003, from http:// www.osisoft.com/ 270.htm.

Olsson H. and K. J. Åström (2001). Friction generated limit cycles. *IEEE Transactions on Control Systems Technology* **9**, 629–636.

Owen, J. G., D. Read, H. Blekkenhorst and A.A. Roche (1996). A mill prototype for automatic monitoring of control loop performance. *In Proceedings of the control systems, 96*, Halifax, Novia Scotia, 171–178

Paulonis, M. A. and J. W. Cox (2003). A practical approach for large-scale controller performance assessment, diagnosis, and improvement. *Journal of Process Control* **13**(2), 155–168.

Pipponen, J. (1996). Controlling processes with non ideal valves: tuning of loops and selection of valves. In: *Control Systems*. Chateau, Halifax, Nova Scotia, Canada. 179–186.

Proakis, J. G. and D. G. Manolakis (1996). *Digital Signal Processing*. 3ed.. Prentice-Hall Inc.

Qin, S. J. (1998). Control performance monitoring – A review and assessment. *Computers and Chemical Engineering*, **23**, 173–186.

Rao, T. S. and M. M. Gabr (1980). A test for linearity and stationarity of time series. *Journal of Time Series Analysis* **1**(1), 145–158.

Rao, T. S. and M. M. Gabr (1984). *An Introduction to Bispectral Analysis and Bilinear Time Series Models*. Vol. 24. Springer-Verlag. New York. Lecture notes in Statistics.

Rengaswamy, R., T. Hägglund and V. Venkatasubramanian (2001). A qualitative shape analysis formalism for monitoring control loop performance.. *Engineering Applied Artificial Intelligence* **14**, 23–33.

Rhodes, C. and M. Morari (1998). Determining the model order of nonlinear input/output systems. *AIChE Journal* **44**, 151–163.

Riggs, J. B. (1999). *Chemical Process Control*. Ferret Publishing. Texas,USA.

Rosenblatt, M. and J. W. Van Ness (1965). Estimation of the bispectrum. *Annals of Mathematical Statistics*, **36**, 420–436.

Rossi, M. and C. Scali (2005). A comparison of techniques for automatic detection of stiction: simulation and application to industrial data. *Journal of Process Control* **15**, 505–514.

Rossi, M., A. K. Tangirala, S. L. Shah and C. Scali (2006). A data-base measure for interactions in multivariate systems. In *Proceedings of the IFAC-ADCHEM* 2006, Gramado, Brazil, April 3–5.

Ruel, M. (2000). Stiction: The hidden menace. *Control Magazine*. http://www.expertune.com/articles/RuelNov2000/stiction.html.

Ruel, M. and J. Gerry (1998). Quebec quandary solved by Fourier transform. *Intech*, **8**, 53–55.

Salsbury, T. I. and A. Singhal (2005). A new approach for ARMA pole estimation using higher-order crossings. In *Proceedings of the ACC 2005*, Portland, USA.

Schreiber, T. (1999). Interdisciplinary application of nonlinear time series methods. *Physics Reports-Review Section of Physics Letters* **308**, 2–64.

Schreiber, T. (2000). Measuring information transfer. *Physical Review Letters* **85**, 461–464.

Schreiber, T. and A. Schmitz (1997). Discrimination power of measures for nonlinearity in a time series. *Physical Review E* **55**(5), 5443–5447.

Schreiber, T. and A. Schmitz (2000). Surrogate time series. *Physica D* **142**, 346–382.

Sharif, M. A. and R. I. Grosvenor (1998). Process plant condition monitoring and fault diagnosis. *Proceedings of the Institution of Mechanical Engineers* (Part E) **212**, 13–30.

Shunta, J. P. (1995). *Achieving World Class Manufacturing Through Process Control*. Prentice-Hall. NJ, USA.

Singhal, A. and T. I. Salsbury (2005). A simple method for detecting valve stiction in oscillating control loops. *Journal of Process Control* **15**, 371–382.

Small, M. and C. K. Tse (2002). Applying the method of surrogate data to cyclic time series. *Physica D* **164**, 187–201.

Smith, S. W. (1998). *The Scientist and Engineer's Guide to Digital Signal Processing*. California Technical Publishing. San Diego, CA.

Srinivasan, R. and R. Rengaswamy (2005). Stiction compensation in process control loops: A framework for integrating stiction measure and compensation. *Industrial and Engineering Chemistry Research* **44**, 9164–9174.

Srinivasan, R., R. Rengaswamy and R. Miller (2005*a*). Control loop performance assessment: 1. a qualitative approach for stiction diagnosis. *Industrial and Engineering Chemistry Research* **44**, 6708–6718.

Srinivasan, R., R. Rengaswamy, S. Narasimhan and R. Miller (2005*b*). Control loop performance assessment: 2. Hammerstein model approach for stiction diagnosis.. *Industrial and Engineering Chemistry Research* **44**, 6719–6728.

Stam, C. J., J. P. M. Pijn and W. S. Pritchard (1998). Reliable detection of nonlinearity in experimental time series with strong periodic components. *Physica D* **112**, 361–380.

Stanfelj, N., T. E. Marlin and J. F. MacGregor (1993). Monitoring and diagnosing process control performance: The single-loop case. *Industrial and Engineering Chemistry Research* **32**, 301–314.

Stenman, A., F. Gustafsson and K. Forsman (2003). A segmentation-based method for detection of stiction in control valves. *International Journal of Adaptive Control and Signal Processing* **17**, 625–634.

Stoffer, D. S., D. E. Tyler and A. J. McDougall (1993). Spectral analysis for categorical time series: Scaling and spectral envelope. *Biometrika* **80**, 611–622.

Stuart, A. and J. K. Ord (1987). *Kendall's Advanced Theory of Statistics*. Vol. 1. 5 ed.. Charles Griffin and Company Limited. London.

Sugihara, G., and R. M. May (1990). Nonlinear forecasting as a way of distinguishing chaos from measurement error in time-series *Nature*, **344**, 734–741.

Swami, A., J. M. Mendel and C. L. Nikias (1993). *Higher-Order Spectral Analysis Toolbox Users Guide*. United Signals & Systems Inc. CA, USA.

Taha, O., G. A. Dumont and M. S. Davies (1996). Detection and diagnosis of oscillations in control loops. In: *Proceedings of the* 35th *Conference on Decision and Control*. Kobe, Japan.

Tangirala, A. K., J. Kanodia and S. L. Shah (2007). Non-negative matrix factorization for detection of plant-wide oscillations. *Industrial and Engineering Chemistry Research* **46**, 801–817.

Tangirala, A. K., S. L. Shah and N. F. Thornhill (2005). PSCMAP: A new tool for plant-wide oscillation detection. *Journal of Process Control* **15**, 931–941.

Tao, G., S. Chen and S. M. Joshi (2002). An adaptive control scheme for systems with unknown actuator failures. *Automatica* **38**, 1027–1034.

Terdik, G. Y. and J. Máth (1998). A new test of linearity of time series based on the bispectrum. *Journal of Time Series Analysis* **19**(6), 737–753.

Theiler, J., S. Eubank, A. Longtin, B. Galdrikian and J. D. Farmer (1992). Testing for nonlinearity in time-series – the method of surrogate data. *Physica D* **58**, 77–94.

Theron, J. P. and C. Aldrich (2004). Identification of nonlinearities in dynamic process systems. *Journal of the South African Institute of Mining and Metallurgy* **104**, 191–200.

Thornhill, N. F. (2005). Finding the source of nonlinearity in a process with plant-wide oscillation. *IEEE Transactions on Control Systems Technology* **13**, 434–443.

Thornhill, N. F. and T. Hägglund (1997). Detection and diagnosis of oscillation in control loops. *Control Engineering Practice* **6**(10), 1343–1354.

Thornhill, N. F., M. A. A. S. Choudhury and S. L. Shah (2004). Impact of compression on data driven process analyses. *Journal of Process Control* **14**, 389–398.

Thornhill, N. F., J. W. Cox and M. A Paulonis (2003b). Diagnosis of plant-wide oscillation through data-driven analysis and process understanding. *Control Engineering Practice* **11**(12), 1481–1490.

Thornhill, N. F. and A. Horch (2007). Advances and new directions in plant-wide disturbance detection and diagnosis. *Control Engineering Practice*, **15**, 1196–1206.

Thornhill, N. F., B. Huang and H. Zhang (2003*a*). Detection of multiple oscillations in control loops. *Journal of Process Control* **13**, 91–100.

Thornhill, N. F. and H. Melbø (2006). Detection of plant-wide disturbance using a spectral classification tree. In: *ADCHEM 2006*. Gramado, Brazil.

Thornhill, N. F., H. Melbø, and J. Wiik (2006). Multi-dimensional visualization and clustering of historical process data. *Industrial and Engineering Chemistry Research* **45**, 5971–5985.

Thornhill, N. F., M. Oettinger and P. Fedenczuk (1999). Refinery-wide control loop performance assessment. *Journal of Process Control* **9**, 109–124.

Thornhill, N. F., R. Sadowski, J. R. Davis, P. Fedenczuk and M. J. Knight (1996). Practical experiences in refinery control loop performance assessment. In: *IEE Conference Publication*. 427/1. 175–180.

Thornhill, N. F., S. L. Shah and B. Huang (2001). Detection of distributed oscillations and root-cause diagnosis. In: *Preprints of CHEMFAS-4 IFAC*. Korea. 167–172.

Thornhill, N. F., S. L. Shah, B. Huang and A. Vishnubhotla (2002). Spectral principal component analysis of dynamic process data. *Control Engineering Practice* **10**, 833–846.

Valle, S., W. H. Li and S. J. Qin (1999). Selection of the number of principal components: The variance of the reconstruction error criterion with acomparison to other methods. *Industrial and Engineering Chemistry Research* **38**, 4389–4401.

Wallén, A. (1997). Valve diagnostics and automatic tuning. In: *Proceedings of the American Control Conference*. Albuquerque, New Mexico. 2930–2934.

Watson, M. J., A. Liakopoulos, D. Brzakovic and C. Georgakis (1998). A practical assessment of process data compression techniques. *Industrial and Engineering Chemistry Research* **37**, 267–274.

Welch, P. D. (1967). The use of fast Fourier transforms for the estimation of power spectra. *IEEE Transaction on Audio and Electroacoustics* **AU-15**, 70–73.

Wetherill, G. B. and D. W. Brown (1991). *Statistical Process Control*. Chapman and Hall. London, UK.

Whalen, B. R. (1983). *Basic Instrumentation*. 3 ed.. Petroleum ExtensionService (PETEX). Austin, TX.

Xia, C. (2003). Control Loop Measurement Based Isolation of Faults and Disturbances in Process Plants. PhD thesis. University of Glasgow. UK.

Xia, C. and J. Howell (2003). Loop status monitoring and fault localisation. *Journal of Process Control*, **13**, 679–691.

Xia, C. and J. Howell (2005). Isolating multiple sources of plant-wide oscillations via independent component analysis. *Control Engineering Practice* **13**, 1027–1035.

Xia, C., J. Howell and N. F. Thornhill (2005). Detecting and isolating multiple plant-wide oscillations via spectral independent component analysis. *Automatica* **41**, 2067–2075.

Xia, C., J. Zheng and J. Howell (2007). Isolation of whole-plant multiple oscillations via non-negative spectral decomposition. *Chinese Journal of Chemical Engineering*, **15**, 353–360.

Yamashita, Y. (2004). Qualitative analysis for detection of stiction in control valves. *Lecture Notes in Computer Science*, **3214**(Part II), 391–397.

Yamashita, Y. (2006*a*). An automatic method for detection of valve stiction in process control loops. *Control Engineering Practice* **14**, 503–510.

Yamashita, Y. (2006*b*). Diagnosis of oscillations in process control loops. In: *Proceedings of ESCAPE-16 and PSE2006*. Garmisch-Partenkirchen, Germany.

Yuan, J. (1999). Testing linearity for stationary time series using the sample interquartile range. *Journal of Time Series Analysis* **21**, 713–722.

Zang, X. and J. Howell (2003). Discrimination between bad turning and non-linearity induced oscillations through bispectral analysis. In *Proceedings of the SICE annual conference*, Fukui, Japan.

Zang, X. and J. Howell (2004). Correlation dimension and Lyapunov exponents based isolation of plant-wide oscillations. In: *DYCOPS*. Boston, USA.

Zang, X. and J. Howell (2005). Isolating the root cause of propagated oscillations in process plants.. *International Journal of Adaptive Control and Signal Processing* **19**, 247–265.

Zang, X. and J. Howell (2007). Isolating the source of whole-plant oscillations through bi-amplitude ratio analysis *Control Engineering Practice*, **15**, 69–76.

Copyright Acknowledgements

Parts of the book have appeared in the archived literature and the authors gratefully acknowledge permissions to re-use material from the following papers in the archived literature.

Chapter 4 is an expanded and revised version of Thornhill, N.F., Choudhury, M.A.A.S., and Shah, S.L., 2004, The impact of compression on data-driven process analysis, *Journal of Process Control*, 14, 389–398, © 2004, Elsevier Ltd., with permission from Elsevier.

Chapter 5 contains excerpts from Thornhill, N.F., Cox, J.W., and Paulonis, M., 2003, Diagnosis of plant-wide oscillation through data-driven analysis and process understanding, *Control Engineering Practice*, 11, 1481–1490, © 2003, Elsevier Ltd., with permission from Elsevier.

Chapter 6 is a revised version of Choudhury, M.A.A.S., Shah, S.L., and Thornhill, N.F., 2004, Diagnosis of poor control loop performance using higher order statistics, *Automatica*, 40, 1719–1728, © 2004, Elsevier Ltd., with permission from Elsevier.

Chapter 7 contains excerpts from Thornhill, N.F., 2005, Finding the source of nonlinearity in a process with plant-wide oscillation, *IEEE Transactions on Control System Technology*, 13, 434–443, © 2005, Institute of Electrical and Electronics Engineers (IEEE), with permission from IEEE.

Chapters 11–14 are reprinted (with revisions) from Choudhury, M.A.A.S., Thornhill, N.F., and Shah, S.L., 2005, Modelling valve stiction, *Control Engineering Practice*, 13, 641–658, © 2005, Elsevier Ltd., with permission from Elsevier.

Chapter 15 is an expanded and revised version of Choudhury, M.A.A.S., Shah, S.L, Thornhill, N.F., and Shook, D.S., 2006, Automatic detection and quantification of stiction in control valves, *Control Engineering Practice*, 14, 1395–1412, © 2006, Elsevier Ltd., with permission from Elsevier.

Chapters 17 and 19 contain excerpts from Choudhury, M.A.A.S, Kariwala, V., Thornhill, N.F., Douke, H., Shah, S.L., Takada, H., and Forbes, J.F., 2007, Detection and diagnosis of plant-wide oscillations, *Canadian Journal of Chemical Engineering*, 85, 208–219, © 2007, with permission from the Canadian Society for Chemical Engineering.

Chapter 18 contains excerpts from Thornhill, N.F., Huang, B., and Zhang, H., 2003, Detection of multiple oscillations in control loops, *Journal of Process Control*, 13, 91–100, © 2003, Elsevier Ltd., with permission from Elsevier.

Chapter 18 also contains excerpts Thornhill, N.F., Melbø, H., and Wiik, J., 2006, Multi-dimensional visualization and clustering of historical process data, *Industrial Engineering and Chemistry Research*, 45, 5971–5985, © 2006, American Chemical Society (ACS), with permission from ACS.

Chapters 18 and 19 contain excerpts from Thornhill, N.F. and Horch, A., 2007, Advances and new directions in plant-wide disturbance detection and diagnosis, *Control Engineering Practice*, 15, 1196–1206, © 2003, Elsevier Ltd., with permission from Elsevier.

Chapters 18 and 19 also contain excerpts from Jiang, H., Choudhury, M.A.A.S., and Shah, S.L., Detection and Diagnosis of Plant-wide Oscillations from Industrial Data using the Spectral Envelope Method, *Journal of Process Control*, 17(2), 143–155, © 2007, Elsevier Ltd., with permission from Elsevier.

Chapters 18 and 19 also contain excerpts from Tangirala, A.K., Shah S.L., and Thornhill, N.F., PSCMAP: A new tool for plant-wide oscillation detection, *Journal of Process Control*, 15, 931–941, © 2005, Elsevier Ltd., with permission from Elsevier.

Index

Other titles published in this Series (continued):

Soft Sensors for Monitoring and Control of Industrial Processes
Luigi Fortuna, Salvatore Graziani, Alessandro Rizzo and Maria Gabriella Xibilia

Advanced Control of Industrial Processes
Piotr Tatjewski
Publication due December 2006

Process Control Performance Assessment
Andrzej Ordys, Damien Uduehi and Michael A. Johnson (Eds.)
Publication due December 2006

Modelling and Analysis of Hybrid Supervisory Systems
Emilia Villani, Paulo E. Miyagi and Robert Valette
Publication due January 2007

Model-based Process Supervision
Belkacem Ould Bouamama and Arun K. Samantaray
Publication due February 2007

Magnetic Control of Tokamak Plasmas
Marco Ariola and Alfredo Pironti
Publication due May 2007

Continuous-time Model Identification from Sampled Data
Hugues Garnier and Liuping Wang (Eds.)
Publication due May 2007
Process Control
Jie Bao, and Peter L. Lee
Publication due June 2007

Distributed Embedded Control Systems
Matjaž Colnarič, Domen Verber and Wolfgang A. Halang
Publication due October 2007

Optimal Control of Wind Energy Systems
Iulian Munteanu, Antoneta Iuliana Bratcu, Nicolas-Antonio Cutululis and Emil Ceanga
Publication due November 2007

Model Predictive Control Design and Implementation Using MATLAB®
Liuping Wang
Publication due November 2007